An Unforgiving Land

Hardscrabble Life in the Trapps, a Vanished Shawangunk Mountain Hamlet

Published by Black Dome Press Corp.
649 Delaware Ave., Delmar, N.Y. 12054
blackdomepress.com
(518) 439-6512

First Edition Paperback 2013

ISBN-13: 978-1-883789-73-2

ISBN-10: 1-883789-73-7

Library of Congress Control Number: 2013944838

Map of Ulster County courtesy of Ulster County Tourism.

Front cover photograph: Mary Enderly, born and raised in the Trapps, is flanked by members of her family ca. 1911. Courtesy of Joan Hayes Wustrau.

Back cover photograph: U.S. Route 44/55 passes through the Trapps Gap (upper right) in the cliff escarpment of the Shawangunk Mountains. The Trapps hamlet once lay beyond to the right, its people isolated from the surrounding valleys for much of its 150-year existence. Courtesy of Bob Larsen.

Back cover illustration: Detail of Lake Minnewaska, E. J. Whitney, illustrator, 1887. Courtesy of Mohonk Mountain House Archives.

Design: Ron Toelke Associates, www.toelkeassociates.com

Printed in the USA

10 9 8 7 6 5 4 3 2

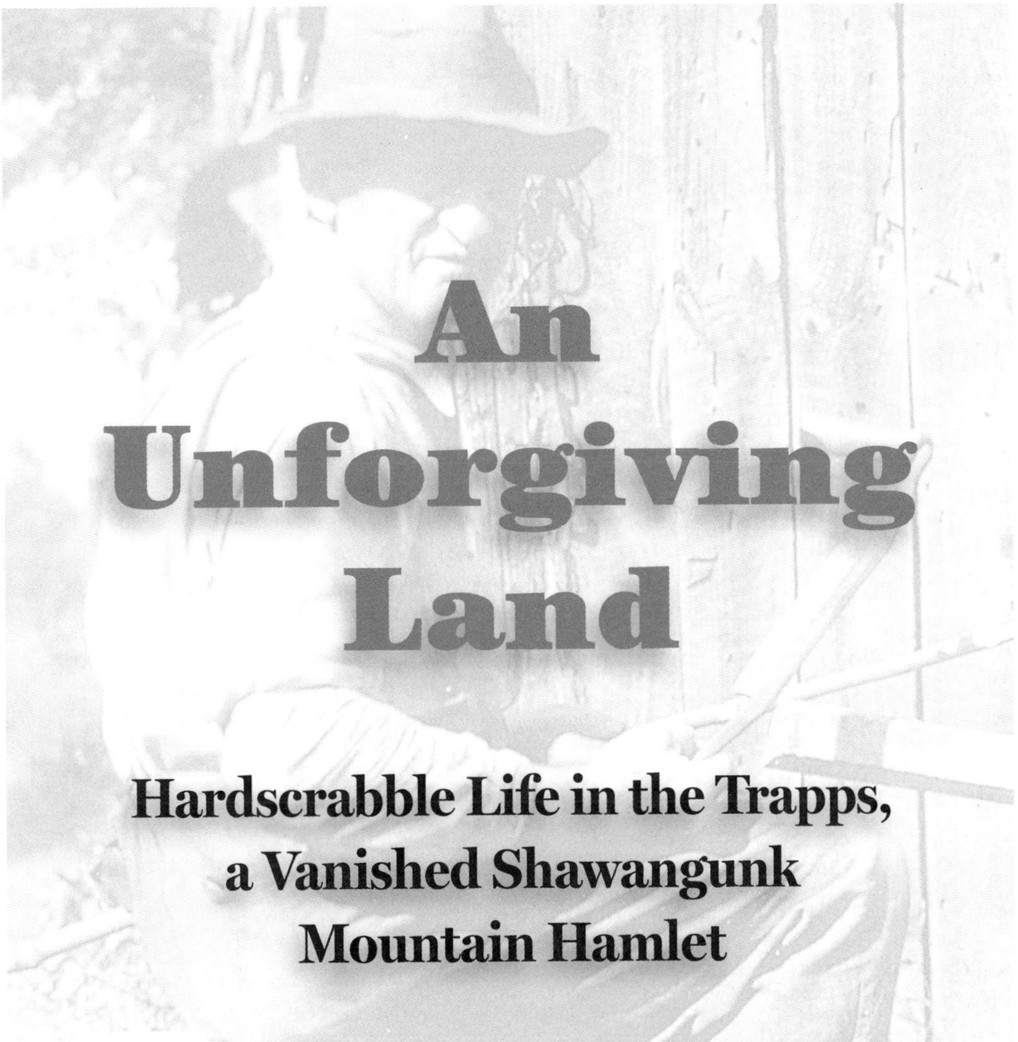

An Unforgiving Land

Hardscrabble Life in the Trapps, a Vanished Shawangunk Mountain Hamlet

Robi Josephson and Bob Larsen

Black·Dome

Dedicated to the Trapps people
for their many contributions
to Shawangunk Mountain history.

We also remember with love
Thelma Atkins, Harold Bittner,
Evelyn Coddington DeWitt, Clarence Kidney, Joe Otis,
Melvin Sharpe, Harold Van Leuven, and John Van Leuven, Sr.,
who gave us the gift of their memories.

"Surely the memory of one's youth is stronger than thought —
that tacit feeling of the shape of things as they were."
Willie Morris, *Taps: A Novel**

Contents

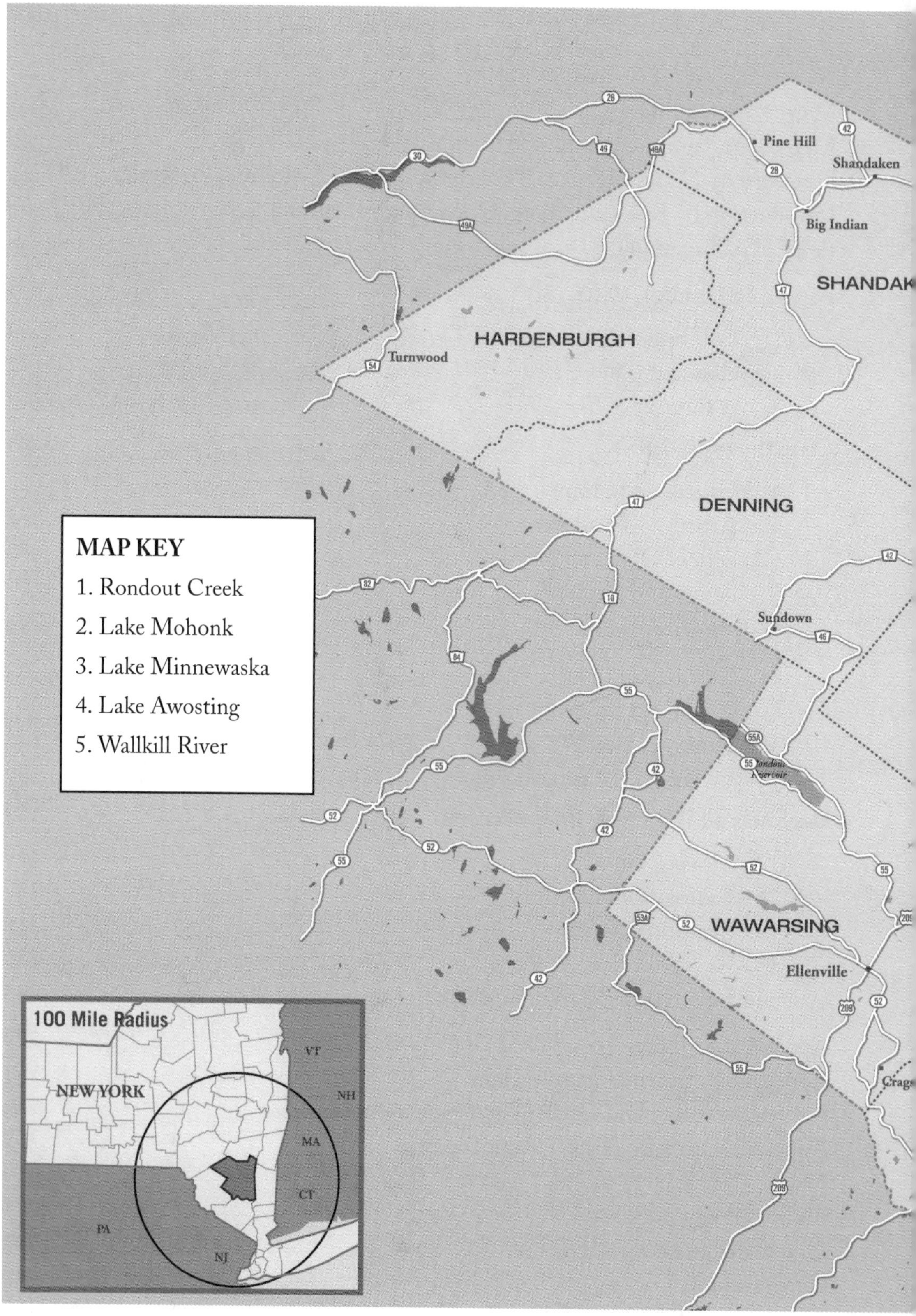

MAP KEY

1. Rondout Creek
2. Lake Mohonk
3. Lake Minnewaska
4. Lake Awosting
5. Wallkill River

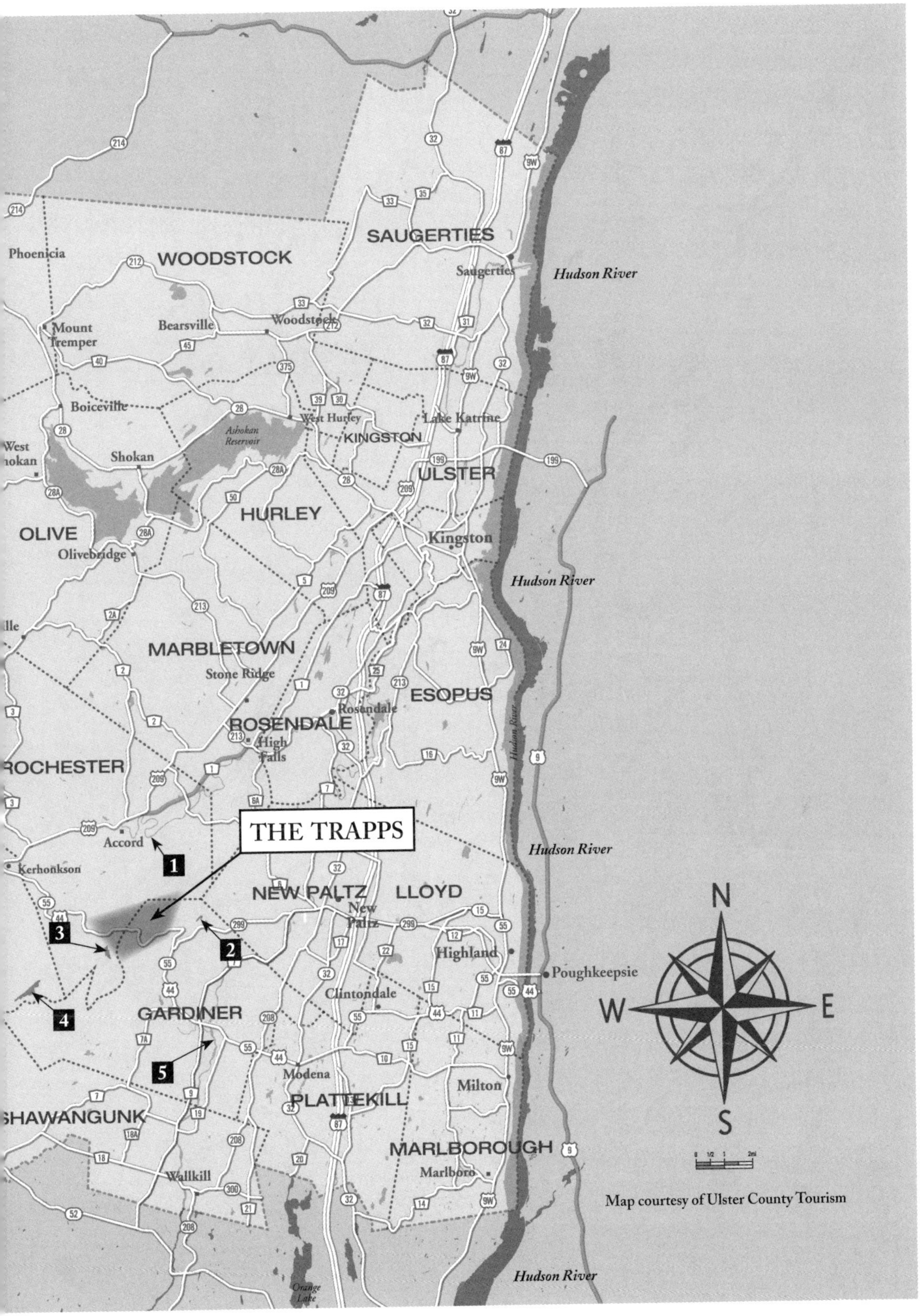

Map courtesy of Ulster County Tourism

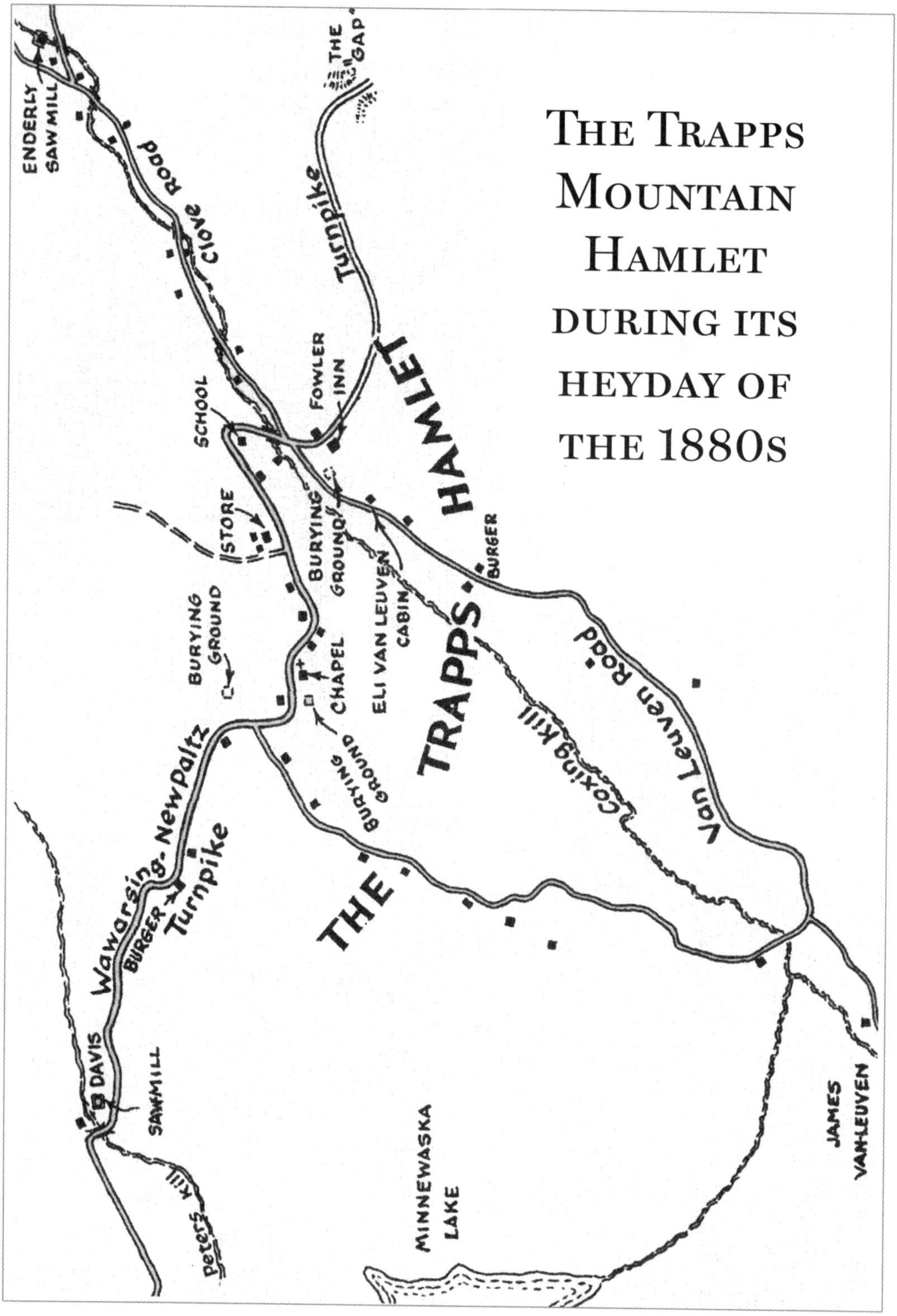

The vanished Trapps hamlet was once located in the northern Shawangunk Mountains, Ulster County, New York. This map by Jack Fagan shows its heyday of the 1880s. Courtesy of Jack Fagan.

Foreword

Life in the Trapps hamlet—high atop the Shawangunk Mountains of Ulster County, New York—was hard but good. Families adapted to the unyielding nature of this harsh and "unforgiving land." The deep knowledge and tireless research of coauthors Robi Josephson and Bob Larsen show how life in the Trapps was "dangling, uncertain." Yet everyone had work to do and did it on his or her own terms while living and breathing the mountain's free spirit. In this new book, these two dedicated cultural historians breathe life into the history and humanity of this now-vanished hamlet.

Settlers, primarily of Dutch, English, and Huguenot heritage, left old lives to start anew on the mountaintop. In pioneering times they bought lots partitioned from the Grote Transport and Nineteen Partners tracts. These two eighteenth-century patents encompassed thousands of acres of cliffs and woodlands, spanning two Shawangunk Mountain streams called the Coxing Kill and Peters Kill.

From the post-Revolutionary period to the mid-twentieth century, the Trapps people harvested what they could from the rocky terrain. They cut trees and grubbed out stumps to work subsistence or scratch farms on thin mountain soil. They hunted game, gathered nuts and berries, shaved barrel hoops, cut millstones and tan bark, burned charcoal, logged, and milled lumber. A cow or two and a handful of chickens provided milk, butter, cheese, eggs, and meat.

Tree saplings, grown from logged stumps, were shaved into strips to make sturdy barrel hoops. Tan bark was stripped from oaks and hemlocks for the tanning industry. Millstones cut from the Trapps and northern Shawangunks—350 tons in 1893 alone—at one time represented 80 percent of U.S. millstone production and were shipped as far away as Sydney, Australia. Cordwood from the Trapps stoked the brick making industry of Kingston. Huckleberries picked on Shawangunk ridgetops were in high demand across the Northeast and made for a lively summer trade.

The Trapps people built homes of log, sawn lumber, or thick, rough-hewn planks. Wood battens and insulation of mud, straw, or newspaper helped protect families against harsh winters. Wood stoves warmed them within. Streams, springs, and hand-dug wells or house cisterns provided water essential for life. Despite rugged conditions, families had virtually everything they needed to survive.

Over time a community infrastructure developed. A one-room school house, tavern, church, general store, dance hall, and two sawmills were the centers of commercial and social life in the isolated mountain hamlet. All year people gathered in the evenings or for special occasions to play music, dance, sing hymns and songs, and tell stories.

By the mid- to late 1800s, city people began to seek green relief in the beautiful Shawangunks. Trapps men served as guides for city folks, and a long-distance walking mania called "pedestrianism" swept the region. Meanwhile, mountain folks themselves walked, in some cases miles, to their places of work at stone quarries, charcoal mounds, woodlots, and the mountain hotels established by Albert and Alfred Smiley at Mohonk and Minnewaska Lakes—hotels that Trapps men helped build in the post–Civil War years when New York State's mountains became summer havens for New York City's burgeoning middle class.

Many enthusiastic individuals, including the last generation to live in the Trapps and descendents of Trapps families, shared their personal reminiscences with Robi Josephson and Bob Larsen, who were able to glean the essence of this unique place and its evolution over time.

The Mohonk Preserve is dedicated to preserving the Shawangunk Ridge. After the Trapps hamlet faded and the last residents left, most of the land they worked and lived their lives on has been preserved, protected from development but open to public enjoyment. I am proud of our work restoring the Eli Van Leuven Cabin, protecting cultural history sites, and interpreting the old trails, woods roads, cellar holes, and burying grounds of the Trapps Mountain Hamlet for each new generation of mountain enthusiasts.

No one has done more than Bob Larsen. Bob's forty-year career as a Mohonk Preserve ranger and cultural historian has brought forth renewed interest in the hardscrabble life of the Trapps and its people. He was instrumental in the designation in 2000 of the Trapps Mountain Hamlet as an entire historic district listed on the State and National Register of Historic Places. He gathered the hamlet's last residents and family descendents to help as advisors, consulting with them on the accurate restoration of the Eli Van Leuven family cabin to U.S. Secretary of Interior standards. Bob developed the "walk back in time," an interpretive trail with brochure, wayside interpretive stops, and a podcast on the Mohonk Preserve's Web site relating the story of the Trapps Mountain Hamlet.

Robi Josephson has been the perfect partner for Bob. She has had a long affiliation with the preserve as a consulting archivist, historian, interpretive guide, writer, and editor. For more than a decade, Robi has been immersed alongside Bob in extracting the real stories of mountain life and this place called the Trapps. Their book is the definitive chronicle.

I urge you to visit the Mohonk Preserve to take a step back in time. Walk through what once was the Trapps Mountain Hamlet and imagine mountain life as it was lived by the generations of people you are about to meet in this book.

Glenn Hoagland
Executive Director, Mohonk Preserve
July 2013

Mary Susan Enderly, right, poses on her wedding day in 1876. She was a daughter of Benjamin and Catharine Fowler, who once operated a tavern in the Trapps hamlet. The woman at left may have been a sister of Mary's. Courtesy of the Daniel Smiley Research Center, Mohonk Preserve (DSRC-MP). Donated by Joan Hayes Wustrau.

Introduction

I first discovered the Trapps Mountain Hamlet in the early 1970s when I was a nineteen-year-old seasonal ranger for the Mohonk Trust (now Mohonk Preserve), which owns much of the land in which the hardscrabble hamlet is located. It was less than twenty years since Irving Van Leuven, the last person to live the old ways in the mountainous hamlet, had died after a life of burning charcoal, shaving barrel hoops, and living without electricity or running water. Already the mountain hamlet was becoming a distant memory, overshadowed by the modern two-lane highway that now ran through it and the bustle of the late twentieth century.

Bob Larsen was a Mohonk Trust ranger and my boss. I lived in a small cabin located on the remnants of the old Enderly farm and sawmill within the Trapps hamlet, and when Bob and I were not repairing carriage roads or patrolling the nearby Trapps rock climbing area he would show me the secrets of the Trapps hamlet. An old foundation, an overgrown graveyard, an abandoned millstone quarry—nothing escaped Bob's interest. Above all I remember the secretive hamlet and its historic remnants as intoxicatingly beautiful, with its rushing mountain streams, gleaming white Shawangunk rock, and deep woods and ravines.

Bob never lost his fascination for the eighteenth-century landscape and the colorful people that inhabited it. He spent the next forty years quietly walking its paths and trails, wondering about the origins and workings of the small and staunchly independent farms, plank houses, sawmills and forges that made up the hamlet. Even more, he spent years single-mindedly pursuing the lives and collecting the memories of "his people"—the descendants of the Enderlys, Van Leuvens, and others who worked the land and survived in a harsh and demanding environment.

Bob Larsen went on to become instrumental in securing the listing of the Trapps hamlet as a National and State Historic District, and lovingly restored and interpreted the last iconic structure in the hamlet, the Van Leuven Cabin. Today a visitor can not only visit the restored cabin but also follow an interpretive trail that brings to life the old stonewalls, rock quarries, and headstones that Bob quietly ruminated over for forty years.

Bob also had the good fortune and foresight to team up with Robi Josephson, a uniquely talented writer with a keen and accurate eye for local history—especially that of the vast Mohonk and Minnewaska estates that sit astride the Shawangunk Ridge of Ulster County and protect much of the diminutive yet strikingly pretty mountain range. Together Mr. Larsen and Ms. Josephson have distilled a remarkably detailed and fascinating history of a lost hamlet and a lost time.

Bob is fond of reminding me that, of the many national and state historic structures and districts in New York, the Trapps Mountain Hamlet District does not preserve gilded-era mansions or the lifestyles of the aristocracy. Rather, the men and women who lived in the Trapps over the last two hundred years were industrious and hardy people who lived at the edge—farming subsistence plots and engaging in whatever local industry the land would afford. This delightful history is a vignette into a disappearing chapter of American history and a window into the life and times of post-colonial settlers who worked the land in the Shawangunk Mountains.

This book is to be savored and read again and again. It is a testament to the love and affection for detail that both Larsen and Josephson bring to the history of this remarkable mountain hamlet and the Shawangunk Ridge, and it is a testament to the generations of settlers who lived and worked in this stark and beautiful landscape until the last of them died over fifty years ago.

Bob Anderberg
Vice President and General Council, Open Space Institute
July 2013

PREFACE
A Beginning

U.S. Route 44/55 is a spectacular drive over the Shawangunk Mountains of Ulster County, New York. As the road climbs west from the Wallkill or east from the Rondout Valley, homes and buildings give way to a realm of forest and stone. The Trapps mountain hamlet was located in this lofty realm extending on both sides of the modern highway.

It is "a region where the Maker of Mountains went mad," novelist Arthur O. Friel wrote about the Shawangunks in the 1920s.[1] This sedimentary range is composed of two layers—a hard conglomerate top and a much thicker bottom of softer shale. From Sky Top above Lake Mohonk south to Sam's Point, a great wall or escarpment of conglomerate is visible from the town of New Paltz and across the Wallkill Valley. From these vantage points, the Shawangunks typically cut a handsome figure in off-white or light gray. At times, though, when the cliff escarpment blushes with the rising sun, the mountains look their dapper best and downright dashing.

Gaps or breaks periodically punctuate the great wall of cliffs. Route 44/55 passes through one such gap called the Trappen. This Dutch word means "stairs" or "staircase," which the gap resembled to those settling the Wallkill Valley more than three centuries ago. The plural suffix "-en" was later anglicized to "-s," with the gap called "The Trapps" even in modern times by the hamlet's last residents.[2]

The gap was such a dramatic feature that it warranted a nod from nineteenth-century gazetteer Hamilton Child, who described it as "a deep gully, 650 feet wide, extending through the mountains" in the Town of Gardiner. For Friel, it was "the Jaws of the Trapps."[3]

The place-name "Shawangunk" is pronounced as written, with three syllables, or as the two-syllable "shon-gum" or "shawn-gum," a long-used regional contraction. Popularly known as "the Gunks," the range extends about fifty miles from Port Jervis (at the tristate corner of New York, New Jersey, and Pennsylvania) northeast to Rosendale. Often incorrectly called "the foothills of the Catskills," the Shawangunks are geologically distinct from its mountain neighbor, more complex in formation, and about 100 million years older. In the Shawangunks, the Martinsburg shale was deposited as clay and silt onto the bottom of a vast inland sea more than 450 million years ago during the Ordovician period.

The overlaying Shawangunk conglomerate or grit was deposited as pebbles, sand, and quartzite into another inland sea millions of years later during the Silurian period. The northern half of the range, where our story takes place, was further shaped by uplifting, erosion, and glacial scouring. Over eons, a mountain range was born.[4]

It was within this rocky madness, nestled in a high hollow between Lake Mohonk and Lake Minnewaska, that the Trapps people forged a living for over 150 years. The American Revolution and World War II were rough bookends of its history.

Today, much of the former hamlet is protected by the Mohonk Preserve, a nature preserve open to the public, and the bordering Minnewaska State Park Preserve. The old hamlet also lies within one of The Nature Conservancy's "Last Great Places on Earth" and along New York State's Shawangunk Mountains Scenic Byway. We're lucky to have such stewards.

Thus we begin our tale of a vanished place and bygone times on the mountain.

Robi Josephson and Bob Larsen
Ulster County, New York
October 2013

Birth of a Hamlet 1730–1830

1

Entering the Wilderness

Late November 1876, Friday afternoon. The Wallkill Valley Railroad screeches to a halt. A traveler from New York City steps off the train at the Rosendale train depot, some ninety miles to the north up the Hudson River Valley. He brushes soot and dust off his suit and satchel, strides down the hill into town, and checks into a hotel on Main Street.

"Welcome, mister. I'm at your service. What brings you to our fair village?" the proprietor asks.

"Heading south," the traveler responds. "Along the spine of the mountain."

"Rough country down there. Full o' wild beasts. You got a good rifle?"

"Much obliged for your concern, my good man. I have my wits."

"You'll need 'em. Sign here, and I'll show you to your room."[1]

Photograph by Ted Reiss.

The traveler from New York began his pedestrian escapade in Rosendale, then a thriving center of the cement industry and busy port on the Delaware & Hudson Canal. After a night's rest, he set forth south into the Rondout Valley and began a two-day tramp traversing an ancient hump of long, craggy ridges rising west of the Hudson River and southeast of the Catskills—the Shawangunk Mountains.

Our metropolitan explorer made his way to a mountaintop lake and hotel, both named Mohonk. A high point above the lake named Sky Top offered visitors sweeping vistas of the Shawangunks, the Catskills, and surrounding valleys.

Later Saturday, he departed this lofty haven of Quaker twins Albert and Alfred Smiley. He found his way down to a mountain glen called the Clove and followed an old road paralleling a stream called the Coxing Kill. He eventually reached the Trapps, a mountain hamlet where he found lodging at the home of sawmill owner George Davis. Sunday morning, our pedestrian left the hospitality of the Davis family. He saw homes made of wooden planks and even more primitive dwellings whose residents he regarded as "the least wealthy" of the Trapps people.[2] These rustic shelters, windowless and floorless with crude wooden doors, were a far cry from the parlors of Manhattan. Yet, he fancied them "quite endurable."

From there, our wayfarer tread higher, threading his way southward through a strange land with strange names: Minnewaska, Palmaghatt, Awosting. He spelled the Palmaghatt "Paulymer Gut," his best guess from local speech.

He identified the Trapps as a "wide hollow or pass ... in an irregular wall or palisades of rock." He continued, "Excepting the few fields in the hollow of the Traps [*sic*], and the farms north of Mohunk [*sic*], the range is a barren, uninhabited wilderness. ... The huge rocks, black with lichens, the heavy forest in deep shadow, the damp thick mosses, and the unbroken silence, make this place very impressive."[3]

Our traveler made his way across the "bleak November" landscape, reentering civilization at Ellenville, another busy canal port located in the Rondout Valley between the Shawangunks and the Catskills. He returned to New York, where "Pedestrianism: The Shawangunk Mountains from End to End" appeared in the Christmas Eve 1876 edition of *The New York Times*.[4]

Today, we can only guess at the reaction of metropolitan readers to his account of an alien place within a day's travel from New York. What odd names! Hardly a sign of life! A place called the Trapps? Do they trap wild animals for a living? Do the people attend church, the children school? Yet he was silent regarding these questions.

The Times reporter stated the Shawangunks "give ample returns for money and time to the vigorous man who likes a short run in good mountain air and bold scenery." His "short run," some twenty miles long, was penned in the spirit of pedestrian sport for urbanites game for a two-day trek close to home.

He all but omitted the Trapps hamlet, home then to some forty to fifty families. During much of its existence, the Trapps was given scant media or literary exposure beyond mid-Hudson Valley towns. In truth, there was no reason to mention those who led a hardscrabble life atop a rocky escarpment far from major metropolitan centers.

Yet theirs was a unique community isolated by its ridgetop location. The Trapps people, inhabiting rocky land poor for farming, led very different lives from those who dwelt in rural valley communities in mainstream America. This seclusion made for closer ties among families through marriage, work, and church, as well as with sister mountain hamlets, including the upper Clove and Lawrence Hill. The Trapps remained in a cultural lag, its residents living a nineteenth-century subsistence way of life well into the twentieth century.

Some families lived in the Trapps for only a short time; others stayed for two or three generations. A handful remained for more than a century—families named Burger, Enderly, Harp, Van Leuven. The day-to-day work of the Trapps people and the colorful aspects of their lives appeared regularly in local newspapers. This local exposure, other public records, and oral histories tell us much about their bygone way of life and how families survived on the high ridges of the Shawangunks.

The men stripped tan bark from mature stands of hemlock and oak for the leather industry. They cut tons of conglomerate rock, shaping them into millstones destined for gristmills and industrial towns across the country and, indeed, around the world. They logged acres of wood lots for lumber and for charcoal.

In winter, the men fanned out onto Shawangunk Mountain lakes to cut massive blocks of ice for refrigeration. They moved into farm sheds to shave innumerable hoops to secure barrel staves. Barrels then bore America's bounty, including cement from nearby Rosendale for major construction projects across the country.

Families churned butter, boiled maple sap into syrup, grew apples and peaches—all for sale. They picked quart after quart of huckleberries, so savored from Buffalo to Boston to New York City and beyond. Girls and boys helped their parents in these varied occupations. Everyone had to pitch in to survive, for the Shawangunk mountaintop was a harsh place, an unforgiving place. It was an unforgiving land.

Change came to the land and the Trapps people after the Civil War when major construction projects in the northern Shawangunks provided steadier employment. Trapps men worked as laborers and stone masons erecting mountain hotels at Lakes Mohonk and Minnewaska, as well as auxiliary structures and an extensive network of carriage roads. Men and women of the hamlet worked to keep these resorts operating at first during summers and then year-round. Trapps men also helped to construct two state highways crossing the mountain range and two aqueducts cut under it during the early 1900s.

In short, the Trapps people were part of the invisible workforce that helped build our region and our nation. Their legacy lives, but their hamlet is no more. Little is left of their farms and homes but for stone foundations or rubble. Yet we can explore this vanished place thanks to the Minnewaska State Park Preserve and the Mohonk Preserve. The latter's restoration efforts are especially noteworthy.

The Mohonk Preserve has restored the Eli Van Leuven Cabin, one of the hamlet's last remaining structures. This plank house, constructed of rough-cut wooden slabs from a Trapps sawmill, was the typical home found in the Trapps not so very long ago. The Mohonk Preserve also created a path leading to the cabin from its West Trapps Trailhead on Route 44/55. Along the path, people can discover rough-cut millstones in an abandoned quarry, old stone walls running through a forest that was once farm fields and pastures, and a burying ground where Trapps families laid their loved ones to rest.

A short drive away, at the Mohonk Preserve's Coxing Trailhead on Clove Road, are the remains of the Enderly farm. Here once stood a farmhouse, three-story barn, sawmill with millpond, and blacksmith forge. In modern times this site has become a popular swimming spot on the Coxing Kill known as Split Rock. The Mohonk Preserve has interpreted this area to show visitors how one Trapps family lived and worked over a century ago.

Furthermore, and with another nod to the Mohonk Preserve, the Trapps Mountain Hamlet, including the sites just noted, is now listed on the National and New York State Registers of Historic Places—the first time an all-but-vanished subsistence community of our state has received such designation.

The Trapps and its people may have nearly vanished, but they have captured our imaginations and won our hearts. It has been our great privilege to meet and interview the last generation to grow up in the Trapps and those whose parents or grandparents lived there. Many of those interviewed have become, much to our delight, cherished friends.

We have tried, in our short history, to recreate this lost community, to give life to its people. We honor their lives and labors, their dreams and endeavors, their struggles and successes.

• 2 •

Colonial Land Grant

The first of June 1730. Philip DuBois, Frederick Schoonmaker, and Moses DePuy in turn dip a quill pen into ink, signing the document before them with a grand flourish. Each is a colonial trustee of the Town of Rochester.

The document grants them in common a great tract from the Rondout Creek east across the Shawangunks to the highest point of the steep cliffs facing the Wallkill Valley. The tract's northeast corner is a break in the wall of cliffs known as the Trappen.

"We'll see what that mountain land brings," Philip comments.

"It's rocky land," Frederick says. "Hard to farm, but promising woodlots."

"I've heard of gold mines and treasure," Moses says, "but who knows?"

Their business concluded, the three men shake hands and take leave.

The story of the Trapps mountain hamlet began with a flourish of quill pens. When that trio of Rochester trustees sat down to business in June 1730, their act set the stage for the hamlet's future settlement. By then, the Hudson Valley corridor from New York to Albany had become an oceanic merging of cultures. Slicing through the high seas, ship after ship carried English, Dutch, French Huguenot, and other European colonists to American shores. Newcomers from afar swept the indigenous peoples from their homelands; those few native people who survived disease and war were physically forced away. American Indians driven from the vicinity of the Shawangunk

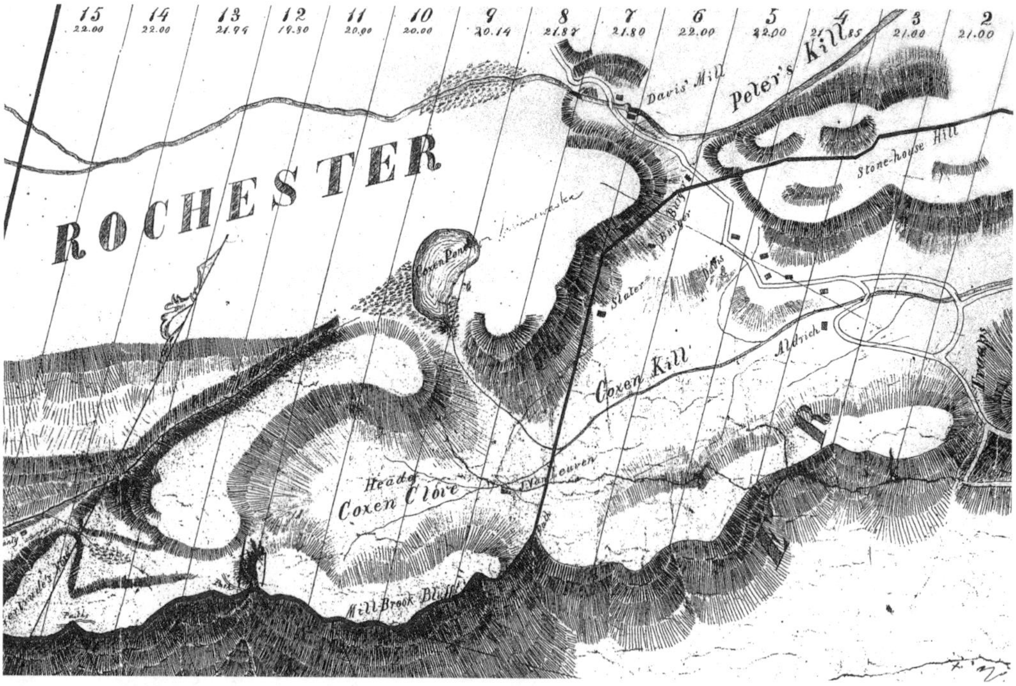

The Grote Transport of 1730 (above) and the Nineteen Partners Tract of 1770 (next page) were two colonial land grants in the Town of Rochester of the northern Shawangunk Mountains. The lines of both maps show how lots were partitioned in the Trapps. Coxen Pond on the Grote Transport map was an earlier name for Lake Minnewaska. Moggunk and Paltz Point in the lower right of the Nineteen Partners Map were earlier names for Sky Top, the promontory above Lake Mohonk and visible from the Wallkill and Rondout Valleys. Detail of an 1860 copy (above) courtesy of the Haviland-Heidgerd Historical Collection, Elting Memorial Library (HHHC-EML). Detail of an 1875 copy (next page), Map #1017, courtesy of the New York State Department of Environmental Conservation, Region 3, New Paltz.

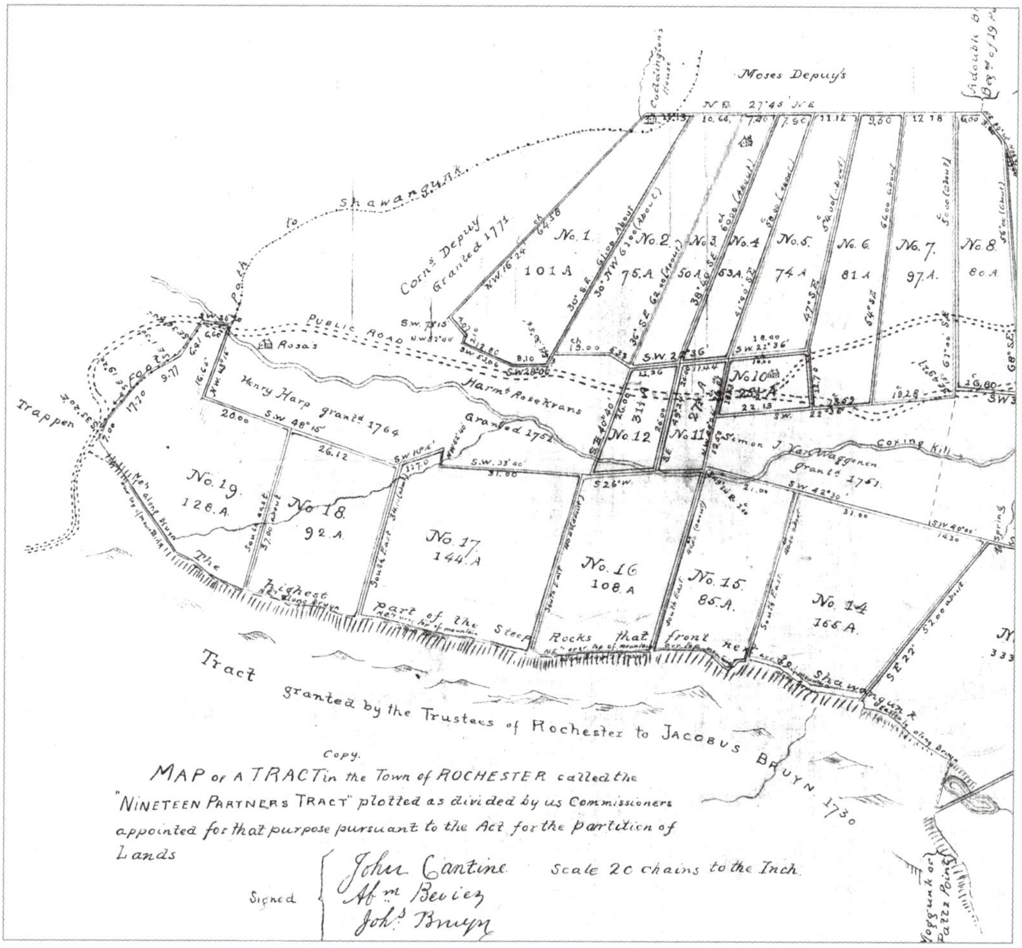

Mountains later became part of the larger Delaware Nation, also called the Lenape.

The race was on for New World riches. The virgin continent riveted the attention of speculators and settlers, who ventured across the Atlantic and then inland via coastal rivers. Yet the Shawangunk Mountains west of Hudson's River held little interest well into the 1700s. Mohonk "was very difficult of access from New Paltz and considered of no value," wrote French Huguenot descendant Ralph Le Fevre. A steep wall of cliffs facing the Wallkill Valley prevented settlement from the east, and the range's rocky soil was deemed unfit for farming or grazing. Mountain ranges, though, were "largely held intact for speculation," noted the Junior League of Kingston. Land speculation in the Shawangunks gained momentum during the 1700s, primarily from the Rondout Valley to the west.

In 1703, Queen Anne of England had conveyed the massive Rochester Patent, replacing the Dutch name of Mombaccus for the region with the name of an English earl.

Anchored in the eastern Catskills, the patent's vast swath of land spanned the Rondout Valley and a huge chunk of the northern Shawangunks, including the site of the future Trapps hamlet, to the brink of the cliffs facing New Paltz and the Wallkill Valley. A land patent is legal recognition of title or interest to a tract of land, either purchased or received freely as a grant; a government might grant land as an award for military service or as an incentive for settlement or development. The lands of the Rochester Patent were conveyed to the town trustees, prominent residents who acted as administrators and land agents.[1]

During the 1700s, two great tracts in the Shawangunks were carved from the Rochester Patent. The first was the Grote (or Groote) Transport of 1730 when the three Town of Rochester trustees—Philip DuBois, Frederick Schoonmaker, and Moses DePuy—became joint owners of more than 20,000 acres. The title "Grote" was well chosen, for it meant a "great transfer" in Dutch. The second was the Nineteen Partners Tract of 1770, when nineteen men, all leading citizens, were granted in common well over 1,800 acres.

The Nineteen Partners Tract's northeast corner was a striking high point known then as Paltz Point and Moggunk, a place-name of the Warranawonkong people of the Esopus Indians. (Moggunk evolved into Mohonk.) This high point, renamed Sky Top, was also the southwest corner of the New Paltz Patent. These two great land tracts met along "[t]he highest part of the steep rocks that front next to Shawangunk," which became an important boundary line. This line formed the eastern boundary of land patents rising up from the Rondout Valley, such as the Rochester Patent, and the western boundary of grants rising up from the Wallkill Valley, such as the New Paltz Patent.

Together, the Grote Transport, extending south, and the Nineteen Partners Tract, extending north, encompassed much of the northern Shawangunks. They shared a common boundary in one section just west of the cliff escarpment facing the Wallkill Valley, along a footpath long used by native peoples to access and cross the mountain range. This vital route joining the Rondout and Wallkill Valleys was an obvious choice for a boundary line between the two land tracts and between private parcels along its way. The path cut through a deep gap between cliffs, by that time known as the Trappen, a Dutch plural noun meaning "steps" or "staircase."[2]

As new settlers fanned out across the Wallkill and Rondout Valleys, pressure intensified for new fertile land. Settlers began inching up from the Rondout Valley onto the Shawangunks' lower western slopes and up into its ravines, including the Clove in the

The Number 18 boundary stone is one of the few remaining from the partitioning of the Nineteen Partners Tract. Roman numerals, enhanced here for better visibility, were used then to mark the stones. Photograph by Robi Josephson.

watershed of the Coxing Kill, a stream tumbling off the mountain from a lake later called Minnewaska. The word *kill*, from the Dutch, means a "channel" or "creek." A "clove," from the Dutch *kloof*, refers to "a cleft or saddle between heights," noted Hudson Valley historian Marc B. Fried.

Individuals also purchased mountain parcels separate from the great tracts. In April 1720, John Beatty of Marbletown was among the earliest to acquire land in the Clove. The Rochester trustees sold Beatty a 250-acre parcel for 6 pounds, 5 shillings "situate on the northwest side of the great mountain near to the high point Called Mogonick

[Sky Top]." In July 1750, Edward Wood of Kingston bought the same parcel from Beatty for 108 pounds. These men likely viewed their Shawangunk mountain land as investments or as new land for farming or for harvesting one highly valued commodity—timber.[3]

As the first rumblings of the Revolution began, the northern Shawangunks had been carved up and claimed. By war's end America had survived a painful birth. Many in the mid-Hudson Valley remained at their homes and farms while others left to try their luck elsewhere. A few turned their sights to the Shawangunks. The mountains were poised for settlement, a course peaceful but perilous for those adventurous enough to begin anew on its rugged ridges.

• 3 •

A Pioneer Settlement

Evening, 1825. After a day's labor on a valley farm, John Burger and his yoke of oxen trudge up the dirt wagon road toward the Trapps gap. Up ahead, a lamp flickers and grows brighter.

"Bless you, Ann, but you take chances to meet me."

"I have the gun, John."

Suddenly, the night itself snaps and snarls. Wolves! Far off, but getting closer. The two hurry on to the safety of their forest cabin.

Back at their mountain farm, the young couple take turns all night keeping pine knots ablaze around the livestock pen. Each is ever alert, ever ready, shotgun gripped, to blast any predators who stalk their mountain homestead, waiting . . . .[1]

After the American Revolution, land ownership was in flux with waves of people on the move. In 1803 the Louisiana Purchase doubled the size of the new nation. A vast wilderness lay west of the Mississippi River. Yet the frontier was still a very real place back East. Those with a pioneer spirit in New York's mid-Hudson Valley had only to gaze as far as two nearby mountain ranges—the Catskills and the Shawangunks.

By 1800 the great Shawangunk land grants of the 1700s were undergoing partitioning. The Grote Transport and the Nineteen Partners Tract were carved into smaller and smaller parcels. Many parcels remained in the families of the original patentees, but over time more were sold to newcomers. Some of the individual lots on the mountain also changed ownership.[2]

Up rutted roads, over stony ridges came new families. Oxen or horses hauled overloaded wagons, creaking in useless protest from the weight of worldly belongings. It was a time when many a young man or family staked out a claim. Homesteaders began reaching high into the northern Shawangunks—a "little frontier" still as wild as the Kentucky or Tennessee hills—arriving at a hidden hollow soon called the Trapps.

The process of clearing land and building a homestead might take two to four summers, noted Tom Wessels. Using "his keen axe, his firebrand, and his trusty ox team," a man cut down and burned out the forest for his home, often sowing his first crops around charred roots and stumps, wrote Jared van Wagenen, Jr.[3] Once a cabin was ready, the man (with his family if married) moved into the wilderness, hoping the land would provide and his luck would hold. The earliest homesteaders in the Trapps were named Roosa, Aldrich, Burger, Enderly, and Van Leuven. The settling of the northern Shawangunks had finally truly begun, well over a century later than in the neighboring valleys.

In June 1785, Henry Harp, age fifty, "being weak in body but of a sound mind and memory," wrote his will. While dipping a quill pen into ink, perhaps he paused to reflect on his life. His German boyhood in Pennsylvania. His wives—Lidia, the mother of his children, and Eve. The year 1764 when he acquired land in the upper Clove at the great split rock on the Coxing Kill. Joining that Nineteen Partners deal in 1770. Fighting with his brothers for freedom. His adult life in the Rondout Valley.

When Henry died in 1788, his land was divided among his three sons, with stipulations for one highly desirable site on his mountain acreage: "Then I give to my son Edward all the land lying on the Anderieses Kill, Except the falls and a good conveniences and

Room to set a Saw Mill and Room enough for the Logs to which I give to my three sons, Edward, Henry, and Jacob." Anderieses (the origin of the name unknown) was another early name for the Coxing Kill. Henry's will is crucial, as it may be the first record of an early American industrial center in the Shawangunks. If Henry's sons followed his wishes, they operated the earliest sawmill in the Trapps. Yet that distinction might have gone to later owners. The first of these was Teunis Klaarwater Roosa, a Marbletown man of Dutch descent.[4]

Within two years of Henry Harp's death, his sons sold or gave the Coxing parcel to Teunis, who had served with Henry during the Revolution. The Rochester census of 1790 places Teunis there, and the 1799 partition map of the Nineteen Partners Tract pinpoints a house marked "Rosa's" within the boundaries of Henry's old parcel along the Coxing. "Rosa's" was the homestead around which the Trapps hamlet grew, higher up the mountain but still part of a Dutch enclave in the upper Clove strung out along the public road. After the British had seized political control in the previous century, those of Dutch descent had clustered up and down the Hudson Valley where they were free to speak their language and practice their traditions. Teunis, however, left this Dutch neighborhood soon after acquiring the Coxing parcel. In 1801 he sold the mill site and 135 acres to a family who would remain in the Trapps for more than a century—the Enderlys. Alas, the name of those who built that first sawmill, whether Harp, Roosa, or Enderly, was lost long ago in the tumbling waters of the Coxing Kill.

The new family at the Roosa place was that of Johannes (John) Enderly. The son of a German colonist, Johannes fought for America during the Revolution. He married a Dutch woman named Marytje Calder, according to descendant Joan Hayes Wustrau. Theirs was a time when land was wealth, when families of an agrarian society did not bank or invest money but to put it into land. Land was the main income producer, and presumably Johannes and his family settled at the mill site in 1801 or soon after.

Johannes found himself laboring on more than just his new farm. For the common good, as well as the benefit of living along a public road (albeit unpaved as all roads then were), he was required to maintain the section bounding his property. Landowners were assigned to work a number of days depending on the length of their road frontage. In 1802 Johannes worked four days, more than any other landowner in the Trapps, known then as the Second Clove District. By 1827 his assigned work days still numbered four, half to be completed by July first.[5]

This typical tavern sign of early America welcomed travelers for rest and refreshment. Courtesy of the Library of Congress, HABS MASS, 9-LIN, 7-4.

As more families settled the Clove, three institutions followed—churches, schools, and taverns. Taverns, also called public or road houses, served spirits with the day's gossip, business, and politics. They often doubled as inns and meeting places. The tavern keeper, or taverner, was licensed and responsible for his clientele. Taverns became important public institutions in the community, noted David M. Ellis. *The Wallkill Valley World* printed these New York State tavern rules for "Ye Olde Taverne" in 1850:

Four pence a night for Bed.
Six pence with Supper.
No more than five to sleep in one bed.
No Boots to be worn in bed.
Organ Grinders to sleep in the Wash house.
No Beer allowed in the Kitchen.
No Razor Grinders or Tinkers taken in.
No dogs allowed upstairs.

A handful of taverns operated in Rochester between 1800 and 1810, noted Nathaniel Bartlett Sylvester. Most were located off the mountain, yet a perfect spot was found along a wagon road crossing the Shawangunks through the Trapps. Peter Aldrich, a young New England Yankee, likely established this earliest tavern in the Trapps. The time was right for a place where pioneers could meet over a pint of ale and where growing numbers of travelers could rest themselves and their horses. While most homesteaders were moving up onto the mountain from the Rondout Valley, Peter was one of the first to do so from the Wallkill Valley.[6]

A Rhode Island lad, Peter Aldrich was born in 1773, the year of the Boston Tea Party. War is a great impetus for population movement, with the Revolution no exception. The tidal wave of Yankee migration surged in the decades after America's independence from Great Britain. New Englanders by the thousands eyed New York and the western territories as places to change their fortune. Peter, seventeen years old, joined this migration about 1790. He arrived in New York, crossed the Hudson River, and soon after married Hannah Polhemus, a young woman of Dutch descent who may have lived north of Newburgh.

Something, though, pulled Peter to the mountain westward. He and Hannah lashed their possessions onto a wagon and with their young children set out for the Shawangunk wilderness. By 1800 they had settled on a rugged tract remote from the nearest village and very different than the pastoral farmlands of Rhode Island or the Wallkill Valley. While isolated, there were Dutch neighbors down the road for Hannah. Peter became a Yankee "schoonzoon," or the Hudson Valley Dutch equivalent spoken then for "son-in-law."

The Aldrich family might have been only the second to settle in the Trapps. On August 12, 1801, Peter officially became a Trapps landowner, acquiring a woodlot, perhaps his first. The property was a portion of Great Lot 2 of the Grote Transport, south of the old boundary line between the Grote Transport and Nineteen Partners Tract, and Peter acquired it from Philip DuBois Bevier, grandson of an original patentee. By 1810, Peter and Hannah had at least seven children, who helped run the mountain farm according to their age and ability. Within those ten years, a Rhode Island Yankee had chiseled a new life for himself and his family up in the rough country of the Shawangunks.

Yet farming was not enough for Peter Aldrich. In 1802 he acquired a license from the Town of Rochester to operate a public house or tavern "near the Clove." Peter's land bordered the unpaved public road up through the Clove, for that same year he was required to work two days maintaining his road frontage. By 1827 Peter's number of days had risen to six, two more than John Enderly and the highest of any Trapps landowner.

Peter never shook off his nickname of "Yankee." Some of his mountain land was called "the Yankee Lot" even years after his death in 1855.[7]

The Burgers, a long-time Hudson Valley family, also arrived on the mountain in the very early years of the nineteenth century. Benjamin Burger, born about 1776, wed Helena ("Lena" or "Laney") Palmateer of New Paltz at the Marbletown Reformed Dutch Church in May 1799. They likely started married life with his parents in Marbletown, where in 1806 Ben was listed as a merchant or taverner. Soon after, Ben and Lena moved onto the mountain east of a small lake named Coxen Pond (Lake Minnewaska). Records and anecdotal accounts verify the Burger homestead as a very early one in the Trapps, settled either by the couple or by one of their sons. A century later, their grandchildren Calvin Burger and Sarah Voorhis relished telling stories of early mountain life. "There were no Indians in the Traps locality when Burger settled there," Calvin told a New Paltz paper, "but wolves were very abundant." One of his sister's tales, the basis for this chapter's opening vignette, was first published in Ellenville then reprinted in New Paltz in 1906:

> Mrs. John Voorhis entertained us recently with an account of the abundance of wolves hereabouts when her father [John] was first married and settled here. She said he had often told her when it was time for him to come home when he had been away with his team of oxen that her mother would meet him on a hill some distance from the house with a lamp and walk beside the cattle [the oxen] with it to keep the wolves away from them. There were so many and they were so fierce

DEADLY ATTACK OF A WOLF UPON A MAN, AND HEROIC CONDUCT OF THE MAN'S WIFE.

Wolves and other predators were constant threats for early Trapps residents. The engraving, "Deadly Attack of a wolf upon a man, and heroic conduct of the man's wife," illustrated a page-one story for Frank Leslie's Boy's & Girl's Weekly, *March 2, 1867. Courtesy of the Library of Congress, LC-DIG-ppmsca-02883.*

with hunger that even then it was difficult, and she said they were compelled to keep fires running at night to frighten them away from the cattle or they would have been devoured by the ferocious animals.

During this period, new families named Fowler, Harp, and Van Leuven arrived in the Trapps. Gilbert Fowler was followed by Samuel Fowler, perhaps a son. Henry Harp, perhaps a son or nephew of the elder Henry, also appeared. By 1820, James Van Leuven was farming on his own with his parents, David and Elizabeth, nearby. James's place "was located near the foot of Millbrook Mountain ... far back in the mountains ... [where] a spring of fresh water was found," noted Ulster County historian Kenneth E. Hasbrouck. Within another ten years, James's younger brother Henry O. started his own place near the Enderly mill site. Henry O. married Ann Eliza Burger, daughter of Ben and Lena.[8]

With families came services. While taverns served spirits, the human spirit needed nourishment from church and school. In January 1807, mountain families could at last worship more conveniently at a Reformed Dutch Church built in the lower Clove rather than having to cross rivers to attend church in the surrounding valleys.[9] The earliest public schools had operated in the Town of Rochester in the 1790s. The first to open near the Trapps was in 1798 in the upper Clove at Pleasant Ridge (later Rock Hill), but attendance, not yet compulsory, was sparse.[10]

In 1812 the United States declared war against Britain, its first such act as an independent nation. As they had during the Revolution, mountain men of Rochester responded to the call of duty. Enlistees from the upper Clove were Samuel T. Jansen and Samuel Fowler, and from the Trapps possibly David Van Leuven. The three returned safely home after the war. By 1820 the two Sams had married and were raising families. Sam Jansen wed neighbor Tryphena Aldrich, daughter of Trapps pioneers Peter and Hannah. "The War of 1812 served to strengthen the Union and a feeling of nationalism and to open up the West," wrote Arthur G. Adams.[11] Veterans returned home on a tide of renewed nationalism. A period of growth, progress, and expansion transformed the young nation. Steam and coal power fueled industry. New railroads and better roads, called turnpikes, improved land transportation, with the building of canals the greatest catalyst for opening up the interior.

In 1825 the "big ditch" of New York State Governor DeWitt Clinton achieved the seemingly impossible. When the Erie Canal linked the Atlantic Ocean and Great Lakes, "canal fever" struck America. In 1828 a new, 108-mile-long waterway of Swiss brothers Maurice and William Wurts joined the coal mines of northeastern Pennsylvania with the Hudson River near Kingston. Their Delaware and Hudson Canal (D&H) ran north through the Rondout Valley, bringing new jobs, new towns (many called "ports"), and new economic bounty to the region. With products and people moving faster and easier than ever, benefits reached well beyond the tow paths up into the Catskills and Shawangunks. The D&H gave work to local men, including those of the

Trapps, and became a lifeline for exporting local commodities, including two critical for sustaining Trapps families during the 1800s—millstones and barrel hoops.[12]

In the decades following the Revolution, pioneering families found sustenance in the Shawangunks' stony landscape. Similarly to the native peoples who had once lived in the surrounding valleys, the Trapps people planted crops, hunted game, and gathered nuts and berries. Dissimilarly, they built permanent homes and milled lumber. They burned charcoal and cut stone. The ridges echoed with new sounds, of axes cracking against wood, of black powder busting out stone.

Their children grew up as neighbors and married each other. With each marriage, family trees became increasingly intertwined, due in large part to the isolation of the Trapps. Mountain families lived closely together through kinship, church, and business, a closeness that remained well into the 1900s.

The Trapps people bought smaller lots carved from the old Grote Transport and Nineteen Partners Tract. During the 1820s, John Enderly added 111 acres to the old Roosa parcel. By 1830, Alex Burger and his brother John, sons of Ben and Lena, had each bought long, narrow parcels, smaller subdivisions of the Grote Transport that ran west over the Shawangunks from the "highest part of the steep rocks." Alex bought a parcel of about 90 acres running down to the Coxing Kill. John paid $200 for about 150 acres stretching over the Dickie Barre ridge to the next stream, the Peters Kill. Families named Aldrich, Burger, Enderly, Fowler, and Van Leuven were transforming a mountain wilderness into mountain farms and homes.[13]

• • • • • •

The early to mid-1800s was a period of great expansion for the young nation. American art and literature emerged, two resplendent butterflies shedding their European chrysalises. Artistic expression fluttered then took wing across the new continent. Writers and artists—Irving, Cooper, Bryant, Cole, and Durand, among others—turned to the grand places in nature as romantic and dramatic settings for their work. They sowed seeds of adventure in the minds of a previously unknown entity, upper-class tourists, who were keen to be awed by the grandeur of nature. Traveling in style and comfort, with trunks and servants in tow, they journeyed north to the Catskill Mountain House, a hotel perched atop an escarpment overlooking the Hudson River, one of the must-see destinations of a Grand American Tour that also included Niagara Falls, the Erie Canal, and the White Mountains. Unbeknownst to those sailing the Hudson River between New York and Albany, however, off to the west in a massive rampart of white cliffs, a torturous dirt wagon road climbed up through a deep gap. There, a cluster of homesteads was becoming a new community called the Trapps.

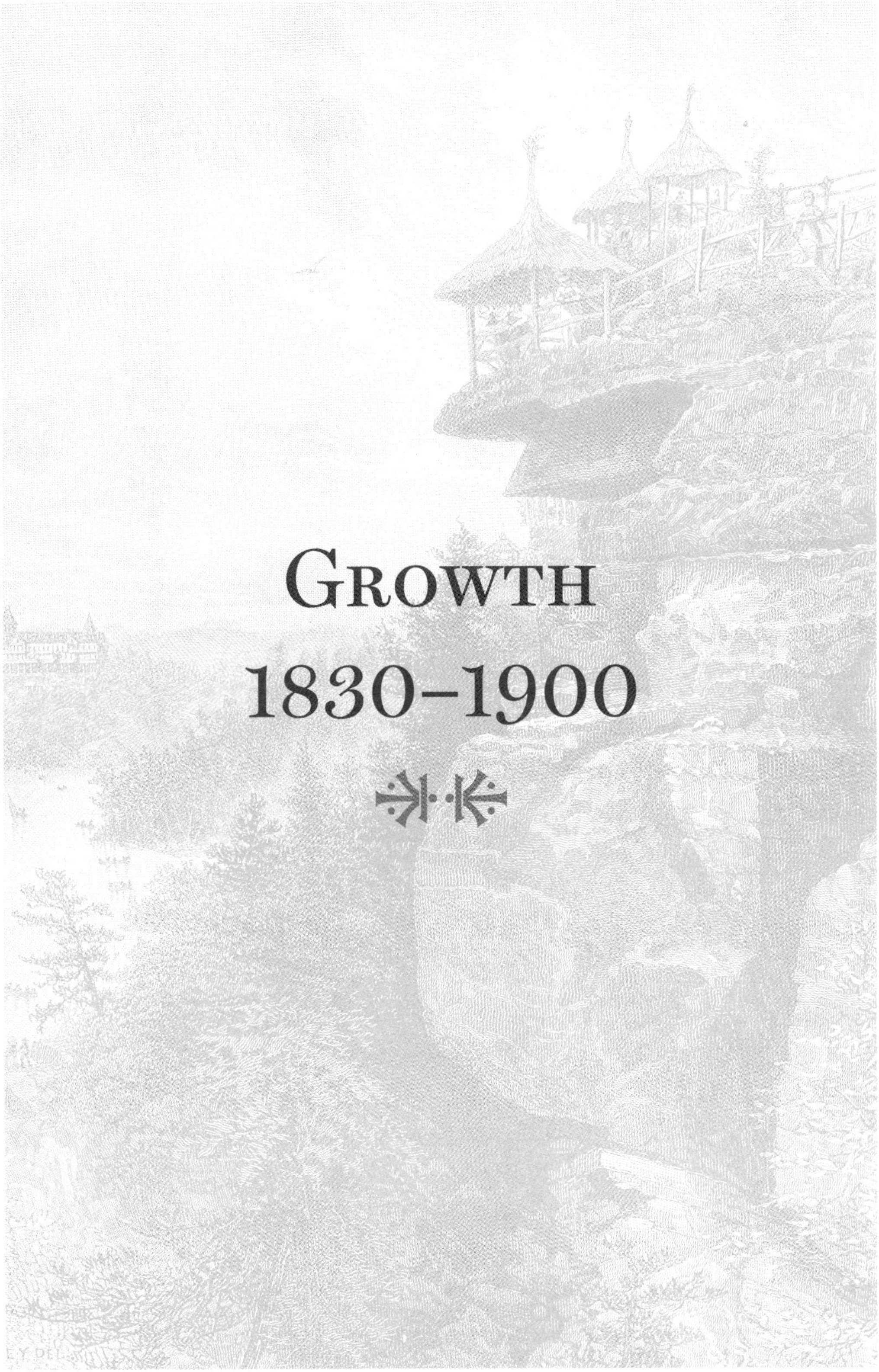

Growth 1830–1900

4

Hearth and Home

Spring, 1868. Thom Lyons scoops up his wife, Lucy, and their baby son Oscar, and whisks them over the threshold of their new home.

"Put me down, you Irish rogue!" Lucy shouts. "You'll muss my dress!"

"Muss, fuss, me sweet colleen!" Thom sings out. "Here's ta us 'n us 'n us!"

Thom spins them 'round 'n 'round 'til all'r dizzy and Oscar squirms.

"Okay, me boyo, yer da is off ta work," he says putting the boy down.

There are stumps to haul out and crops to put in. Thom smiles as broad as the sky, content with his new life and family in America up on Trapp's Mountain.[1]

Little is known about Thomas Lyons, the first European immigrant to settle in the Trapps. He journeyed from Ireland sometime about 1841, probably entering America through New York harbor. Thom found his way to the Trapps, where during the 1860s he was a farm laborer, living on the farms of James Van Leuven near Millbrook Mountain and of George Davis, who ran the sawmill on the Peters Kill.

By the early 1870s, Thom was a naturalized citizen, headed his own household, was married and raising a family, and earned enough to lease twenty-five acres for farming. He likely leased from George Davis, given their relationship and George's stature as the largest landowner in the Trapps.[2] Thom married neighbor Lucinda Burger, who had one son, John B. with her first husband, neighbor Alexander Wynkoop. Thom and Lucy had two sons, Oscar, born in 1867, and Charles B., known as "Charlie," born in 1879.[3]

Yet, from these scant facts emerges the picture of an immigrant who left the old sod for America, found work and a family on an Ulster County ridgetop, laboring first for others but then for himself and his own. The Lyons family's gay entrance into their new home is, like other chapter vignettes, a construction of the mind. Constructing hearth and home, however, was a very real undertaking for settlers.

Three types of homes were constructed in the Trapps: log was the earliest; plank, the most common; and frame, the most modern. It may seem curious that no stone houses remain in the Trapps given their abundance in the valleys. On the mountain, stone was found for the taking, but lumber was easier and cheaper to use.

Philip H. Smith described characteristics of log cabins built by mountain pioneers—bark roofs, paper windows glazed with hog's lard, and jambless fireplaces "into which a back-log ten feet in length could be rolled." A feature of many Dutch homes was the jambless or open fireplace without sides. Floors were dirt; ones of wood were expensive and therefore rare. In 1855 several Trapps families lived in log cabins—those of James Van Leuven, Alex Burger, and Ann Burger, widow of Alex's brother, John. Mountain pioneers "rolled up" their log cabins with the help of family and neighbors. They cut down tall straight chestnuts, resistant to rot, and other mixed timber and dragged the logs to the site by horses or oxen. They cleared away brush and rocks and perhaps put down a layer of stone for a foundation (or perhaps not). They notched the ends of each log so that each would nest into the one above and below.

Walls grew log by log. The men raised a pitched roof of bark or rough-cut shakes. They cut out a door and a window, then added a gabled attic space for storage and sleeping.

This log cabin in the Trapps was one of two built in the mid-nineteenth century that stood into the twentieth century. Rough-cut boards covered the logs. Courtesy of Daniel Smiley Research Center, Mohonk Preserve (DSRC-MP). Donated by descendant Clarence Kidney.

They chinked the logs with wattle, made of mud or clay and sticks, to close gaps against the wind and rain. They erected a chimney, with heat provided by iron stove or fireplace. Window shutters and a door were fashioned and hinged in place. A ladder nailed to the wall gave access to the attic. The one-room cabin was ready.

No log cabins remain in the Trapps. Several scenarios likely doomed them: they rotted if built without foundations or cellars; they were abandoned and left to disintegrate; or they were dismantled to build a better house.[4] Mountain families built better homes of plank or frame as they could afford to. A plank house was an earlier version of today's prefabricated house, but without completed rooms. A more accurate term for them is "plank-and-beam" house.

Plank slabs, cut at a Trapps sawmill, were the bark-covered outer sections left over after squaring a log. These sawmill discards were once a common and cheap material for roofing and siding. Rougher cut than milled boards, planks had to be thick enough to support the house on their own. Ralph Le Fevre, owner of *The New Paltz Independent*, published an account of an 1878 excursion to Sullivan County. After passing through the Trapps, his party began their descent: "Down the mountain we go ... past a few cabins roofed with slabs, and sided up and down with rough boards."

Before a plank house was built, a cellar might be dug and a dry stone foundation, without mortar, put in. Next came a chimney. Then, two hand-hewn beams were set horizontally on the ground about six feet apart with the planks nailed vertically to the beams to form a wall. (The bottom beam was the sill; the top one, the header.) The four walls were raised over the foundation, their ends simply nailed together. On the narrow sides of the house, the planks were cut longer, extending to the peak to support the roof.

Once the walls were up, boards called floor joists were laid down as a base for the first floor and attic. A square opening was cut in the first-floor ceiling, and a ladder hung for access. Next, small trees were cut, the bark left on for rafters. The ends were butted against each other to form a peaked frame for the roof. As was the custom, the rafters were not nailed to the ridge pole. Thick planks were set vertically over the rafters and the whole was covered with cedar shakes. A space was left between the planks to let the shakes breathe. Larger spaces for a door and windows on the first floor were also cut.

Thick horizontal battens might be nailed inside or out over the cracks between the planks to keep out the worst drafts. Down in the valleys, where style counted, formal board-and-batten exteriors became the rage during the American Gothic Revival of the 1800s. Yet up in the Trapps, where practicality ruled, battens were rough cut like planks.

Inside walls might be coated with whitewash, a mixture of water and readily available substances such as lime, milk, flour, or eggs; hide glue made the whitewash adhere better. The house then was ready for the family to move in. Needs were few—a stove, table, chairs, dishware, pots, and cutlery. Wall pegs to hang items such as garments. Rifles, tools, axes, and knives. Beds or pallets, a cradle, a trunk for clothes, linens, and bedding. Mattresses stuffed with poultry feathers or pine needles. Quilts woven or patched together from old scraps. Sewing notions. The family Bible.

The "first dwelling houses" of early European settlements in Ulster County were "typically bark-lined dugouts and rough, wood-framed structures covered with crude boards," noted Marc B. Fried. The latter sounds similar to plank homes in the Trapps, where most families never acquired enough wealth to build fancier frame ones.[5]

Rough, vertical planks became the common building material in the Trapps. Thick horizontal battens overlaid the planks to keep out drafts. This Trapps home was abandoned and torn down about 2010. Photograph by Robi Josephson.

A plank house differs from a framed "post and beam" house in one key way—the former has only horizontal beams, but the latter adds vertical studs and corner posts. In 1865 the hamlet's three wealthiest men lived in frame houses—George Davis, sawmill owner, Benjamin Fowler, tavern keeper, and John Shaffer, hotel owner. In 1855, John Burger's widow, Ann, lived in a log cabin; ten years later she lived in a frame house, perhaps built for her by their sons, while John's brother, Alex, remained in a log cabin.

Life in these early homes, whether log, plank, or frame, was no picnic. The structures were cold and drafty with little or no insulation. The Junior League of Kingston noted one pre-1850 farmhouse, "located in the 'old Clove' settlement" west of Lake Mohonk and extensively remodeled, had "walls laid up with mud and straw." Natural materials,

This abandoned well once provided fresh water for a Trapps home and farm. Photograph by Robi Josephson.

including cloth and newspaper, were used for insulation as they were handy and cheap.[6]

The Trapps people continued to build log and plank homes well into the 1800s, after frame homes became more common elsewhere, reflecting the poverty of these families. Yet some survived under even harsher circumstances. In 1892, M. H. Pike wrote a three-part series about Shawangunk flora for *Garden and Forest*, in which he noted "squatters hidden among the rocks." These residents, along with the small farmers, "make a poor living by the harvest of Blueberries and other kindred fruit."[7]

Regardless of abode, water was essential for survival. People hauled water from streams, springs, wells, or house cisterns. Even during the coldest months, stream water can flow quickly over rocks or collect in deep pools that never completely freeze. At times, though, ice might need to be broken or snow melted for water. Families also used "spring runs," overflows from underground sources. Some springs acquired names, such as Hugo's Spring named for Trapps resident Hugo Slater.

A time-honored tradition was to offer water to any traveler and his horse, but in 1874 Catharine "Laney" Fowler refused "a pail of water" when Kaiser, the New Paltz village barber, was in the Trapps and needed water for his horse. "She wanted the water to wash with," stated the village paper. Such a refusal from the owner of a public house—or from anyone in the Trapps hamlet—created gossip for the local papers.

Wells reached water running along rock ledges or seams below the surface. The surface-to-bedrock depth is shallow in the Shawangunks, a geologic condition allowing relatively easy

Wooden yokes helped with the chore of hauling water and maple syrup. Yokes were sized for adults or children, yet this chore was often a child's job. Illustration by, and courtesy of, Jack Fagan.

access to water. Enderly descendant Arthur Hayes noted that his family well "was dug by hand, stone lined the sides and a chain on a pulley was used to pull up the pail of water." Wells had to be cleaned out periodically. Joe Otis recalled his father, Ray, climbing down into the stone well on their Trapps farm and using the stones as steps. While simply constructed, some Trapps wells remain intact to this day.

Children were given the chore of lugging water back to the house, hauling buckets by hand or perhaps hanging buckets from wooden neck yokes. Trapps descendant Evelyn Coddington DeWitt grew up in the early 1900s at Lawrence Hill, a sister hamlet to the west. Evelyn said it was her job to haul water from the spring each morning. Children were assigned daily chores to help keep the farm and household going. Major chores were assigned to days of the week. Monday was typically wash day, and Tuesday was ironing day. Saturday was often bath night. Whether the Trapps people followed this ritual or not, they likely washed and

This old sanitary facility, or outhouse, had three seats to accommodate a large mountain family. Photograph by Robi Josephson. Courtesy of Bob Larsen and Barbara Rubin.

bathed in large metal tubs or in local streams and ponds depending on the weather and season.

Waste water was dumped right on the ground. Families used outhouses, called privies, for human waste. What little garbage there was, such as broken stone- or dishware, might be thrown into the privy hole or a separate trash heap. Recycling was a way of life—food scraps to the pigs, broken tools or scraps of iron to the blacksmith. Old clothes were reworked into rugs or quilts or ripped for toilet paper, rags, or insulation. (That's why buttons can be found at former privy sites.)

Privies, erected near the house, provided minimal privacy. Before toilet paper, people used whatever was available—leaves, mosses, corn cobs, rags, books, paper, and those perennial favorites, the *Farmers' Almanac* (1818), with convenient nail hole, and the Sears, Roebuck catalog (1888). Chamber pots were kept under beds for use overnight, especially in bad weather. Summer or winter, chamber pots were emptied each morning, and children often had the chore of emptying the "night water."[8]

The Trapps people used wood stoves for heating and cooking. Some families "ran two stoves" at their mountain homes, as did the Hayes and Coddington families, Trapps descendants who settled nearby. The Hayes house had a kitchen, living room, two bedrooms, and attic, all heated by a wood stove in the kitchen and a potbelly stove in the living room. "We had no electricity," Arthur Hayes wrote. "We used oil lamps to see. My mother would iron and do practically everything in the kitchen. She would take a steel iron and put it on the wood stove to heat it up, then iron for a bit, then put it on the stove again to heat up."

Life then was messy and dirty without hot running water, modern appliances, or cleansers. Many women alive today remember the long days needed to keep a house and to feed the family hot, home-cooked meals three times a day. The main occupation for women at the time was "keeping house." Yet this role was equal to the husband's work outside the home. A wife was responsible for the needs of her family—for feeding them, clothing them, caring for them, while also tending livestock, crops, or the kitchen garden. With large families, "keeping house" was an arduous job even with children pitching in. The saying "a woman's work is never done" was too true. There was no such thing as free

Crocks once were essential storage containers in American homes. These crocks were produced in Ellenville, a major center of stoneware production in New York State during the 1800s and the closest such center for Trapps families. Photograph by Ted Reiss. Courtesy of Barbara Rubin.

time or taking a vacation. Many a woman only "found her first real rest in the arms of death," wrote Arthur O. Friel in a 1924 short story set in the Trapps.

Despite rugged conditions, the mountain gave people virtually everything they needed to live. Wood and stone were plentiful for building, as was wood for heating and cooking. Water ran above and below ground. Nearly all homes in the Trapps are long gone, yet their short service was not in vain. While rough-hewn and transient, they provided shelter for several generations. Some foundations still stand, their stones even now holding back the earth. Today, we can peer down into these cellar holes, imagine the homes built over them, and reflect on the families who lived there. We can hear streams gurgling nearby and discern, to quote the Bard, "books in running brooks."[9]

• 5 •

Artists and Builders

September 1875. George Davis hobbles into his Trapps home.

"Sarah! Alfred Smiley wants to buy land, including parcels around Coxen Pond! I'll sell him some of ours and more if I can get the Deckers, Jim Van Leuven, and the others to sell. He plans to build a road and another hotel like at Mohonk."

"Those are some plans, George. Be careful."

"With the sawmill gone, Sarah, it's a good deal for us. What's bothering you?"

"Your leg. It worries me. You'll be on your feet day and night."

"That sure smells good. When will that pie be ready?"

"When you promise to be careful."

"Okay. Okay. I promise," George said, his mouth watering.[1]

When George Davis first ventured up the mountain with his family in the mid-1850s, the Trapps and nearby mountain hamlets bustled with human endeavor. Mills hummed up and down the Coxing Kill and lower Peters Kill. Families labored on their farms and woodlots. Three decades later, historian Philip H. Smith intrigued readers with the secluded place he had found: "There is a singular and romantic formation on the top of the Shawangunk mountain known as The Traps [a frequent alternate spelling]. Quite a village has sprung up within its sheltering bosom, and boasts of a hotel, store and chapel."

From the 1850s to the 1880s, progress marched into the Trapps, linking it as never before with the outside world. A graded, dirt wagon road called a turnpike was built over the mountain through the hamlet. New business and steadier work came. An immense mill rose, and small hotels opened. Philip H. Smith visited a high hollow echoing with the energies of some forty to fifty families eking what they could from the mountain and working at the expanding mountain hotels nearby at Lake Mohonk and Lake Minnewaska.

This postcard looks south from high above Lake Mohonk to "The Traps" cliffs and cleared fields of the Trapps hamlet. This lofty scene became one of the most popular Shawangunk Mountain views painted by nineteenth-century landscape artists. Courtesy of the Mohonk Mountain House Archives (MMHA).

While the Trapps people bore a practical vision of life, others arrived in the Shawangunks bearing a far different one. These visitors came not with axe and hammer but with brush and paint, pen and ink, canvas and notebook. Their landscapes, essays, and poems portrayed nature in a new light. At their hands, a gnarled tree, erstwhile ignored or cleared away, might become a picturesque *objet d'art*. A woodland scene, its trees still measured for board feet and cord wood, might become an intimate portrait. Nature, heretofore exploited, might be admired and revered. Wildlife that had been fair game for the table and trophy wall might be coveted for lines artistic and poetic. Nature's minutiae—its flowers, insects, and grasses meticulously copied—soon would appear in popular new books called "field guides." These words and images became powerful voices and advertisements for an emerging conservation movement.

Landscape painters explored hill and dale, seeking their own private Shangri-las, gravitating to the ethereal peaks of New York, New England, and the Western territories. Among the most acclaimed to search for wilderness up in the Shawangunks were Thomas Cole and Asher B. Durand. In August 1838, Cole wrote to his friend: "Do let me hear from you when you get among the Gunks. I hope you will find everything there your heart can wish." Cole wended his way up into the range, probably more than once. Packing sketch pads and pencils, Cole arrived in August 1846 with a pupil and two friends, one of whom was likely his biographer, Louis Legrand Noble.

Noble wrote that the party spent two days at a "remarkable lake, which sleeps in a high fissure of the mountain ... [with] ... singularly wild shores of perpendicular precipices and huge crags." They stayed at a small mountain inn nearby, their host "a short-spoken Dutchman" who smoked "a rank old pipe" and unnerved his guests with snake stories. Cole made "several choice studies" and "a half finished picture, exquisitely beautiful in its sky" of Lake Mohonk. The original of Thomas Cole's half-finished painting survives today in private hands; the inn's location remains a mystery.[2]

While mountain ranges such as the Shawangunks became short-term oases for artists and writers, some families pulled up their valley stakes for the long haul and moved onto higher ground. In 1850, George Davis of Marbletown was a farmer, butcher, and family man with Rondout Valley property valued at a very respectable $5,000. Yet, George must have found the Trapps to his liking. Good lumber was there for the cutting. Why not set up a sawmill on the Peters Kill like the one Enderly operated on the Coxing Kill? George found just the spot below a roaring waterfall. By 1856 he had established his mill and a home along the main wagon road. George's was no run-of-the-mill operation; it was the most massive structure built in the Trapps hamlet. William D. Jones's 1856 survey map of the New Paltz and Wawarsing Turnpike confirmed George's sawmill along the "Peters Creek" and his house nearby. Sawyer Davis was already busy filling orders when Will lugged his equipment up the mountain to survey the route of a proposed turnpike.[3]

A turnpike was an unpaved, graded toll road built to connect rural areas with city markets. Peter H. Harp, a descendant of Henry Harp, explained: "Turnpikes were owned and operated by individuals who financed their venture through the sale of stock with the approval of the State." The New Paltz Turnpike, charted in 1807, was built about 1830 to join that growing Huguenot town with the Hudson River. This toll road "proved a great blessing to the farmers of the Wallkill Valley," who then sent their grains, dairy, and meats down the pike for delivery to New York, wrote Ralph Le Fevre.

Local investors had wanted to cash in on the success of the 1828 D&H Canal, Harp noted. In 1834 the New York State Legislature passed an act incorporating a turnpike company joining Ulster and Sullivan Counties, but things did not take off until the mid-1850s. Impressive volumes of goods shipped from the Rondout Valley and Catskills enticed investors. Harp wrote that "the Ulster Tannery at Lackawack processed 225 tons of hides a year; 14 saw mills produced 20,000,000 board feet of lumber a year; the Napanoch Ax Factory made 9,910 axes in the month of January; the Ellenville Glass Factory produced 80,000 units per year, and the Napanoch Iron Co. produced 800 tons of iron per year with coal, 350 tons with charcoal and 500 tons of Salisbury Iron."

Based on such promising figures, the New Paltz and Wawarsing Turnpike Road Company was born. Construction costs were figured at $1,000 per mile, and the shortest route was calculated from New Paltz through the Trapps to Wawarsing. Will Jones's survey route, partially hopscotching the older dirt wagon road, was approved, and the turnpike began inching its way west over the mountain. Ralph Le Fevre of New Paltz recalled the adventure of traveling at about age fourteen through the Trapps on the new turnpike. It was a big event for many, but most of all for the Trapps people.

A tollgate was set up at the gap in the cliffs facing New Paltz, where a stile blocked the road with a pole or pike. Once the toll was paid, the pike was turned or lifted to allow passage. Tolls varied with type and size of vehicles, and number and type of animals. An 1834 act specified 12 1/2¢ for wagons, carts, and wheeled carriages drawn by two horses, mules, or oxen; 3¢ more for each additional animal; 6¢ for wheeled vehicles drawn by one animal; 6¢ for sleighs or sleds drawn by two animals; 3¢ more for each additional animal; 20¢ for every twenty horses, mules, or cattle; 6¢ for every twenty sheep or swine; and 4¢ for every horse and rider or led horse. Empty vehicles might pay half. Walkers and funeral processions passed free.

The New Paltz and Wawarsing Turnpike Company hired Trapps man Hiram Van Leuven to be "gate keeper" collecting tolls at the gap. In 1860, Hiram and his wife, cousin Anna Maria (pronounced Mariah) Van Leuven, in their early twenties, were raising two young children and owned property worth $200. Hiram was luckier than many, for he had landed one of the first steady paying jobs on the mountain.

The state regulated the construction and operation of turnpikes. Frederic J. Wood noted that in Massachusetts "it was directed that the turnpike should be built in as straight a line as possible, and this was nearly always done with unfortunate results, as the resultant location led up and down hill regardless of grades, and disregarding centers of population, usually rendered the road of little practical use."

Truer words were never spoken. Business over the mountain never materialized. The turnpike was saddled with taxes while the nearby D&H Canal was not. "The old company couldn't make it pay," *The Kingston Democratic Journal* reported in August 1860. "The road was sold under the hammer, and Wm. DeGarmo, of New Paltz, bought it for a few hundred dollars." Perhaps DeGarmo, an original investor, intended to reorganize, but the turnpike stayed public, young Hiram Van Leuven was out of a job, and the D&H remained transportation king of the region for much of the century.[4]

Elsewhere along the Shawangunks, new roads crossing the range had greater success. South of the Trapps, the Newburgh-Cochecton Turnpike opened in 1809, followed in 1851 by the toll Ellenville and Newburgh Plank Road.[5] North of the Trapps, the section of the New Paltz Turnpike over the Shawangunks to the Rondout Creek was never built, but a road existed over that part of the mountain. The grade there was so steep that drivers needed to rest their horses at the crest of the ridge. Thus began a new summit neighborhood aptly called Mountain Rest, where a small hotel was operating by the early 1850s. To the south in the Trapps, the Aldrich Tavern was followed by two small hotels run by the Fowlers and Shaffers.[6]

Ben Fowler and his wife, Catharine "Laney" Gorsline, had made a home in the Trapps next door to pioneers Peter and Hannah Aldrich. Ben was the son of War of 1812 veteran Sam Fowler; Laney was the daughter of Sam Gorsline. Peter and Hannah died in 1855 in their early eighties, and the Fowlers, both in their mid-twenties, saw an opportunity in the Aldrich land, once part of Lot 3 of the old Grote Transport. The Aldrich children accepted the Fowlers' offer for the 99-acre parcel, and on December 29, 1859, Ben Fowler became its new owner. In 1862, Congress enacted the first federal income tax to help finance mounting expenses for the Civil War. Ben began paying taxes, $5 for keeping a hotel and $20 for retailing liquor.

In November 1854, John C. and Lydia Shaffer arrived in the Trapps from Greene County. The family settled along the old dirt wagon road west of the Peters Kill. John worked as a laborer, but then watched the new turnpike go through with enough traffic to support a second hotel. By 1861 he was open for business. In February, John catered an oyster supper for about 150 people, and in April 1862 he was reported "busily engaged in enlarging and painting his hotel at 'The Traps' with a view of accommodating summer boarders." John also paid the new federal taxes for his hotel and liquor license, yet something changed. Whether trade fell off or mountain life was too harsh, in February 1866 John auctioned everything off, including dairy fixtures, and moved down to Springtown northwest of New Paltz, where he bought another hotel.

Ben and John's neighbor, George Davis, was busy farming, running his mill, and buying up property. In 1860 he owned real estate worth $3,500; by 1865 it was up to $5,000. He had amassed 2,000 acres, or about 3.12 square miles, with 60 acres improved as crops, pasture, and meadow, and 1,940 acres in unimproved woodlots. The largest landowner and wealthiest businessman in the Trapps raised Indian corn, oats, buckwheat, rye, and potatoes, while keeping three cows, three pigs, and some chickens. In the 1860s, George was taxed $15 as a peddler and $1.20 for three cows. His woodlots supplied a seemingly endless supply of trees for his business. By then there was an added benefit—the turnpike was public, and his valley clients no longer had to pay tolls.[7]

Meanwhile, artists continued to journey up into the Shawangunks. Jasper F. Cropsey, Jervis McEntee, Sanford R. Gifford, and T. Worthington Whittredge were among those who followed Cole and Durand to seek out Lake Mohonk and environs. Promontories above the lake, soon called Sky Top and Eagle Cliff, attracted these artists for their dramatic and accessible vantage points. During an 1864 visit, Gifford made several studies or sketches of the range, including "View of the Trapps, Shawangunk Mountains." Back in his New York City studio, Gifford painted *The Shawangunk Mountains, 1864*, which he considered "one of 'his chief pictures,'" noted Avery and Kelly. In 1865 he displayed this 30-by-48-inch

Albert K. Smiley, right, and Alfred H. Smiley established the Shawangunk Mountain hotels at Lake Mohonk and Lake Minnewaska, respectively, after the Civil War. These resorts provided year-round work for Trapps families who lived near each lake. Courtesy of MMHA.

landscape at the National Academy in New York, where viewers could admire this striking range north of the city and peer at wilderness farms deep within its forests.

Artists coming to the Shawangunks before 1860 helped set the stage for a new endeavor at Lake Mohonk. John Festus Stokes, a farmer and wagon maker by trade, lived west of Mohonk in the Clove. He bought land at that summit lake in the late 1850s and first opened the place for picnics. In July 1860, John was constructing the first wing of a hotel with ten rooms, similar to the modest beginnings of the Catskill Mountain House he wished to emulate. "The Mountain House at Lake Mohonk or Paltz Point" quickly became a popular spot, with day parties and boarders of that first decade coming from as far away as Pennsylvania and Long Island. Visitors were attracted to the lake's picturesque setting and John's hotel with tavern and dance hall. That August, *The Ellenville Journal* stated that Sam's Point, another Shawangunk Mountain promontory to the south, was as popular with locals as Central Park was to New York City dwellers. *The Journal* also reported that the names of 300 area residents "had been put on the register" atop Sam's Point on one day that month alone. By the time Gifford painted "his chief picture" of the Shawangunks in 1864, the range from Lake Mohonk to Sam's Point was already a tourist destination.[8]

Meanwhile, providence—both the New England city and that mystifying principle called "destiny"—played a role in bringing Quaker twins Albert K. and Alfred H. Smiley to the mountain. The brothers' arrival, in turn, helped change the destiny of the Trapps people. In 1869, Albert was principal of Friends' Boarding School, a Quaker college preparatory school in Providence, Rhode Island. The year before, Alfred had left his administrative position there to operate a farm in Poughkeepsie that both brothers owned. One fall day, Alfred ventured up to Lake Mohonk and the rest, as they say, is history. John F. Stokes sold just over 300 acres, including the lake, to Albert for about $28,000, with the twins quickly expanding the modest hotel for their first spring opening.[9]

The Smileys' timing was perfect. The iron horse was transforming the nation, and in 1868 a new railroad began inching north through the Wallkill Valley's Orange and Ulster Counties. In May 1869, only months before the Smileys arrived at Mohonk, the Transcontinental Railroad forever linked eastern and western America. In December 1870, the Wallkill Valley Railroad reached New Paltz to the cheers of its citizens and the delight of Mohonk's new proprietors and their guests, whose journey from places such as New York and Philadelphia was made far shorter and easier.[10]

Alfred Smiley soon became the full-time manager at Mohonk and took to exploring his new surroundings. Picnics with his family were a particular pleasure. In the summer of 1875, his curiosity took him south through the Trapps to Lake Minnewaska, then

The Wallkill Valley Railroad reached New Paltz in December 1870. The train depot there was a travel hub for guests of Lake Mohonk and Lake Minnewaska. Courtesy of MMHA.

called Coxen Pond. As Lake Mohonk had six years earlier, the beauty of this second lake impressed Alfred, and that fall he purchased several parcels around the lake from various owners. Alfred also met Trapps resident George Davis, another owner of lake property, and proposed a business plan to him.

If George could persuade the owners of the remaining lake parcels to sell, Alfred would buy the whole from George acting as his agent. George's mill on the Peters Kill had burned down one year earlier. He had no insurance, nor rebuilt as far as we know, and might have seen Alfred's proposal as a very welcome financial opportunity. On Alfred's behalf, George negotiated with the owners, including three from the Trapps—brothers Samuel and Alexander Decker and James Van Leuven, who lived out Millbrook Mountain way.

On October 3, Alfred wrote to Albert: "I went yesterday to see Mr. Davis & look at land. He was too lame to walk much but went to the Lake with us—the first time he had seen the Lake in 14 yrs. & the second time he had ever seen it though he has owned it 18 yrs!!" It may seem astounding to us today, as it did to Alfred, that George had been to the lake only once before in so many years. Yet, life then and there demanded a practical outlook. A lake with such natural and wild beauty was largely useless for generating cash income, or perhaps George was too busy at his sawmill, while it operated, to enjoy the lake. (It is unknown how or when George had become disabled.)

On October 22, 1875, George's wife, Sarah C. Davis, and Severyn Hasbrouck, a Marbletown merchant, sold the largest parcel (of unknown acreage) worth $2,000 to Alfred Smiley. As agreed, the parcel was composed of lots combined by George Davis. Alfred also bought land along the old turnpike, from where he planned a road up to the lake. The new road started near the former Shaffer Hotel, bought by John H. McGibney after Shaffer sold out. Born in New York City, John had made his way north to the Trapps and in 1865, eighteen-year-old John "Magivny," as listed on the census, boarded at the home of Henry O. Van Leuven. Young John worked as a boatman on the D&H Canal, possibly getting the job through Henry O.'s neighbor, Charles Aldrich, thirty-five, also a canal boater. By the early 1870s, John farmed twenty acres and still worked for the canal, noted Hamilton Child.

John McGibney met Martha Wynkoop, who came from a large, orphaned Trapps family. Her mother, Maria, had died in the 1850s; her father, Joshua, had hanged himself, beset over "family troubles" in 1866. In 1870, Martha, eighteen, and an older sister worked as domestics for another Trapps woman with four daughters. Upon marrying in 1872, John and Martha became hosts of a popular mountain spot. "The ballroom was quite large and it is said that ten sets might be formed for dancing," wrote Kenneth E. Hasbrouck. The McGibney Hotel was in full swing while Alfred Smiley acquired Lake Minnewaska.

As Alfred did so, he began work on a new entrance road from the turnpike to the lake. Alfred brought workers and horse teams from Mohonk, also hiring George Davis's three sons aged sixteen to twenty-four years. While driving back and forth between Lakes Mohonk and Minnewaska, Alfred saw the need for another carriage road to connect the two lakes and began laying out possible routes. To build the road, Alfred needed rights-of-way, including one crossing the Enderly farm and mill, which by this time was likely operated by a grandson of Johannes. Alfred spoke to the mill owner and on November 14 wrote to Albert that Mr. Enderly "had no objection to such a road."

As significant as Alfred's purchase of the new lake was the appearance of its new name, "Lake Minnewaska," in local papers as early as October 1875. While researching old deeds up in Kingston, Alfred had seen the name Minnewaska associated with Coxen Pond, wrote author Nan Knickerbocker. Alfred may have made the name a public one, but a relative of the Trapps Burgers might have been the bestower. In September 1912, Ralph Le Fevre stated in his New Paltz paper: "Cal Burger told me his uncle gave this lake the name of Minnewaska. He had probably been reading some dime novel and thought Minnewaska a pretty name." Cal's uncle was Wessel Davenport, his mother's brother. Cal had told Ralph that the name Minnewaska was given "in memory of an Indian maiden, bearing that name, who drowned herself in the lake through grief at the faithlessness of her white lover."

With a more romantic name and new roads opening, construction commenced on a mountain hotel at Lake Minnewaska in fall 1877, noted Knickerbocker. With Davis out of business, lumber—and lots of it, more than the Enderly mill could handle—had to be found elsewhere. The Honesdale sawmills in Pennsylvania met that need, with 200,000 feet of lumber shipped via the D&H Canal to Kerhonkson and then dragged up the mountain via the old turnpike. Stone for the foundation was harvested locally using dynamite and cut by hand, wrote Knickerbocker. Sand for plaster came from a bank west of Kerhonkson. Alfred found a built-in labor force of skilled workers right in the Trapps and its sister hamlets, workers who found much-welcome work and steadier income.

To be on hand, Alfred and family moved into the old mill house near the ruined Davis mill on the Peters Kill, noted descendant Alfred B. Smiley. Whether George Davis

The Smiley families of Lake Mohonk and Lake Minnewaska provided telegraph service for the convenience of guests. Railroads and steamboats improved travel for citizens across America, including those headed for the Shawangunks and Catskills. Detail of Currier & Ives lithograph "Progress of the Century," ca. 1876. Courtesy of the Library of Congress, LC-DIG-ppmsca-17563.

rented to Alfred or let him stay there free is unknown. For the first year of construction, Alfred "drove a horse and buggy between Lakes Mohonk and Minnewaska daily," while tending to his duties as manager at Mohonk, author Larry Burgess wrote. In 1878, Alfred and family moved to a completed section of the new mountain hotel. The next spring he oversaw the finishing touches for the grand opening of Minnewaska, even running the first telegraph line to that lake from Mohonk for the convenience of his guests.

Yet, Old Man Winter unleashed one final blast, and with a blizzard came death. George Davis suddenly took ill and died on April 17 "while a furious snowstorm was raging," wrote Knickerbocker. "On the day of the funeral, the snow was so deep in the Trapps Village, where the Davis family lived, that no minister from the valley was able to get to the house, and the head engineer who was in temporary quarters at the unfinished hotel, accompanied by Edward Smiley [Alfred's oldest son], waded down the mountain in snow waist-deep, and conducted the funeral services at Mr. Davis' home."

The New Paltz Independent ran a one-sentence obituary in May 1879 for the Trapps sawyer who had helped broker one of the most critical land sales on the mountain. Sarah Davis laid her husband, aged sixty-nine, to rest in the family burying ground. Had George lived another two months, he would have witnessed Alfred's dream come true, one that he had helped further. On June 14, 1879, a large bell pealed from the tower of Alfred's new mountain hotel, proclaiming that Lake Minnewaska was open for business. Yet, that

Artist E. J. Whitney captured the dramatic landscape of Lake Minnewaska for an 1887 brochure. Wildmere, Alfred Smiley's second of two mountain hotels at Minnewaska, perches in the background. Courtesy of MMHA.

bell broadcast another message to Albert—Mohonk no longer had a manager. To fill the gap left by Alfred at Mohonk, Albert resigned from the Friends' Boarding School after twenty years of service and moved his family to Mohonk. At Albert's urging, the twins' younger half-brother Daniel joined him as Mohonk's new manager in 1881. That same year, the Davis house was "greatly enlarged," presumably by his widow and sons, to take in summer boarders.[11]

The ventures at Lakes Mohonk and Minnewaska set the stage for a rosier future for mountain families of the Trapps and its sister hamlets of the Clove, St. Josen, Lawrence Hill, Rock Hill, and Granite. The mountain's dramatic backdrop inspired the creative energies of the three Smiley brothers. Their visions, in turn, resulted in steady work and

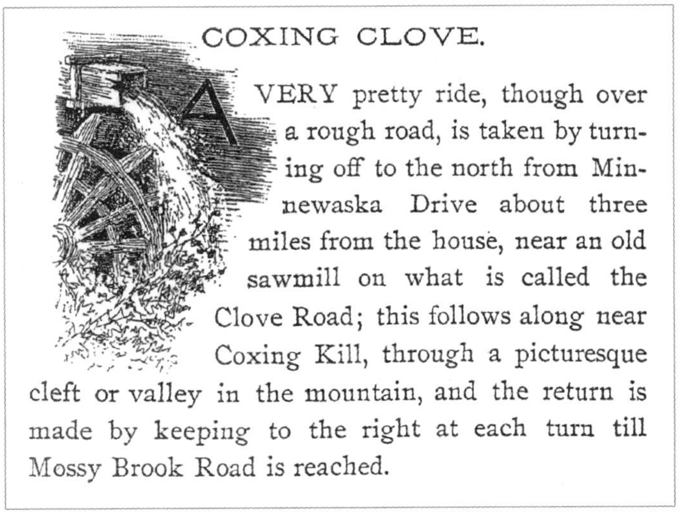

COXING CLOVE.

A VERY pretty ride, though over a rough road, is taken by turning off to the north from Minnewaska Drive about three miles from the house, near an old sawmill on what is called the Clove Road; this follows along near Coxing Kill, through a picturesque cleft or valley in the mountain, and the return is made by keeping to the right at each turn till Mossy Brook Road is reached.

Artist E. J. Whitney illustrated the mill wheel of the Enderly sawmill in the Trapps for an 1888 Mohonk Mountain House guidebook. Courtesy of MMHA.

wages for those dependent on subsistence farming and seasonal harvests. At Mohonk, Albert and Daniel built additions only to tear them down and build new ones. Eight years after Alfred opened Cliff House, his first mountain hotel at Minnewaska, he opened a second one named Wildmere. By that year, 1887, the Wallkill Valley Railroad added the "Mohonk and Minnewaska Express" train to its schedule, and the Shawangunk Mountains were a bright star on New York's upstate tourist horizon.

Land stewardship was an integral part of the Smileys' Quaker upbringing, as well as a broader conservation movement slowly emerging during the nineteenth century. In 1872, two years after the Smiley brothers' first full season at Mohonk, tourists journeying to the American West gaped at the strange, volcanic wonders of Yellowstone, the world's first national park. The Shawangunks, too, had their wonders, but on a more intimate scale.[12]

Once Alfred's mountain hotels at Minnewaska opened, guests enjoyed reciprocal hospitality at the two lakes. Journeying between the two, they rode on Alfred's new carriage road through the Trapps, passing through the Enderly farm. A Mohonk guidebook of 1875 noted that guests could travel "through the wild gorge of the Traps Mountains." Coxen Pond, not yet called Lake Minnewaska, was a secluded destination: "No house, or clearing, or road, or boat even, disturbs its primitive wilderness." An 1888 Mohonk guidebook described the drive through Coxing Clove as "very pretty" but "rough," passing "near an old sawmill on what is called Clove Road; this follows along near Coxing Kill, through a picturesque cleft or valley in the mountain." E. J. Whitney

Guests of Lake Minnewaska strolled to Pine Cliff to view Sky Top to the north. Below were the cleared fields and farms of the Trapps hamlet. The Detroit Publishing Company, detail of photo ca. 1900–1906. Courtesy of the Library of Congress, LC-D4-16853.

illustrated this description with a rough-cut wooden sluiceway and waterwheel still in operation. Thus, the Enderly mill entered the realm of nineteenth-century picturesque. Two pages later, the Trapps Road was portrayed as going "down through the jagged cleft in The Traps Mountain and is considered very interesting."

The 1888 guidebook extolled Eagle Cliff at Mohonk Lake as having what was "often considered the finest single view on the place." Eagle Cliff became so popular with landscape painters that "its western verge" became known as Artists' Rock, noted the same guide. From Eagle Cliff and Sky Top, guests and artists alike took in the sweeping vistas to the south of Minnewaska, of surrounding mountains and valleys, and of spacious skies above. Pine Cliff at Minnewaska provided similar dramatic views looking north to Sky Top and beyond.

Yet, there was much more within this lofty scene. Smoke swirled from farmhouse chimneys, blacksmith forges, and charcoal mounds. Cows lowed, horses neighed, roosters crowed. Horses, called to pull their weight and more, clopped along dirt roads. Ox teams snorted and strained under heavier loads. Black powder busted out rock. Axes cracked against timber. An old up-and-down saw whined and whirred. These sights and sounds spoke of families struggling from predawn to after dark in a mountain hamlet called the Trapps.[13]

• 6 •

School Days and Heydays

October 9, 1856. Kids line up across the dirt road, each steadying a stick on an upright wooden hoop. One calls out, "Winner makes it to the creek! Ready, set, go!"

Boots and bare feet pound the dirt. Hoops and sticks fly. Only the best make it to the finish line. Two rivals only make it a few feet.

"You pushed me!" six-year-old Willie Shaffer cries.

"Did not!" Hiram Slater, also six, yells back.

"Children! Recess is over," Miss Van Leuven calls out. "Right now."

"Watch out after school," Willie whispers, poking his rival.

"You watch out." Hiram gets in the last word for now. Others follow in, racers gasping and tossing their hoops and sticks in the Trapps schoolyard.[1]

School Days

October 9, 1856, was the first day of fall term when Sarah Van Leuven gathered her charges for afternoon instruction. New York State had established the earliest public schools, called common schools, in 1812, but the Trapps population was too small then to warrant its own school. Instead, any Trapps youngsters attending school would have traveled by "shank's mare" (on foot), by horse, or by wagon a few miles north to Rock Hill or, after 1840, to the Clove. By mid-century, however, the Trapps hamlet had grown, and a one-room schoolhouse, designated District School Number 8 of the Town of Gardiner, was built in 1850. Because the Trapps community spanned the mountain regions of Gardiner and neighboring Rochester, children from both towns attended.[2]

On January 1, 1857, the three elected school trustees met to fill out the annual state report. Henry O. Harp, Henry O. Van Leuven, and Hugo Slater filled in the blanks: erected 1850; a frame building in good repair; one room about eighteen feet square; two rows of desks; one blackboard; heated by a stove in winter (not a fireplace or furnace). The next question asked, "How are the rooms supplied with pure air, so important to the progress, comfort, health and life of the children in your district?" The answer stated, "It is well ventilated by windows."

The questions went on regarding the fenced, ¼-acre site.

Have you provided separate privies for the sexes? "None."

Are the school grounds planted with shade trees? "Yes."

Is the school located in a convenient and pleasant part of the district? "Yes."

At what do you value the house and school? "Two hundred dollars."

How long has school been taught during the year starting January 1 and ending December 31? "Six months."

The next section concerned the teacher. Miss Sarah E. Van Leuven was hired in May 1856 to finish out the school year and teach the fall term beginning October 9. She closed school, or her services were dispensed with (which is unknown), on January 2, 1857. Miss Van Leuven was paid $37.00 for her services.

A section asking about a separate "colored" school was left blank. The next asked the amount spent for upkeep, repairs, fuel, furniture, and insurance. None was given, only that funds were "furnished by the inhabitants sending to school." Textbooks were *Cobb's*

Spelling Book & Reader, *Davies Arithmetic*, and *Smith's Geography*. The last question on the annual report asked for suggestions on other matters. The answer: "Our school is situated in a mountainous District which is the cause of our school being so small and it is with difficulty we can keep a school, but a turnpike is going through our neighborhood and will facilitate the way of children getting to school another year."

The annual report was done. Henry O. Harp signed his full name. Below his signature, Henry O. Van Leuven and Hugo Slater each scratched out an "x". Neither knew how to write, but both valued education for the Trapps children, including their own. The handwriting suggests someone other than Henry O. Harp filled out the report. It was likely Schoolma'am Van Leuven.

The Trapps school census of 1856 listed forty-two students of school age between four and twenty-one years. Yet, only twenty children were registered or attended school that year, some for only a few months. Reasons reflected the times. Children were needed at home to help with chores and harvests. Older ones might be employed as domestics or laborers or even be married. New York State passed the first compulsory education laws in the later 1800s, but enforcement was difficult. Survival in remote places, such as the Trapps, was at times more important than book learning.[3]

Two schoolhouses were built in the Trapps, likely at the same location on the turnpike. French's Ulster County map of 1858 placed the school on the Clove road to the northeast, well past the Enderly mill at the Gardiner-Rochester line. Kenneth E. Hasbrouck noted, however, "The first school stood on the same site as the present one and was built before the Civil War." In September 1867 the school trustees paid $1.00 to Ben and Laney Fowler for a .27-acre parcel for "school purposes and no other." Was the school always on the Fowler parcel, or had it been relocated there?[4]

Another question is just as hard to answer. Winslow Homer immortalized the little red schoolhouse in *Snap the Whip*, his 1872 painting of a one-room school to the north in the Town of Hurley.[5] The neighboring Clove School was called "The Little Red Schoolhouse," noted Louise Yeaple. Was the first school in the Trapps red, white, or left bare of paint?[6]

In 1871–72, Gardiner's population was just under 2,000. The sole church was to the east in the Walkill Valley at Guilford with sixty-nine members. There were nine school districts, each with one teacher. Hamilton Child wrote, "The number of children of school age [four to twenty-one] is 813; the number attending school, 465; the average attendance, 198; the value

This one-room schoolhouse was the second one built in the Trapps hamlet ca. 1890–91. It was 620 square feet, nearly twice the size of the first school built in 1850. Courtesy of DSRC-MP.

of school houses and sites, $4,730." A Professor S. R. Shear described conditions of circa 1800 that were to remain virtually unchanged into the next century for many rural districts, including the Trapps:

> About this early period of the district schools, little is known. The school buildings were primitive; the teachers were in the main, poorly prepared for their work and illy paid, and the school year for the older pupils at least, was very short. The early school rooms were unique; on either side were rows of desks facing the middle of the room; injured dignity breathed from every desk. … As the children sat in their seats, they resembled partly closed jack-knives; the angle between the seat and its back being keenly acute. From the teacher would come the order: "Sit up straight and give attention"; the order was obeyed as fully as possible, as are all such orders.

Men taught in winter, women in summer, wrote Byron "Bide" Snyder, recalling school days of the 1870s for *The Woodstock Weekly* in 1924. Discipline was maintained "by brute force," and male teachers did so or retired. "It was useless to try to rule by moral force," he wrote. "We had no morals." The one-room school held sixty pupils and was heated by "an antiquated, big box stove ... kept nearly red hot." Pupils roasted or froze depending on where they sat. The teacher changed their seats during cold days, "sort of stirring us around so that we would only half freeze or half roast," he noted.[7]

The level of schooling was not universal at first, but the quality of education and teacher training steadily improved throughout the nineteenth century. In the 1840s the first state normal school for the training of teachers was established in Albany, and teachers' institutes began to form, including one in Ulster County. Trapps teacher Sarah E. Van Leuven earned a teaching certificate, from where is unknown. In 1886 a state normal school for the training of teachers opened at New Paltz, with Albert K. Smiley of Lake Mohonk its first president.[8]

Once hired, an unmarried teacher might room with a local family or in a boarding house. Single women normally left teaching upon marriage, but by 1900 this was changing. Catherine Ellsworth and Wilson Tinney of the nearby Town of Esopus described teacher housing in rural districts: "It was the custom in those days for the teacher to live in with one of the farm families in the area. It was not unusual for the teacher to have a lunch made up by the farmer's wife consisting of cold pancakes spread with lard." While this meal might seem meager, families shared what they could. We wonder what Miss Van Leuven ate during lunch recess at the Trapps School.

Walter Needham of Vermont spoke about his grandfather's winter lunch of the 1840s. The log cabin was so cold he had to "thaw out his cowhide boots in front of the fireplace before he went to school." Not owning gloves, he baked two potatoes and held one in each hand on the way to school. The potatoes warmed his hands, and running warmed his feet. He ate the cold potatoes for lunch but "got pretty tired of them, for a steady diet." With potatoes a staple in the Trapps, did a gloveless child there tote hot potatoes to school, eat them cold for lunch, and grow weary of them?

Much like a rural church, a one-room schoolhouse became "the heart of the community," wrote Sharon S. Palmer. She noted that today's peer teaching, where students help each other learn, dates back to this period when students listened to lessons for all grades, when older children helped younger. Subjects included reading, 'riting, and 'rithmetic (ciphering), spelling, history, and geography. Other subjects deemed important for a child's education included penmanship, recitation, and elocution or the art of speaking, noted Palmer, adding that subjects reflected the rural society of the times. For example, the class might be asked to cipher how many bushels of wheat a farmer could expect from a certain number of acres if one acre yielded "x" number of bushels, the kind of figuring needed at home.[9]

County commissioners visited district schools several times a year and sent their reports to local newspapers. It took them some months to visit all schools in the county. The number of weeks school was open varied by district and by year. In January 1875 a published report listed at the Trapps School "50 children of school age, 38 have attended school, average attendance 16, valuation $7,000, teachers' wages $347, public money $108. John H. Voorhis has taught the school for 38 weeks. Hugo Slater, Wm. H. Decker and Wm. H. Chase trustees."

In July 1878 a New Paltz school report stated: "So many of the children of our district school are trying to earn an honest penny by picking currants and berries that the attendance has fallen off very much. This is to be regretted [*sic*] on account of the examination exercises on Friday, since the school appear [*sic*] to a much better advantage with full ranks."

John Harry Voorhis hailed from New Jersey. By 1859 he had arrived on the mountain and married Sarah Burger, a daughter of Trapps pioneers. John and Sarah left for Poughkeepsie and Kingston, where John worked for city newspapers, but then returned to the Trapps. By the end of January 1875, John, whose teacher training is unknown, was no longer welcome at the school. It was reported locally:

> There is no school at present in the Traps district in Gardiner. Mr. John H. Voorhis has been teaching for over a year past, and claims that he has been employed to teach for several months longer. Two out of the three trustees have locked the school house door against him, refusing to allow him to continue longer, and have engaged Mr. Alexander McGibney as teacher.

Alex McGibney had just arrived from Texas, where he had lived for many years teaching school. "In coming to our office for a certificate to teach he took his first sleigh ride in fifteen years," *The New Paltz Independent* stated. "Mr. McGibney has a brother residing in the Traps district." Alexander's brother was hotel owner John H. McGibney, who likely had written Alex about the opening. Alex, probably single, was a master penman in an age when society held that skill in high esteem. In 1875, the same paper reported, "The best penmanship we have ever noticed among our correspondents comes from the schoolmaster

Ben and Catharine Fowler of the Trapps hired J. W. Owen, for pay or trade, to record family events in their Bible. Penmanship, a highly regarded skill then, was taught in schools. Courtesy of descendant Joan Hayes Wustrau.

at the Traps." Alex also established an evening school "at the 'Traps' two nights during each week, and bids fair to meet with success, as many in that neighborhood have long felt the need of even a primary education."

Alex's evening classes did not last long and neither did he, for the school advertised for a new teacher in November. John Voorhis was hired back one month later, and with him came new trouble, but the school report in January 1876 made no mention of difficulties: "Twenty-five students, average age 13, attended District No. 8, valued at $7,000. Wages for John Voorhis and Alex McGibney: $269."[10]

John was let go again and a new teacher was hired, but conditions remained poor. Commissioner H. H. Holden led his July 1876 report with School District Number 8; *The Independent* printed his report on page one:

> The school at the Traps, Gardiner, is taught by Miss Ella Splann. On register 21 names, present 17 scholars. This school was taught the past winter by John

> Voorhis, average eleven scholars. We found the school very backward, not well supplied with books. A school taught twenty eight weeks in each and every year, by competent teachers, with but few books, and a blackboard should be father [*sic*] advanced. One girl was ciphering in reduction [subtraction]; two in elementary geography, all trying to read, spelling did very well.

By then, Alex McGibney had left for Illinois, and things finally seemed to settle down at the one-room mountain school. In 1880, John Voorhis listed "schoolteacher" as his occupation on the census; perhaps the third time was the charm. By 1890, however, the trustees grappled with a different problem—District #8 needed a larger school. Trustee Edward Smiley, son of Alfred, "led the effort to build a new school," noted Carleton Mabee. That same year, Edward advertized for bids for a new school thirty feet by twenty-four feet by twelve feet high and for a foundation. Who won the bid or when work began is unknown, but based on school reports, the second school likely was built ca. 1890–91.

At 720 square feet, the new school was more than twice the size of the first at 324 square feet. Yet, since 1856, the number of registered school-age children, aged four to twenty-one, had only increased from forty-two to fifty-one. (Average daily attendance in 1892–93 was half that.) Perhaps the old school was run-down or the state required it to be modernized. The second school was a frame building painted white with wainscoting inside. Larger windows, then in vogue, permitted more natural light. Louise Yeaple, the last teacher of the one-room Clove School before it closed in 1961, recalled that her school had been rebuilt with larger windows and painted white. Elizabeth Burroughs Kelley of West Park wrote of a new school there in the 1890s that had windows above eye-level to discourage distractions.[11]

In 1902 the second Trapps school, of wood, was valued at $500, the site at $150. The blackboard, measuring eighty square feet, was placed along the shorter wall opposite the door. New state mandates required a globe, dictionary, New York and United States maps, an encyclopedia, and modern desks. At one time or another, the Trapps School owned some of these items, usually in poor condition. Students had double desks or single and double. The teacher's desk was properly outfitted with lock and key.

Annual reports reveal the school and hamlet's financial status. In 1892–93, the library had 15 volumes worth $10, but no bookcase to hold them. By the late 1890s it had 123 books worth $55.00, with 58 added in one year. In 1893 the assessed value of taxable

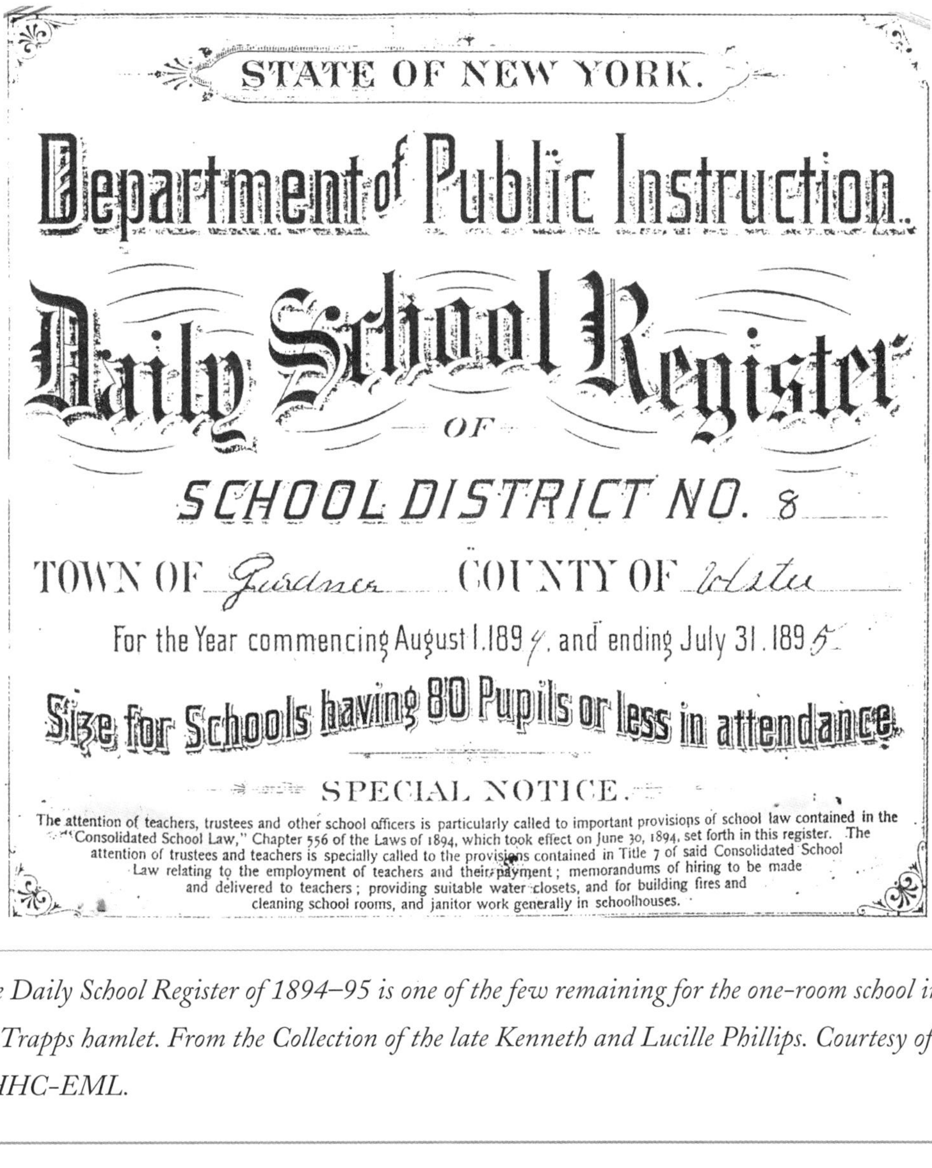

STATE OF NEW YORK.

Department of Public Instruction.

Daily School Register

OF

SCHOOL DISTRICT NO. 8

TOWN OF Gardner COUNTY OF Ulster

For the Year commencing August 1, 189 4, and ending July 31, 189 5.

Size for Schools having 80 Pupils or less in attendance.

SPECIAL NOTICE.

The attention of teachers, trustees and other school officers is particularly called to important provisions of school law contained in the "Consolidated School Law," Chapter 556 of the Laws of 1894, which took effect on June 30, 1894, set forth in this register. The attention of trustees and teachers is specially called to the provisions contained in Title 7 of said Consolidated School Law relating to the employment of teachers and their payment; memorandums of hiring to be made and delivered to teachers; providing suitable water closets, and for building fires and cleaning school rooms, and janitor work generally in schoolhouses.

The Daily School Register of 1894–95 is one of the few remaining for the one-room school in the Trapps hamlet. From the Collection of the late Kenneth and Lucille Phillips. Courtesy of HHHC-EML.

properties in the school district, both real estate and personal, was $26,600. In 1902 that figure rose to $58,995.

Teacher salaries at the mountain school inched higher. In 1856, Miss Sarah E. Van Leuven earned $37 for the year. By 1902 salaries increased to about $44 per month, although figures varied slightly. The educational level of trustees improved greatly as well. Whereas in 1856 two of the three could not read or write, by century's end all could. In July 1893, trustees Edward A. Smiley, William H. Van Leuven, and Alonzo Miller signed their names on the annual report. Edward was the son of Alfred, a college graduate, teacher, and former school administrator. William had attended the Trapps School, as his own father had.

Some of the biggest changes occurred in the areas of student health and welfare. Physiology and hygiene became part of the curriculum, and physicians began paying annual visits. During the 1910s, Dr. William P. Fuller of Kerhonkson vaccinated children at area schools, including the Trapps. While the school form did not specify the kinds of vaccines given, smallpox was the most likely. For the 1918–19 school year, the most devastating year of the Spanish flu epidemic, four children were vaccinated, but eight were not. Dr. Fuller was compensated $5 for this service. It is unknown if any sanctions resulted if districts did not vaccinate their children.

New strict rules governed privy conditions. In 1856 there were no separate privies for boys and girls at the Trapps School, but a new law in 1887 required at least two privies at each school in the district, "which shall be entirely separated from each other and having separate means of access, and the approaches thereto shall be separated by a substantial close fence not less than seven feet in height." Yet the footage separating the boys and girls privies at the mountain school shrunk over decades. In 1893, it was ten feet. In 1898, four feet. In 1902, a yard. No peeking around the fence.

Schools closed for national holidays, including the Fourth of July, Labor Day, Thanksgiving, and Christmas. Washington's Birthday was declared a federal holiday in 1870. Lincoln's Birthday was added in 1892. New York set aside the Friday following May first "to encourage arboriculture" in 1888. In 1895 the Trapps School closed from May 3 to May 10 for Arbor Day exercises. Teachers reported the number of trees planted either that day or over the past year. In 1893, three to four trees were planted in the Trapps; in 1899, none; in 1903, seventeen.

Schools also closed for teacher training. These sessions, usually held at the local teacher's institute, ranged from one day to one week. Teachers were paid full wages to attend. The Trapps School closed for one week in September 1901 when the teacher died. Classes were dismissed to observe her passing and enable the trustees to find a replacement. School doors were shut tight during epidemics, such as diphtheria in 1896 and 1901 and scarlet fever in 1901. The longest period was five weeks from mid-October to mid-November 1918 when the Spanish flu raged across the nation. Lastly, severe winter weather brought joy to kids then just as it does now. The one-room mountain school remained empty for a week after a February blizzard raged in 1893.[12]

Heydays

As summer mountain hotels rose in popularity so, too, did the numbers of people passing through the Trapps. Residents there began to see a new type of visitor, dubbed "the pedestrian." Pedestrianism, or long-distance walking for sport, became a post–Civil War rage. In the 1870s local newspapers announced, "The pedestrian excursions to Paltz Point have commenced" and, playing on a Thomas Paine phrase, "The time that tried men's soles—Pedestrian contests." Another read, "The walking mania has not reached New Paltz, but Kingston, Highland and Walden have been affected in that way." And, in 1891: "The summer boarders in this vicinity and at the mountain houses, likewise, are great pedestrians. It has been no uncommon occurrence for them to walk from Lake Mohonk to Lake Minnewaska, a distance of about seven miles, and the Minnewaska guests are seen daily in the Traps neighborhood, which must be climbed, over the rocks, the whole distance being about two miles, from the Traps to Lake Minnewaska." *The New York Times* correspondent who walked the length of the northern Shawangunks in 1876 truly tried his soles for metropolitan readers.[13]

While pedestrianism was a luxury of the middle and upper classes, mountain families walked to their places of work—stone quarries, woodlots, charcoal mounds, the mountain hotels—which explains the web of woods roads and paths then crisscrossing the Shawangunk range and joining the Wallkill and Rondout Valleys. When families did have free time, they gravitated by foot or by horse and wagon to traditional centers of social life—the church, school, tavern, inn or boarding house, general store, blacksmith shop, and each other's homes.

Social parties and picnics, often open to all, were hosted by the church, local ladies' or men's group, or individuals. The nearest church socials were held down in Guilford. Kenneth E. Hasbrouck noted fund-raising festivals at the Guilford Reformed Church ca. 1860: "Fires from pine-knots illuminated the grounds, a brass band was at one time added and a substantial supper was sometimes served to 200 people. One of the festivals would net around $400." In October 1869, Alexander P. Bush held a social party at his Trapps home. In February 1885, twenty-one partygoers attended a social at John McGibney's, where "All had a pleasant time."[14]

The excerpt below is from an invitation dated December 10, 1876. Whether it was for a social party in the Trapps is unknown, but its origin is local to the area, and we include it for its charm.

> Friend George,
> as we are going to have a Surprise Party Thursday Evening Dec. the 14th. and we would like to have yourself and as many more as you will ask to come ... Please ask the Boys to finde Cheese Candies or Lemon Crackers and the ladies they need not finde anything we will finde for all and a Plenty. it will be at the house of Abram Wynkoop we will have good Music and Plenty of danceing and hope for a grand time do not fail to come and have as many more to come as possible you will all be welcome we have sent out 35 invitations down here. ...
>
> Perhaps it will be late before you get this but if it is to late to ask others come yourself but I haven't had any opportunity to send it to the ofice befor so you will excuse Me for waiting so late I will close for this time hopeing this may finde you enjoying as good health as it leaves us, if we can not see you soon as we hope write and let us hear from you.
>
> No more at present from your friends
> Janey and Phebe Van Leuvan[15]

Food and dancing made "a grand time," as the Van Leuven girls promised. Social events were held year-round, despite bad weather or impassable roads. Bide Snyder explained: "In the winter the town 'holed up.' The so-called roads were blocked with snow for months. In the spring they were merely seas of mud." Conditions were the same in remote places such as the Trapps.

Two other centers of social life were the local general store and tavern. Bide recalled the general store was a "club room and general meeting place," especially for men. "People did not go to the store to do their trading and then go right back," he wrote. "They 'sot around' and visited." In the Trapps there was Ben Fowler's tavern and store, and later Abe Billingham's general store. Bide described Woodstock taverns, perhaps true for the

Picnics became increasingly popular during the nineteenth century. Picnickers might journey through the Trapps to spend a day at Shawangunk Mountain destinations. "The Picnic Excursion," lithograph by Winslow Homer, Appleton's Journal of Popular Literature, Science, and Art, *August 14, 1869. Courtesy of the New York State Library.*

Shawangunks, as having "long tables made of heavy planks" and "a great wood stove that stood in the center with a copper boiler for hot water on top."

By far the greatest commotion occurred during elections. Bide wrote that candidates treated all to free drinks, elections were often held at the local hotels, and fights were typical among residents. Perhaps Shaffer's and later McGibney's Hotel saw the following election-day tradition that Bide described: by mid-afternoon, many voters were "pie-eyed," resulting at night in a "grand glorious free for all fight." Each Ulster County town had champions who fought it out. "If a man got licked one election day," Bide wrote, "he would nurse his grudge until the next one and try again." Votes were bought in plain sight. At times, Bide added, pollsters working for opposing candidates "would grab the same man, then there would be a struggle until one or the other landed him."[16]

Skating was a popular winter sport as shown in this engraving by Elias James Whitney, Harper's Weekly, *February 12, 1859. Courtesy of the Library of Congress, LC-USZ62-89868.*

When they could spare a moment in good weather, people enjoyed bathing, swimming, fishing, boating, and picnicking. All year people gathered in the evenings to play music, dance, sing hymns and songs, and tell stories. People played banjos, fiddles, accordions, and mouth harps. Some had pianos or even an organ, as did the Coddingtons, where father Simon played fiddle to daughter Evelyn's organ music. Evelyn said people took out the furniture and took up the rug for square dancing.[17]

In winter people skated and sleighed. Bide Snyder made his own skates "of wood with a steel runner." Skating boots had a hole in the heels (bored with a gimlet, a small tool with a screw-like tip for boring holes) to fit a screw for fastening them to skates. Straps also helped secure boot and skates, but got wet and froze stiff. "The straps would be coated thickly with ice," he wrote, "and you couldn't get your skates off and you would have to hobble home and thaw them out."

People traveled by sleigh for work, for visiting, and simply for fun. Catskills historian Alf Evers wrote that snow "was packed down whenever possible to form a smooth and sleek surface over which the runners of sleighs and farm bobsleds might move almost effortlessly. Horses which had spent the milder months at mundane farm tasks seemed to take delight in pulling a light cutter over the smooth surface of a well-packed winter road." Evelyn DeWitt, whose mother, Laura Van Leuven Coddington, was born in the Trapps, recalled traveling by sleigh from their home at Lawrence Hill to school and several miles south to her Uncle Andreas Van Leuven's home in Gardiner for Sunday dinner. The Coddingtons rode through the Trapps to get to her uncle's house and back again.[18]

Auctions were another social event, a chance to gawk at the private lives of others, gain a much-needed item, or meet up with family and neighbors. Hotelier John C. Shaffer left the Trapps in 1866 to run hotels in Springtown and New Paltz. His ad announced: "An auction sale will take place at John C. Shaffer's Hotel, at the traps [*sic*], town of Gardiner, on Thursday, Feb. 8th, at 10 o-clock, a.m. Stock, wagons, household furniture, dairy fixtures, etc., will then be sold. See bills. Jas. Ketcham, auctioneer."

More trouble befell John Voorhis after school trustees locked the school doors against him; John's property went on the auction block in fall 1879. In October, sale items at another auction in the Trapps included his "one good work horse, a wagon, harness, shoats [young pigs], bay [a reddish-brown horse], household furniture, &c."

In June 1892, George Enderly of the Trapps was forced to auction off a fifty-acre farm that he likely had acquired from his family, who owned other mountain lots. George went into debt, for the county sheriff seized the property to pay for the lien. On June 3 a "sheriff's sale" was held at the County Court House in Kingston with Abe Billingham, husband of George's cousin Rachel Enderly, buying up the parcel.

On the 24th, the 99-acre farm of George's late in-laws, Ben and Laney Fowler, was sold at an auction held at the Callahan Hotel in Gardiner. Laney, who outlived Ben by some years, bequeathed their property to the couple's heirs. Daughter Rachel, administratrix of her mother's estate, and her husband George Coddington entered into a legal wrangle with Rachel's two sisters, their husbands, and the widower of a third sister. Their differences unresolved, the farm was put up for auction. Widower Edwin Davis, son of George, bought the 99-acre Fowler farm for $800. *The New Paltz Times* noted, "Only $8 an acre for land in the town of Gardiner."[19]

Auctions generally got little press, but elopements might garner much attention. One elopement involved Sarah Hedden, thirteen, of Gardiner, and David Williams, about twenty, a laborer living in or near the Trapps. The story ran in July 1880:

> A RUNAWAY MATCH AND ITS ENDING—The Traps region in Gardiner furnishes more than its full quota of news for the newspapers. The latest excitement in

"Country Courtship," engraving by S. G. McCutcheon, Harper's Weekly, *May 20, 1882. Courtesy of the Library of Congress, LC-USZ61-1815.*

> that region is in regard to a runaway match, the heroine being Sarah E. Hedden, daughter of Daniel Hedden, and the young man David Williams. Sarah was only 13 years of age—not legally old enough to contract matrimony. They left home Sunday before last, proceeded to High Falls, where they were married, and then went to Fishkill where the young man had a sister. The cruel parent employing officer Elias Coe, of this village, started in pursuit, found and arrested Williams, and after a preliminary examination before Justice Sears he was sent to jail to await the action of the Grand Jury. What a crushing blow to two young hearts that beat as one.

The marriage was likely annulled, one heart again beating as two.[20]

Weddings received scant press. Most announcements were, like obituaries, bare bones, although longer descriptions might appear for more prominent families. Mountain weddings were modest affairs. In the earliest decades with limited social contact, a girl often married the boy next door. The ceremony was held at the local church or a private home, where a simple reception might be held. In 1876, Mary Fowler married Trapps neighbor George Enderly at the home of "Mr. Fowlers" and the "Bride's mother," noted records of the Alligerville Methodist Church. Mary's father, taverner Ben, had died, but the house was still referred to as his.

Town of Rochester historian Milford Ebert listed familial terms of yesteryear: "Boyfriends were beaus, dating was courtin' or keepin' company, anything more was

Trapps neighbors Mary Fowler and George Enderly pose on their wedding day, two days after St. Valentine's Day in 1876. Courtesy of their great-granddaughter Joan Hayes Wustrau.

known as sparkin'. An engaged girl was said to have been spoke fer, the wedding ceremony was gettin' hitched." He added that people used the terms "in the family way" or "bein' like for a baby," or even "she's like," for being pregnant. Ebert also described a rite of passage called a "skimelton" or "skimmington," where the town folk "would sneak up to the honeymooners' house and surprise them with roaring shotgun blasts, banging of pots and pans, and ringing of cowbells."[21]

Newlyweds often stayed with one set of parents until they could afford to set up their own households. George and Mary Fowler moved in with his widowed mother Laney, likely to help her run the tavern, which enabled her to remain there.

As children arrived, baptisms might depend on the parson's rotational schedule. Two mass baptisms were held at the Clove Reformed Dutch Church in February 1809 and March 1816. The first was held for fourteen children, including two four-month-old sons born a day apart in October 1808. One was the son of Daniel Carson and his wife Sarah Frost of the upper Clove, the other of Ben and Laney Burger, early Trapps settlers.

At the second baptism, twenty-seven children were anointed. Three were children, aged one to seven, of George Frost and his second wife, Agnes Green; two, aged three months and three, of Daniel Carson and Sarah Frost, George's daughter with his first wife, Elizabeth. We can imagine one family after another, dressed in their Sunday best, arriving at the church by horse and buggy. Perhaps those closely related, like the Frosts and Carsons, returned to one home or the other to celebrate these momentous events in their lives. Finally, the day's celebrations over, they changed back into regular clothes and went back to their daily chores.[22]

• 7 •

Saints and Sinners

December 1871. The Trapps School. Some forty adults and children crowd into the one-room schoolhouse for baptism into the Alligerville Methodist Church. Trapps families include Billingham, Burger, Bush, Coddington, Davis, Enderly, Fowler, Van Leuven, and Wynkoop, with Enoch Dingee the sole member of his family to attend.

Seven years later in April, an officer of the law arrests Enoch's son, Alonzo, for stealing hides from the McKinstry Tannery in Gardiner. Under arrest for the same crime is William Enderly, Jr., brother of Enoch's wife, Mariah.

"Lonnie and Will are off to jail," Mariah cries. "Our son. My baby brother."

"Let us look to the Lord for his help and wisdom," Enoch says, folding his hands over Mariah's as the two kneel in prayer.[1]

"Sure, mister, there were the sinners.

But we had a share of saints, too."

So wrote resident Ray Veres about the Trapps of the 1920s. The hamlet had its share of both, much like any other community. Local newspapers, though, relished telling the more colorful stories about the Trapps, and locals consumed them with equal relish. *The New Paltz Independent* defended the hamlet's reputation in March 1875 when it responded to an earlier piece published in the *Kingston Rondout Freeman*.

> THE TRAPS NOT A "HARD" PLACE.—A correspondent writing from the Traps comments upon a recent statement in the *Rondout Freeman* that the section in which he resides in [is] a hard place. He says: There are at least thirty families who are brought in relation with the Traps, by attending its meetings and school. The majority of "heads" of families here have but little education, or none at all.
>
> Though uneducated they are not ignorant, nor idle, nor dissipated, with few exceptions. They are a more or less contented people, with their simple life. During the residence of persons who have lived here a long number of years, such a startling calamity as a murder has not been known. But one case is on record of a person being tried by a district court for theft. There are no cases in court for indebtedness by parties whose home is in the Traps. There has been but one instance of a person from here having to be supported as a pauper. We have no taverns or public houses, therefore intoxication is rarely seen here, except as it comes from outside places. The women stay away from Mrs. Woodhull's lectures, and therefore do not give curtain lectures. We always pay *The Freeman* for his paper.

The Freeman writer reinforced prevalent opinions held then by valley residents. Yet, *The Independent* writer peered through glasses perhaps a little too rosy. Many Trapps residents then could not read or write, but they were hard working and a majority were members of local churches. Most mountain families were poor, and some activities, though illegal, might have stemmed from a desire to improve their lot in life.[2]

Where better to begin than at the beginning, with a snake-in-the-garden story? America then was overwhelmingly Christian and much more culturally homogeneous than now, especially in rural areas. The owner and publisher of a small town newspaper might wear the hats of reporter, writer, and editor. As such, a nineteenth-century "town crier" was freer to be politically partisan, mix news with opinion, and express religious sentiment. In 1860, Mrs. John J. Dubois of Gardiner killed a large rattlesnake while picking berries on the mountain and "returned with it as a trophy of womanly courage." The phrasing in the local newspaper excerpt below would not appear today in the mainstream press:

> We think she deserves credit for the slaughter of his snakeship. She no doubt bears in common with "rest of mankind" a hatred of the whole snaky tribe, as one of them served her mother Eve a rather scaly trick "one upon a time." We fully justify her in the destruction of this descendant of the "Old Serpent."[3]

The Reformed Dutch Church was the most established one in the Hudson Valley. In early years, Trapps families had a long, steep journey to services down the mountain to Rochester, Marbletown, New Paltz, or Kingston. A Reformed church opened in the Clove in 1807 and then in Guilford southwest of New Paltz in 1833. As the closest Reformed church, Guilford became an important spiritual center for mountain families. Two Guilford residents of the original congregation were Andries and Phebe Van Leuven. Andries was likely related to the Trapps Van Leuvens; Phebe was born in the Trapps, the daughter of pioneers Peter and Hannah Aldrich. The couple paid about $22 for an assigned pew to help raise funds for the new church, noted Kenneth E. Hasbrouck.

Rev. John N. Jansen was the Guilford pastor from 1852 to 1863. His labors included Sunday morning services and, from May to November, Sunday afternoon lectures. Initially, pastors traveled around the district. The Sunday lectures alternated among five locations, including "across the Traps near Mr. William Enderly's," Jansen wrote. William and his wife Catharine, like Phebe Van Leuven, were founding members of the Guilford church and children of Trapps pioneers. Rev. Jansen added, "All these services were well attended." The Trapps School had doubled as a place of worship, much as the Clove School had in that neighborhood. The school was the only building then large enough to hold a sizable gathering in these hamlets.[4]

The Reformed Church, however, was not prepared for the great awakening spurred by the Methodist revival. Here was a chance to break away from the religious regime of the past two centuries. A Methodist church was established in 1858 at Alligerville in the Town of Rochester, with congregations added in High Falls, the Clove, and the Trapps.

The largest Methodist baptism of Trapps residents was held in December 1871 for some forty people, including nine children. Well over a dozen Trapps families came, including ten Van Leuvens. Revival news became standard fare of the press. This 1871 brief ran in New Paltz: "Revival meetings have been held in the Traps on the Alligerville and High Falls charge, for six weeks past. Fifty-two have joined the M. E. [Methodist Episcopal] church on probation. The greater part of them are heads of families."[5]

In 1876 the Clove people erected their own church on land bequeathed for that purpose by the family of the late John F. Stokes and financed in part by the Smiley family of Mohonk and their guests.[6] Alfred Smiley of Lake Minnewaska "deplored" the lack of a church in the Trapps and with his guests "contributed a goodly sum of money" to build one there, noted Nan Knickerbocker. Alfred organized fund-raisers, such as "fairs and other entertainments," and his guests contributed about $400, reported a New Paltz paper.

In December 1884, Sarah Davis donated a small parcel from her farm for a chapel in consideration of one dollar. The parcel was conveyed to three men who became the chapel's first trustees—Alfred's son, Edward A. Smiley, and two Trapps men, Edwin Davis, son of Sarah and George, and Abram Billingham. The deed stipulated the parcel be used "For the uses and purposes of Erecting and, having a chapel for a place of Meeting for Religous [*sic*] worship and for such other purposes as the parties of the second part shall deem proper."

In 1885, Minnewaska employees and the Trapps people erected a one-room chapel. Services were a "union" of the Reformed and Methodist faiths, hence the name Union Chapel. A rural chapel, much like the school, anchored a hamlet, and so it was in the Trapps. Down in the valleys, churches vied for members, but in isolated areas like the Trapps, chapels tended to be interdenominational. The chapel, white with green shutters on the outside and carpeting on the inside, was described as "handsome and commodious" by Ralph Le Fevre and "picturesque" by Nan Knickerbocker.

The Smileys of Lake Minnewaska and their guests actively supported the chapel and its ministers. Alfred H. Smiley paid a $12 premium to the U. E. & E. N. Terwilliger insurance company of Ellenville. (Uriah E. was the proprietor of Mount Meenahga, a hotel north of the hamlet of Cragsmoor atop the Shawangunk Mountains above Ellenville.) Minnewaska guests also attended the Trapps Chapel. The Smiley family held services there on Sunday afternoons into the 1900s, remembered Evelyn Coddington DeWitt. The Smileys often supervised Sunday school classes and exercises,

The Trapps people at last had a church of their own in 1885. The Union Chapel was so-called because alternating Methodist and Reformed services were held there. Photos courtesy of DSRC-MP. Exterior photo donated by Alfred B. Smiley, interior photo by Evelyn Coddington DeWitt.

attended by upwards of forty children. One Sunday in June 1901, Alfred's son George led more than two dozen mountain youngsters, including George's sons Fletcher and Charles, Bertha Van Leuven, Florence Lyons, and Lulu Harp in singing hymns and performing inspirational pieces.[7]

Ministers from the Alligerville Methodist and Guilford Reformed churches preached at the Trapps Chapel on alternating Sundays. Rev. Calvin E. Lasher was the Guilford minister from 1873 until his death in 1923. Rev. Lasher's obituary noted, "Mr. Lasher used to go quite frequently to hold services at Minnewaska [The Traps] where he was very highly esteemed, as he was by everyone with whom he became acquainted."[8] Rotating Methodist ministers from Alligerville, including William Whitfield, also served the Trapps people. The mountain congregation honored "his earnest and faithful work" in January 1888, when about 150 members squeezed into the Trapps Chapel for the event. Rev. Whitfield was given a bonus of over $40 and the "ladies served a bountiful repast which was a model of cookery."[9]

Evangelists also worked the region. In 1900, announcements in New Paltz read: "There have been lately four evangelists at work in the Traps neighborhood and they are now working at Butterville," and "Rev. Mr. Felton, of Dry Brook, who has been holding religious meetings at Butterville and in the

THE TRAPS SUNDAY-SCHOOL CHILDREN'S DAY EXERCISES JUNE 2 1901.

George H. Smiley, Superintendent.

PROGRAM—PART I.

Organ Voluntary..................
March...............By the school
Scripture reading. Wilson Vandemark
Anthem.........Praise ye the Lord
Prayer...........Rev. C. E. Lasher
Opening.............William Harp
Address.............Lillian Brooks
Greeting Song..................
The First Children's Day..........
Bertha Vanleuven.
Recitation and Song...............
By the Juvenile Class.
The Squirrel Home....Chas. Wixon
A Cheery Word...Lizzie Vandemark
My Country......Grover Vanleuven
The Pansy.......Reginald Bloomer
The Young Robins..Isaac Armstrong
Happy Bird Song................
The Dandelion........Maud Wixon
Freedom's Day....Charles Gardiner
The Boy and the Sparrow..........
Fletcher Smiley.
Beautiful Flowers.....Annie Connor
The Lilac...........Oscar Bloomer
The Daisy.........Susie Vanleuven
The Naughty Cat.Allen B. Coddington
The Roses. ......Walter Vanleuven
Song.................Lovely River
Welcome.........Delford Vanleuven
The Barley Loaves.Wilson Vandemark
Woodland Prairie........Lulu Harp
Six Times Nine...Lillian Armstrong
The Promise Wreath of Roses......
eight girls.
Lullaby..........Lizzie Vandemark
A Cheery Heart.........May Walker
Echo Song......................
To the Flag........Edward Bunton
The Squirrel's Arithmetic.........
Charles Smiley.

Children attending the Trapps Sunday school presented "day exercises" for their families and other guests. This detail of a June 2, 1901, program was published a few days ahead in The New Paltz Independent. *Courtesy of HHHC-EML.*

Traps, baptized about twenty men and women by immersion in the Coxing Creek at the Traps neighborhood on Sunday morning." Dry Brook was a hamlet near Margaretville, in Delaware County; Butterville was a Quaker community west of New Paltz on the way to Mohonk.

In October 1889, locals read of a "remarkable incident" at the Trapps Chapel:

> Services are held there by Rev. W. H. Van Hosen of the Alligerville M. E. church and Rev. C. E. Lasher of the Guilford Reformed church on alternate Sabbaths. A week ago last Sunday was dominie Van Hosen's turn to preach and he took for his text a portion of a chapter in St. Matthew, where it refers to Peter walking on the water to meet the Saviour. Last Saturday dominie Lasher preached at the same place and had for his text the identical passage. But what was the most remarkable about the coincidence was that both ministers in the introductory started off with the same words, and all through both sermons were sentences just alike. ... Such was not the case at least circumstances forbid such a thought.[10]

• • • • • •

While saints found the Lord in the swirling streams of the Shawangunks, sinners found Lucifer in the range's nether regions. Thieves, arsonists, drunken brawlers, murderers, even a counterfeiter—the mountain had them all. With only about 200 to 250 souls in the Trapps during its heyday of the 1880s, those few miscreants outshone law-abiding citizens. "The Traps region in Gardiner furnishes more than its full quota of news for the newspapers," began one 1880 article about underage elopement. Arthur O. Friel penned, perhaps a bit too pointedly, in his 1925 novel *Hard Wood*: "Like every community, the Traps had its quota of mean-minded men and of snake-tongued women."

We begin with the 1861 "Grass Widow" trial of two Trapps women. A grass widow was a polite expression for a woman in difficult circumstances. In this case, the two were separated from their husbands. "J. Curious" contributed a full account of the summer trial for *The New Paltz Times*. The "Old War Horse" was prominent Gardiner citizen Joseph O. Hasbrouck. The supper, with unexpected sideshow, and later trial were held at the Shaffer Hotel.

> Mr. Editor:—Permit me to inform you, and the readers of your paper, that the result of the oyster supper, the dance, the speech of the war-horse, and the fight

> which took place at that time, between two GRASS WIDOWS, in February last, ended in the trial of the accused, about the middle of July, at J. C. Shaffer's, for the outrageous crime of assault and battery.
>
> Mr. Editor,—You missed a pile of fun ...

As the story went on, G. Margarette Bush, wife of Alexander Bush, assaulted Catharine Ann Abrams while the two attended the oyster supper on February 19. Catharine was the wife of Albert Abrams and daughter of Henry O. Van Leuven. She filed a complaint at Tuthilltown, with a warrant issued for Margarette's arrest. Margarette let out for the hills until arrested "in the berry woods, and the prisoner refused to walk out—a distance of some three miles—and a very rough road. Officer was compelled to go for his wagon, and carry her out."

At the trial, Catharine testified that Margarette "came up to me, unawares, and struck me in the face; it went through it like a piece of lightning. Clinched me in the hair, and choked me. She was very angry." Upon cross-examination, Catharine revealed that she separated from her husband two years ago, that there was nothing "betwixt" her and Margarette to cause the attack, and that the only reason might have been Alexander Bush's relationship with her father, Henry O. Van Leuven.

Several eye-witnesses testified that Margarette slapped Catharine, but that Catharine also scratched Margarette on the neck. Another eye-witness "contradicted flatly some statements made by Mrs. Abrams [Catharine], in relation to helping or being with [Mr.] Bush, peeling bark or piling bark in woods, &c."

In his closing arguments, Margarette's lawyer argued that she was "robbed of the affections of her husband by the alluring snares of the complainant, her family broken up, and she exposed to the jilts and scoffing of the complainant, and the menaces of the community, were all to be taken in mitigation of the charge, &c."

Margarette added that Catharine "had called her hard names, and underwitted all before the affray. Then she told complainant if she repeated the same, she would slap her face. Also that complainant and accused's husband were too familiar, to her knowledge, and that she had endured all sorts of aggravation before that time."

The trial was rare entertainment for locals. Shaffer's Hotel was "crammed" with the families and friends of the parties. The paper reported: "Old men and women, young men and maidens, and children and babies, all there. The company enjoyed themselves in another apartment by singing hymns, in alternation with some of the old-time plays and kissing, and other jollifications."

The jury returned a verdict of guilty, and Alexander Bush agreed to "pay the fine if it was reasonable." The court imposed a fine of three dollars, which Alexander paid, thus securing Margarette's release. The reporter completed his story to wit:

> Bush and his wife Margarette, buried the hatchet, and were reunited, and the complainant lost her claim to the object of her fancy, proving the old adage, that true love does not always run smooth. So, Mr. Editor, the supper of oysters at the Traps ended, unless some of them that were spilled on the way home from an overloaded stomach should propagate. If so, more anon.[11]

While the Grass Widow trial had elements of broad comedy, some attacks were deadly. William Bunten, nearly fifty years old, attacked Josiah Slater, thirty-four, with an axe in 1869. The story below is one of two that appeared on the same day, one in each New Paltz paper. William's "old grudge" or whether he was arrested is unknown.

> About half-past eight on Saturday evening last, William Bunten, on the strength of an old grudge, attacked Josiah Slater with an axe, cutting his head open badly, and taking one ear entirely off. Mr. Slater is not expected to live. The transaction took place about half a mile beyond the tavern of Benj. Fowler, in the "Traps." The consternation of the neighbors was such on Sunday, that it was almost impossible to induce a farmer to open his door.
>
> Both of these men were under the influence of intoxicating liquors, at the time of the fracas.[12]

In 1877 a young Trapps man named Bush passed counterfeit half-dollars in New Paltz. He was arrested at his Trapps home but was set free because of insufficient evidence.

A fresh grave discovered three years later fostered a rumor mill of murder:

> Some travelers in the Traps neighborhood the past week discovered a grave which had been recently dug. As there is considerable excitement in that locality at present, owing to the fact that pistols have been repeatedly spoken of, who knows but what it was the intention of the grave digger to "make some one scarce."[13]

One Trapps trio couldn't stay on the right side of the law—William Enderly, his nephew Alonzo A. Dingee, and Jacob Wynkoop, related to both men by marriage. William, born ca. 1852, was the youngest of William and Catharine. Alonzo, born in 1861, was a son of Enoch Dingee and first wife, Mariah Enderly, eldest of William and Catherine. Jacob, born in 1837, was a son of Joshua and Anna Maria Wynkoop.[14] Their movements were regularly reported, as in this November 1875 article out of New Paltz:

> Lawsuit.—Elias Coe and A.J. Abrams, agt. J. Wynkoop and Wm. Enderly defend-ants, is an action for trespass, tried before Justice Briggs, at Steen's hotel, Tuesday evening. Plaintiffs own a wood lot at the Traps. Defendants were accused of taking bark, timber, and railroad ties from plaintiffs' lot. Sixty dollars damages were given by the jury. A considerable portion of the population of the Traps were [*sic*] present as witnesses for the plaintiffs, and the trial was a lively one. Lewis Hasbrouck appeared as counsel for the plaintiff, and "Broddy" Hasbrouck for defendants.
>
> Our village people manifest a great aversion to serving on a jury, nevertheless, we saw some of those same people willing to stand up or sit on a hard bench without compensation, listening to the trial, for the whole length of time, while the jurymen were occupying comfortable chairs and earning 25 cents.

Jacob Wynkoop and William Enderly were sent to prison. In April 1878, William and Alonzo Dingee were arrested in Gardiner and jailed, a local paper reported, "for having stolen thirty four green, salted and tanned hides, of the value of $125 from the tannery of McKinstry." The two tried to sell the hides to McKinstry's buyer or agent in Poughkeepsie, but one of McKinstry's men crossed the river to identify them. Some of the hides and other stolen items were found at Alonzo's house hidden under hoop poles.[15]

William left jail, how is unknown, only to be rearrested in February 1879. A constable carted him off to the Kingston jail, or so he tried. When his shackles were removed as they neared the railroad station, William "jumped the train and escaped." Branded "a notorious criminal," William hid out for ten years in the wilds of the Shawangunk Mountains and Pike County, Pennsylvania, as well as the Albany area of New York State.

The law once cornered William in New York's capital city, but he jumped again, this time from a house into the Hudson River. A New Paltz paper reported that "the officers fired after him, but he managed to get away by swimming." William's daring escapades came to an end in late October 1886 when the law finally caught up with him in Ellenville, where he was "engaged in drawing wood" (perhaps moving it or shaving wooden barrel hoops with a draw-knife). William served his time in Kingston. His debt to society paid, he returned to Ellenville, where he lived out his days.[16]

• • • • • •

On April 10, 1853, an event most brutal occurred in the vicinity of the Trapps.

Philip H. Smith unfolded the "startling tragedy" in his 1887 *Legends of the Shawangunk*: "Ben Gosline [sic], a man of middle age and married, became intimate with a young mulatto girl by the name of Maria Cross." One day, Ben brought Maria to "the brink of one of the dizzy precipices with which the locality abounds" to show her an eagle's nest. There, Ben told Maria he would hold her while she leaned over to see the eaglets. A trusting Maria did so, but Ben "loosened his hold, gave her a gentle push, and, with a piercing shriek, the girl went over the precipice."

The same year Smith's *Legends* came out, Town of Lloyd farmer Nathan Williams wrote a slightly different version for *The New Paltz Times*. Williams did not mention Maria's name, Gorsline's first name, or Maria's pregnancy. A hemlock tree near the bottom saved Maria's life, but she was so badly hurt she could only drag herself a few yards. During the night she saw a light moving among the rocks. She lay quietly, presuming it was Ben returning to make sure she was dead. The light finally went away, but she remained all night in agony. The next day Maria crawled down to a farm alleged to have been occupied by Richard Vanderlyn, whose family took her in and cared for her. (The Vanderlyns lived below the cliffs in the newly established Town of Gardiner.) There, "faint from hunger and the loss of blood ... she told the story of her wrongs," Williams wrote.

The two storytellers diverged as to Ben's fate. Smith wrote that Ben fled, never to be seen again; Williams that he was convicted, but "what measure of punishment was merited out to him I hardly remember." Smith thought Maria gave birth soon after, with mother and child still alive, but their later whereabouts were unknown. He ended, "The incidents of the attempted murder, and her miraculous escape from instant death, form themes yet fresh in the minds of the residents of the locality."[17]

The story might have stayed fresh in the minds of many, but after more than three decades, many of the details were very, very stale.

Ben was the eldest of Sam and Sarah Gorsline, a Trapps couple with ten or more children, including Jacob and Laney, wife of Ben Fowler. In 1850, Ben Gorsline was a canal boatman living with his family at St. Josen to the west. Jacob Gorsline, a laborer, lived with his wife and mother-in-law in the Trapps.[18]

The Rondout Courier and *The Ellenville Journal* broke the story on April 22, 1853. *The Journal* hoped that the story, titled "Painful Rumor," was not true. *The Courier* printed facts from a deposition Maria gave in New Paltz. Five days later, Kingston's *Ulster Republican* and *The Kingston Democratic Journal* joined the media frenzy of the day. By the 30th, the story jumped the Hudson River to Poughkeepsie. The press reported that Maria was fourteen to seventeen years old and the illegitimate daughter of a white mother and black father. Her last name was listed alternately as Cross or Schoonmaker. Maria accused Jacob and his wife of trying to kill her. Maria and Jacob lived in the same neighborhood, either the Trapps or the upper Clove. Jacob had worked for the D&H Canal the year before.

The Poughkeepsie Eagle reported that Jacob's wife, unnamed in the other papers, had "strongly hinted to her husband that he must 'clear up' the affair ... [and] ... the guilty pair proceeded to clear it up." On Sunday, April 10, the two allegedly took Maria to the highest point of the Trapps (or between the Trapps and Paltz Point) facing New Paltz. They enticed her with a groundhog, not an eagle's nest. A birch tree, not a hemlock, saved her life. Nothing was said about Maria hiding from a light that night, but over the next two days she dragged herself using two short sticks. Vanderlyn found Maria on Wednesday and took her to the Ulster County Poorhouse outside of New Paltz, where he "called upon Doctor Sears to administer all the aid of his power." Maria did not break any bones, but her body was "dreadfully mangled." At first, the story seemed improbable, but a party found her shoe at the foot of the cliff, the broken tree, and fragments of her clothing and person.

The distance Maria fell was measured "by good judges." It was about 240 feet.

Justice Lefever issued a warrant for the arrest of Jacob and his wife. Jacob slipped the grasp of the law and "fled into a dense swamp in the vicinity." At press time, Jacob's wife had not yet been arrested, "as she should be by all means." All three Kingston papers printed the following: "The causes for this extraordinary crime are well known. Maria is a girl of loose character and a connection existed between her and Goslin [*sic*], which she not only spoke of to others, but had taunted Mrs. G. with.—This at once gives a clue to the motives for the horrible revenge suggested by brutal minds by embittered feelings. It also gives weight to the supposition that Mrs. G. was a full party to the act." The "connection" was a decorous reference to Maria's pregnancy.

On May 6, *The Rondout Courier* reported Maria was "recovering rapidly, being able to walk about already." Jacob, however, had "not yet been taken."

The case took on a new dimension in June and July 1853 when *The Ellenville Journal* reported that a showman named Blake had "engaged" Maria to exhibit her around the country. *The Journal* thought it would be "doubly interesting" if "Mr. Blake would procure a good painting of the spot where she fell, to exhibit with her." (*The Kingston Democratic Journal* and *The Courier* did not print Blake's name.) Plans were canceled when Maria was moved from the poorhouse into protective custody as a witness in her case. *The Ellenville Journal* stated that Maria "will therefore not have the pleasure of going about the country on exhibition."

By July, Jacob was still at large, but his wife was arrested after Maria filed a complaint. Police Justice Martin G. Hayes questioned Jacob's wife, named Jane in federal census records. Jane claimed she was "sick in bed" the day of the attack and imminently due to give birth. Jane's mother and others swore that Jane had been confined for a month previous and could not have climbed the mountain as Maria said. Based on insufficient evidence, Jane was released.

The Gorsline name next appeared in the papers in March 1854 when a man named Ben Gorsline and several others were indicted for assault and battery of a High Falls man the previous June. The indictment was "quoshed" on a technicality, *The Ulster Republican* reported. Jane gave birth to a son, George, in 1854, but nothing is known about the child born the year before.

That April, Jacob's indictment came before a jury in Ulster Circuit Court, reported *The Republican*, the only Kingston area paper to carry the story. By then, Jacob apparently had been apprehended, for his lawyer acted for the "Prisoner." The paper reminded the public that Maria had "experimented with the laws of gravitation by her frightful descent from the Shawangunk mountain" with Jacob allegedly providing the "projectile force." Both sides presented evidence, but the "jury failed to agree" with Maria. Jacob was not indicted and presumably released.[19]

The Gorsline family, except for Laney (as far as is known), was gone from the mountain within a few years. Their father, Samuel, died in August 1853; he was buried in the

Frost Burying Ground in the Clove near the place of his birth. His widow, Sarah, joined three of their other sons who had gone west to Wisconsin, where she later died. Jacob and Jane vanished from New York, only to reappear farming in Michigan. Ben moved to Rosendale. His July 1890 obituary in *The Kingston Argus* described him as "a veteran cement man ... a well known and respected resident ... popular among his fellow-workmen and friends, for many generous manly qualities." By 1860, Richard Vanderlyn and his family had moved to the Town of Shawangunk.

The site of Maria's fall attracted curiosity-seekers for years after. In 1887 Nathan Williams wrote, "The precise spot is discernible from a long distance and is easily identified, from the peculiar reddish color of the mountain rock, and for some time afterwards was much visited by the morbidly curious as a point of great interest; but of late it has almost fallen into 'innocuous desuetude' [disuse]."[20]

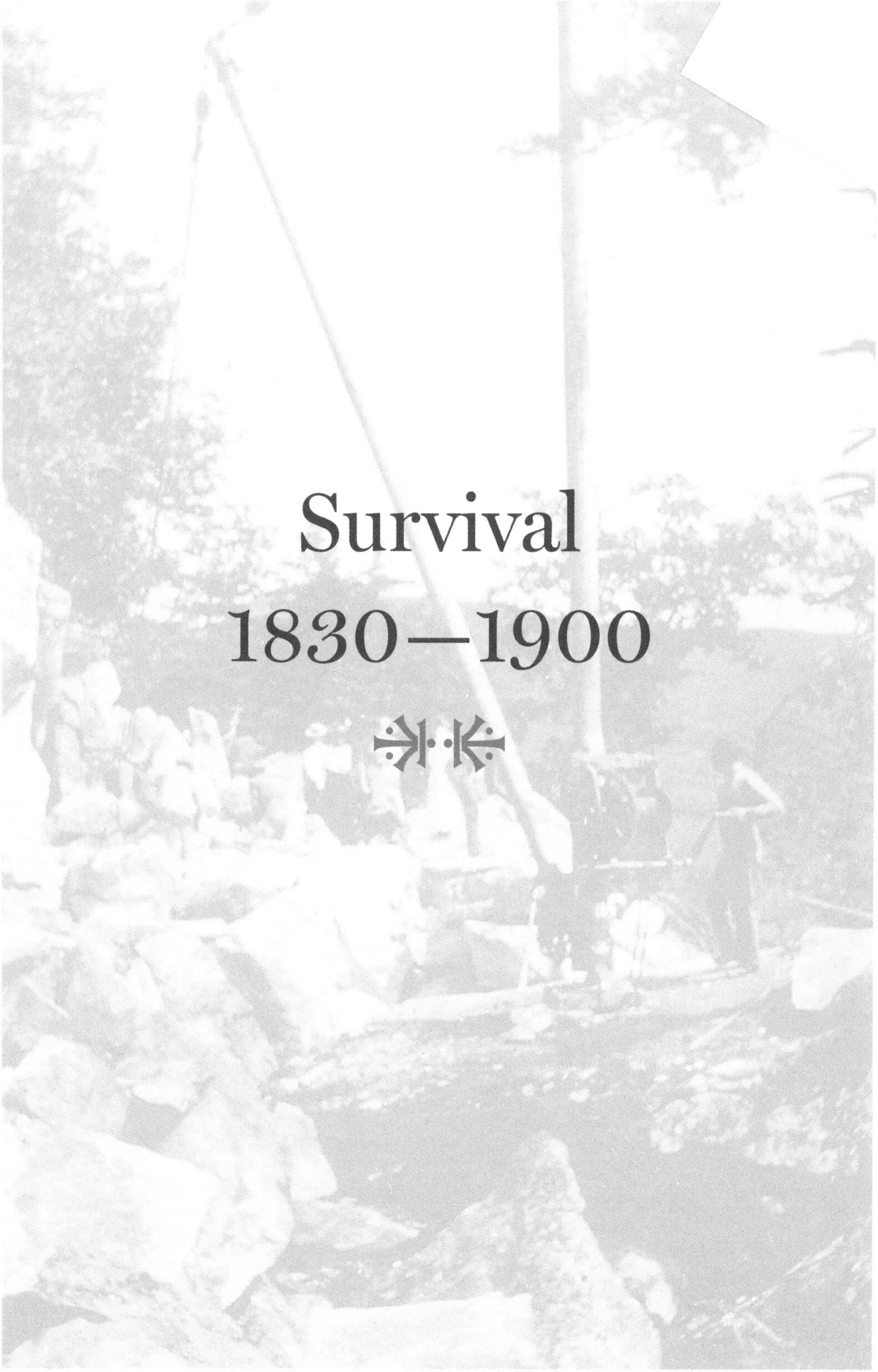

Survival 1830—1900

• 8 •

Farming and Gathering

Eleven PM, August 1897, the Trapps. Abe Billingham, forty-eight, hitches up the team to the wagon, which creaks under a heavy load of berries.

The door squeaks open behind him. "All set, Abe?" His wife, Rachel, forms a dark silhouette in the doorway against candlelight flickering within.

"All set, Rachie. Aiming to catch the first ferry to Po'kipsie."

"I packed some pork, hard eggs, and corn bread. Should hold ya."

Abe hoists himself onto the wagon seat. He grabs the reins and begins his nightly journey. It ends only when the berries run out. "Soon," Abe repeats to the team's clip-clop, clip-clop. "Soon." Once down out of the gap, he'll let the team take over. He'll watch the stars, then drift off to sleep, for the horses know the way by heart, by heart.[1]

Hardscrabble Farming

Trapps farms were "scratch farms," noted Art Lapp of Alligerville. "Scratch" or subsistence farms produced just enough to feed the family. Trapps land was first plowed shortly after the Revolution, yet some disputed the idea. Nineteenth-century New York State gazettes stated Ulster County's "mountain region consists of irregular ridges and isolated peaks with rocky sides and summits too steep and rough for cultivation." Tillson & Brink's 1853 Map of Ulster County even depicted the Trapps as a no-man's-land, showing only the public road joining Alligerville and the newly formed Town of Gardiner. No people, no houses, no school, not even the Coxing Kill.[2]

The gazettes were correct on one thing, though. The uplands were rough, in some places downright inhospitable. Once trees were cut down, a farmer "grubbed out" defiant stumps, noted Jared van Wagenen, Jr. Once stumps were out, rocks had to be dug out. And dug out. And dug out. Norman Van Valkenburgh wrote about this mountain ritual: "The best crop seemed to be stones. No matter how many were carried from the fields in the spring and used to build walls along the boundaries of the farm and between the numerous fields ... an equal number appeared to sprout over the following winter to cover the fields again the next spring." Farmers employed oxen, horses, chains, stoneboats (a type of flat sled), spades, and bare hands to drag away the rocks and boulders that yearly surfaced to vex them.[3]

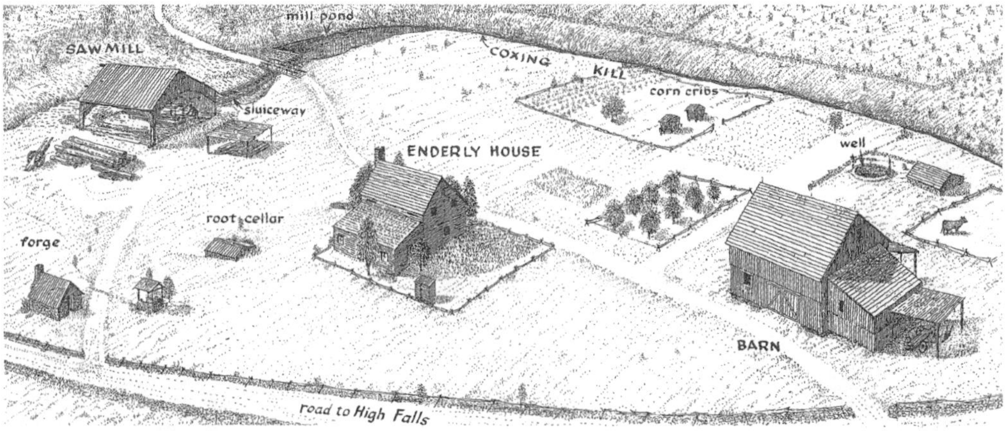

The Enderly farm and sawmill, situated on the Coxing Kill in the Trapps hamlet, as it might have looked during the 1880s. Illustration by, and courtesy of, Jack Fagan.

An ox or horse team was essential for performing heavy labor around the farm. At right is Delvan Smith, whose son Joseph married Myrtle Otis of the Trapps. The man at left is unidentified. Courtesy of Myrtle's nephew, Don Otis.

Once the land was cleared, a home built, and the family moved in, the real work began. People of the Trapps and other remote regions were engaged in a daily struggle to set food on the table and store it for the coming winter. The growing season was shorter and tougher than down in the valleys. Mountain families raised crops able to bear the harsh land and weather—oats, Indian corn, potatoes, buckwheat, and rye instead of wheat. These crops were grown in small quantities, primarily for family use. While men labored at different kinds of work, such as charcoal burning, hoop shaving, wood and tan bark cutting, and stonecutting, most identified themselves as farmers.

Trapps properties ranged from a few acres to a few thousand. Most were a few hundred acres with poor, rocky soil worth a few hundred dollars. A family typically might own an ox, a horse or two, a cow, a handful of chickens, a few pigs. Oxen and horses, the John Deere tractors of the day, performed the heavy work. A family had at least one "milch" cow for milk, cream, and butter. (If only one cow was owned, it was a milch cow.) Three families raised sheep. These few animals and those trapped or hunted gave Trapps

The task of making butter often went to women and children. Butter was stored in stone crocks, and any extra was sold to help support the family. Photo by Ted Reiss. Stoneware courtesy of Barbara Rubin; butter churn courtesy of Robi Josephson.

people the protein and fat they needed, plus wool, hides, and feathers for clothing and bedding. A few acres of hardy crops, including apples, carrots, and potatoes, filled root cellars for the coming year. Peter H. Harp of New Paltz observed farmers "could live for a long time without going to the store for fruit, vegetables, meat and supplies."[4]

Yet, a cash crop could make all the difference. Butter, eggs, fruit, maple syrup or sugar, wild berries—all could be bartered among families or to pay the landlord, tax collector, shopkeeper, peddler, and tradesman such as the blacksmith. Families also bought store items on credit and settled up in cash or barter. Store-bought items included salt, sugar or molasses, tobacco, linseed oil, shoes, cloth, combs, buttons, and other sewing notions. Jared van Wagenen, Jr. noted spices, tea, coffee, saltpeter to cure meat, gunpowder, and medicinal drugs were also purchased. "Enoch Dingee, the old [Civil War] veteran, has traded one of his horses for a cow and a large quantity of hoops," an April 1890 brief stated. We're not sure who got the better deal, but barrel hoops fetched good money and a cow could provide milk, butter, and extra cash for Enoch's family.[5]

Life in the Trapps hamlet was tough. By the mid-1800s the Aldrich family was trying to hold on. In 1850 pioneers Peter Aldrich, seventy-seven, and his wife Hannah, eighty, were still farming on their own on a small place worth $500. Their son Peter P., thirty-one, lived nearby with wife Margret and son Charles. In May 1855 the elder Peter buried his wife of some sixty years and moved in with granddaughter Eliza Harp and family. Eliza, wife of Henry O., was the daughter of Peter P.'s sister, Tryphena Jansen, then fifty-six, a widow and another member of the Harp household. Four generations lived together under one roof for a short time.

By 1855, Peter P. and Margret had separated—but for good reason. Margret moved in with her widowed father, David Van Leuven, eighty-two. Peter, a laborer, went to live with Hugo Slater. Peter P. and son Charles likely supported Margret as she took care of her elderly father, a weaver. The War of 1812 veteran was no longer able to perform farm labor, but weaving, perhaps of rag carpets, could help support the household. By 1860, David had died, Peter P. and Margret were back together running a farm, and Tryphena, sixty-eight, had moved next door into the household of a granddaughter.

Margret Aldrich died in May 1864 around planting time. Her duty as wife, mother, and daughter done, she was interred in a small burying ground behind the farm of Henry O. Harp. Charles's father, Peter P., then fifty-six, boarded with brother-in-law Henry O. Van Leuven, but Peter P. retained ownership of his farm.[6] By 1865, Charles, thirty, was a first-time head of household. That year he owned 118 acres worth $500 and worked as a boatman on the D&H Canal. His family harvested buckwheat, Indian corn, and potatoes, churned 500 pounds of butter, harvested 200 bushels of apples, and made 4 gallons of cider.

Census records of 1865 reveal that the four wealthiest families in the Trapps were hotel keeper John C. Shaffer with real estate valued at $8,000, sawmill owner George Davis at $5,000, tavern keeper Ben Fowler at $3,000, and sawmill owner Catharine Enderly, widow of William, at $2,500. Besides small grain crops, Shaffer grew apples and raised pigs, Fowler had sheep and pigs, and Enderly had apples, pigs, bee hives, and milk cows, from whose milk she churned 400 pounds of butter. (In 1875 her son William produced 700 pounds of butter from seven cows.) George Davis also made butter, but much of his income came from nearly 2,000 acres of woodlots. These families were luckier than most, for their non-farm incomes supplemented their farm production. Most Trapps families barely scraped by.

Some farmers made a go of it in the most inhospitable parts of the Trapps. James Van Leuven, older brother of Henry O., lived behind Millbrook Mountain at the far southern end of the hamlet where the Coxing Kill spills away from its source, Lake Minnewaska.

A dirt road looped south from the turnpike and ended at James's house. A nearby spring gave him water to work the land. In 1855, James, wife Phlinda, and their children ran a 200-acre farm worth $600 with four cows, two oxen, one horse, and four pigs. They raised corn, buckwheat, and potatoes, and sold maple sugar, used as table sugar.

At the northern edge of the Trapps, two farms lay high up on a rocky saddle, the highest one could live and make a go of it. High up where trees grow stunted and gnarled, where storms blow hard, where a dirt road from Rock Hill ended, lived John C. Washburn and David B. Osterhoudt, neighbors from at least 1855 to 1880. A shallow well provided the only water around and made life possible on the all but barren knolls called Dickie Barre and Ronde Barre. The two owned their farms and lived in log cabins. To the east, far below, was the Coxing Kill; to the west was the Peters Kill. Enough topsoil existed when Washburns and Osterhoudts had settled there, but years of plowing and grazing had eroded the poor soil. Still, for some time, these families made do at one of the harshest and most unforgiving places on the mountain.[7]

Trapps farmers found two hardy crops that did well on the mountain's stony haunches—rye and buckwheat. "Rye grows in poor soils—the 'gravel ground' described in local

Laborers harvest buckwheat in a boulder-strewn field. The photograph appeared in Richard Lionel De Lisser's Picturesque Ulster. *Courtesy of HHHC-EML.*

Women and girls of the Enderly family climbed atop a massive hay bale during a harvest of the 1910s. Courtesy of descendant Joan Hayes Wustrau.

deeds," Michael Swift observed. "In fact rye needs poor soil or it 'lodges'—fails to stand and is hard to cut." Rye was scythed, then stored in the barn, where during the winter it was beaten to remove the chaff or husks from the grain. Rye was ground into flour, fed to the animals, or distilled into whiskey, Swift noted. Rye straw made the best bedding for horses, Jared van Wagenen, Jr. wrote.

Buckwheat is a high-altitude crop that also seems to grow right out of the rocks. Karl L. Brooks noted its usefulness in regions such as the Trapps: "Buckwheat is sown in May or June and ripens rapidly, thriving on the poorest soil. Mixed with bran, chaff, or grain, its seeds are sometimes given to horses, either whole or broken, and it is especially good for fattening pigs." The plant's fragrant flowers attract bees, which make a rich, dark honey from its nectar. While buckwheat is not a grass or cereal, it is harvested and milled like wheat. De Lisser's *Picturesque Ulster* shows a photograph of "buckwheaters" humped over like the boulders littering the Catskill Mountain field they harvest.[8]

More extensive orchards existed elsewhere in the northern Shawangunks, but rockier ground in the Trapps supported fewer fruit trees. Still, families raised apples, pears, and peaches for their own consumption or to sell, with apples the number-one fruit grown. In 1864 three Trapps farmers harvested apples, and Catharine Enderly was one of two who raised bees. The tiny honey makers would have pollinated many of the fruit trees and some crops in the hamlet. A decade later at least eleven Trapps families grew apples, with a total of 192 trees yielding well over 500 bushels. Most had about a dozen trees yielding 25 to 50 bushels, with Ben Fowler's 60 trees yielding 300 bushels.

The Smileys of Minnewaska cultivated strawberries and raspberries, and harvested cranberries growing wild near the lake. At times the cranberry harvest was large enough to warrant help. In 1901 a local article stated that employees there "pick the berries and the people residing in the neighborhood also take a hand at the work."[9]

One Trapps farmer tried growing the more unusual crop of ginseng, prized for centuries as a healing plant around the world. Some 17 million pounds of wild ginseng were shipped from America to Asia between 1820 and 1903, noted Lucile P. Macera. Five thousand ginseng farms existed in New York State by 1900, Dr. Robert Beyfuss said. In April 1903 former schoolteacher John H. Voorhis returned to the Trapps after ten years living in New Paltz and working as a printer for Ralph Le Fevre, who reported in his newspaper that John would "make a speciality of raising poultry and ginseng." John might have built outdoor cold frames called lathe houses, under which ginseng was grown, explained Jennifer May. John, though, might not have raised ginseng for long. Macera noted that ginseng has "a long maturing period," taking years to realize any profit. She outlined other problems faced by ginseng farmers of 1900: "mildew, blight, root rot, snails and eelworms"; thieves from New York City, who made night raids on upstate ginseng farms; and a widespread fungus that killed many crops during a "very wet summer of 1904."[10]

The Coddington homestead farm at Lawrence Hill, located on the southwest corner of Lot 1 of the old Nineteen Partners tract, was typical of nearby Trapps farms of the early 1900s. Evelyn Coddington DeWitt sketched her family's farm with a farmhouse and outbuildings, including granary, wood house, wagon shed, barn, shop (wood shop or hoop shop), root cellar, chicken coop, smokehouse, and outhouse. Other separate buildings might be the summer kitchen, which kept heat away from the main house, and the corn crib, built several feet above ground to keep out rodents and dampness.[11]

Mountain farmers struggled to grow enough grain for their own families. If they had any to spare, they might use it as barter. It was not worth going down the mountain to pay a miller to grind the small amounts of corn, oats, rye, and buckwheat grown. Small amounts of these crops might have been ground at home using small mills powered by hand, horse, or water. For the most part, Trapps farms had a half acre of this or an acre or two of that. Three or four acres of one crop were not large enough to consider a cash crop. Crops were grown to be stored for family use—grain in the granary, corn in the crib, and vegetables or fruit, such as potatoes or apples, down in the root cellar.

The Mountain's Wild Harvest

When out in the woods beyond the farm, mountain people saw opportunity in nature's bounty. The wild harvest was marked by the seasons. In spring, tapping maple sap. In summer, picking berries and wintergreen. In fall, gathering nuts. Each wild crop brought additional food and welcome income.

Sugar bush operations were run in early spring when snow still blanketed the land. Daniel Smiley II of Lake Mohonk described the time when maple sap runs, when temperatures dip below 25° F. at night and rise to about 40° during the day. This perfect "sap weather" occurs from February to mid-March, some years as late as early April. As a boy during the 1910s, Daniel learned the traditional method of collecting sap. Gather staghorn sumac or joe-pye weed. Cut the stalk, then burn out the inside with a hot stick, wire, or ice pick. Tap the spouts or stiles into trees and hang buckets from them. Wait for the sap to run clear.

Famed American nature writer John Burroughs knew well the annual ritual of maple-sugaring. He grew up on a Catskill Mountain farm and spent much of his adult life at Riverby, his home and vineyard on the west shore of the Hudson River south of Kingston. Burroughs described what happened after the maple sap was collected:

> In my sugar-making days, the sap was carried to the boiling-place in pails by the aid of a neck-yoke and stored in hogsheads [barrels holding about sixty-three gallons], and boiled or evaporated in immense kettles or caldrons [*sic*] set in huge stone arches; now, the hogshead goes to the trees hauled upon a sled by a team, and the sap is evaporated in broad, shallow, sheet-iron pans,—a great saving of fuel and of labor. Many a farmer sits up all night boiling his sap, when the run has been an extra good one, and a lonely vigil he has of it amid the silent trees and beside his wild hearth. If he has a sap-house, as is now common, he may make himself comfortable; and if a companion, he may have a good time or a glorious wake.[12]

The sap or sugar house was an open shed where sap was boiled down. Boiling a certain percentage of water away makes maple syrup. Boiling it all off makes maple sugar. Burroughs wrote that maple sugar was clarified "with milk or the white of an egg.

Above: This yoke and bucket were used to haul maple sap. The bucket was made in Ulster County. Photo by, and yoke courtesy of, Robi Josephson. Bucket courtesy of Barbara Rubin. Right: Maple trees were usually tapped from February to mid-March when snow might still blanket the ground. Once collected, maple sap was boiled down into syrup or sugar for home use or sale. Courtesy of MMHA.

The product is virgin syrup, or sugar worth the table of the gods." We don't know how much Trapps families earned for maple sugar (Bide Snyder of Woodstock cited 8¢ a pound after the Civil War), but we do know it takes some 40 gallons of sap to make 1 gallon of syrup, and about 1 gallon of syrup to make 7 pounds of sugar. In 1855, James and Phlinda Van Leuven made nearly 51 pounds of maple sugar from about 7 1/4 gallons of syrup from 290 gallons of sap. In 1875, John H. Voorhis made 10 pounds of maple sugar and William Enderly made 15 gallons of maple molasses, which was sap boiled to the consistency of molasses.[13]

Come summer, wintergreen was gathered for distilling into oil and medicine, and for making garlands. The plant had many medicinal uses, including as a pain reliever and cleanser. Dennis Williams picked wintergreen as a boy west of Lake Mohonk during the 1860s. He helped his mother pick the berries and take them to a nearby still, where the distiller paid them 1¢ a pound. It took many a light-weight plant to make up that pound. Dennis wrote, "Many mountaineer women folks picked wintergreens for this still." Mother and son used a two-foot-long, homemade rake to pull fallen leaves off the wintergreen, sometimes uncovering the unwelcome sight of a snake.

Wintergreen was a cash crop, similar to butter and eggs, for women and children. In the Trapps, James Van Leuven's three daughters picked wintergreen. Wintergreen stills, though, were a man's purview. Before his death in 1907, German immigrant Johan Otz operated a still near his Trapps home. Johan made a kind of medicine or cough syrup from the plants and sold it, said his great-grandson Don Otis.[14]

Theodore Ruckle described two kinds of wintergreen plants processed at his father's distillery in Ellenville: "the long-spindly type," only good for eating; and the short, wide-leaved type used for its oil. Local pickers brought the short type to his father in burlap bags over 6 feet high, which held about 100 pounds of stuffed leaves. Pickers might earn as high as $2 to $3 a bag. The leaves were dumped into copper kettles that could hold on average between 900 and 1,100 pounds. The distilling process resulted in little white beads that were "pure wintergreen oil and very strong to the taste," Ruckle wrote. Pure oil was placed in 5-gallon cans and taken personally to buyers, as "people were afraid to ship it for fear the cans might be punctured en route," he added. Ruckle explained "that concentrated wintergreen is very powerful and poisonous and that it takes only a very few drops of wintergreen oil to flavor a gallon of water."[15]

Picking nuts was an annual fall ritual across the region. No written records exist of Trapps people selling nuts, but families would have gathered them for household use and sold any extra. Nuts are a good source of protein and fat, which helped people stay warm in winter while living in drafty, uninsulated homes. Black walnuts and butternuts were also used for dyes. Nuts were once part of the social scene, set out in bowls with nutcrackers for company to enjoy. They were a treat especially during holidays, when youngsters tested their strength against "hard nuts to crack."

Five nut trees are native to the Shawangunks—chestnuts, hickories, walnuts, butternut, and beeches. The first was "said to have covered the ground like marbles," wrote

John and Richard Polhemus. "Animals and people browsed the woods for chestnuts, and the nuts were a staple in the diets of hogs." Of the walnuts, the most prized is the black walnut, a rare native on the range that was sometimes planted.[16]

The Mountain's Blue-Black Gold

"Nothing stops a huckleberry picker," said Ellenville native Katharine T. Terwilliger. Not heat, nor storms, nor snakes kept berry pickers from their appointed rounds. For centuries, "the little black treasures" have beckoned berry pickers to Shawangunk Mountain ridges from June to September.[17] From the 1800s to the mid-1900s, berry picking was one of the chief cash crops for Trapps families.[18] For many, money from berry picking meant survival for another year.

Mountain ranges with thin, acidic soil, like the Shawangunks, are perfect homes for berries, noted Bradley Snyder. The local berry region once spanned from the New Jersey Highlands to the Catskills. "The great berry country of the main Shawangunk range," noted *The New York Times* in May 1889, "stretches from Lake Mohonk on the north to the Delaware River on the south. … Ellenville, in Ulster County, is the heart of this district." During that century, "itinerant pickers followed the ripening berries from the South to Canada," noted Daniel Smiley II.[19]

Typical of local berry news was this August 1895 brief from New Paltz: "A number of our village people who went to the Traps region for huckleberries on Saturday last got caught in the shower and were thoroughly wet. There was also quite a fall of hail near the Traps."[20]

Thunderstorms and poisonous snakes posed very real risks for pickers. Mohonk guidebooks warned guests to stay away from Eagle Cliff during storms. Low or lack of vegetation there and at other exposed spots offered little or no cover. The July 1886 *New York Times* noted the area around Lake Awosting, south of Lake Minnewaska "produces nothing but huckleberries and rattlesnakes." When a family met up with a rattler, the paper said, either parent "grabs a stick and kills it in a business way, and the picking goes on." The July 1897 *Times* noted "every great camp of huckleberry pickers is also more or less a slaughtering party for snakes." Pickers protected themselves with high boots or sheepskin leggings with the wool side out. If a snake struck the leggings, its fangs hooked onto the wool and it was unable to escape. "Then it is easily trod on and killed," the story ended.[21]

~~Susie~~ David
263 w 19th st
~~Dave~~ Susie
29th east 3rd Ave
T8961502 1

n age and ability, with two types of expert pickers. ay, clearing all the berries from one bush with "two d," a June 1885 *New York Times* stated. A "'crack' day. A family of five or six, a typical group size, fter filling their boxes and pails, pickers needed to r speculators. These middlemen sold the fruit in or shipped it directly from local depots.

ry woods, or pickers went down to the peddlers. May 1889 *Times* described, "which they blow as notify any party of berry pickers who may be near pickers had berries for sale, they replied with their and pickers thus connected to make a deal.[22]

More typically, pickers hauled the berries down the mountain at the end of each day. "Most of the berries are carried down the mountain several miles in wooden boxes on the backs of the pickers and fastened with strips similar to a peddler's pack," a news article reported in August 1906. "These huckleberry pickers are very expert in going up and down the mountains. Each person is armed with a long stout pole or staff, and comes down the mountain at a most astonishing rate of speed, springing from rock to rock with the agility of an alpine guide."

The June 1885 *New York Times* astonished readers with an account of "the greatest burden on record ... carried down one of the narrowest and steepest passes of the Shawangunks, near Sam's Point, by a middle-aged German woman. She carried a large tub full of berries on her head. To her back was strapped one of the bushel boxes filled to overflowing. In each hand, she carried a 12-quart pail full of berries. She 'toted' her load six miles without taking a rest, and besides her own burden of berries four of her children trailed after her, each bearing either two pails or two baskets filled with the day's picking. The party reached the village [of Ellenville] with five bushels of huckleberries, which netted them about $7.50."

It was a good day's pay for the times.[23]

While not as commercially active as Ellenville, the Trapps became a busy collection spot for berries, especially after the turnpike went through in the 1850s. Excerpts from the following two articles appeared in August 1878 and 1886 in New Paltz.

> The huckleberry wagons, some two dozen in number, pass through our village almost every night, between 1 o'clock and daylight. They come from the Traps region. The berries are ... [sent to] ... New York. The berries are in boxes holding about half a bushel each and the teams take forty or fifty boxes each.

> THE HUCKLEBERRY BUSINESS—Several huckleberry wagons, a portion drawn by teams and the balance by single horses, come to New Paltz about 6 a.m., every morning. These wagons come down the New Paltz and Wawarsing Turnpike, from the Minnewaska region. The berries are purchased by Josiah J. Hasbrouck and shipped to New York. A small portion are peddled out about this village every Tuesday and Friday. ... The single wagons bring about 20 boxes, holding usually 12 quarts each. The teams bring double that quantity. ... The men who go to Poughkeepsie with huckleberries pass through our village about midnight. They drive a long distance before they get to our village, and are on the road all night. Sometimes they manage to catch a nap while on the road. It is rather a hard life on account of the loss of sleep. Otherwise it might be pleasant enough.

Five known Trapps men became berry peddlers—sawyer George Davis, George Coddington, a Mr. Decker, Abram Billingham, and Lewis Wynkoop. One of the earliest records of a "huckleberry shanty" in the Trapps was noted in February 1869 belonging to George Davis. George Coddington had married Rachel Fowler, daughter of taverners Ben and Laney.[24] Mr. Decker was most likely Howard, whose family lived at the western edge of the Trapps hamlet center (in an area today marked by Jenny Lane, an unpaved road). A New Paltz paper noted, "Abm. Billingham and Mr. Decker go through our village at [*sic*] with loads of huckleberries for the Poughkeepsie market." Howard married Abe's daughter, Elizabeth or Eliza. Abe and Howard also became Trapps School trustees.

Abe Billingham was crowned "The King of the Trapps," noted Peter H. Harp. Abe was born ca. 1846–49, likely in Neversink, Sullivan County, with Abe or his father first arriving in the Trapps by the mid- to late 1860s. Abe married Rachel Enderly of the Trapps, with whom he later ran a general store, small hotel, and post office in the hamlet. Two New Paltz articles about Abe ran in February 1885 and in August 1897:

> Abm. Billingham, our mountain groceryman is doing quite a business this winter in hoops and trade. Go in [*sic*] Abe you are a clever fellow—hoops in the winter and huckleberries in the summer.

> The people who take loads of huckleberries to Poughkeepsie at this season of the year have a lonely night ride, and the farmer who raises potatoes and disposes of them in the same market, should start before daylight if he wishes to dispose of them to good advantage. At the first crossing of the ferry boat at six o'clock a few mornings since we found Mr. Hiram Minard who had come from Springtown with a load of potatoes. Here were also two teams with loads of huckleberries, which had traveled a great part of the night, one having come from Abm. Billingham's in the Traps neighborhood and the other from a much farther distance in the mountains. Now that we have a direct train from Poughkeepsie, farmers can dispose of a considerable amount of produce in that city at a much less cost for transportation than to take it by team and wagon.[25]

One berry peddler of the Trapps became so renowned he acquired a trade nickname. Lewis Eamon "Huckleberry" Wynkoop was a member of one large Trapps family and married into another. He was the son of Joshua Wynkoop and husband of Sarah Enderly. (Sarah was the sister of Rachel Billingham.) Their eldest son, Melvin, joined him in the business, with the two known as "The Peddlers." Perhaps Lewis's berry money enabled him to build the grandest house in the hamlet, a two-story house of plank. Lewis was sought after for his opinions, shown below in two local articles of 1880 and 1890. The Trapps was becoming known as Minnewaska, and Lewis's middle initial was given as "H." His nickname had become his middle name! Cantonville was a hamlet along the D&H Canal near Ellenville.

> Mr. Ean Wynkoop informs us that the huckleberries on the mountains are drying up on account of the drouth [drought], and that there will not be more than a third of a full crop. Mr. Wynkoop suggests that a greater profit would result to people engaged in the huckleberry business, if a couple of good teams would be employed in transporting the berries that are brought down on this side of the mountain instead of taking a score of horses, and their drivers, each going on his own hook. Cooperation would pay in the berry business as well as elsewhere.

> On Friday last Lewis H. Wynkoop of Minnewaska, came to Ellenville for a load of berries. On his return in the afternoon, when near Cantonville, his horse kicked over the dashboard and striking Wynkoop broke his right leg above the ankle. The injured man was removed to the hotel of A. Jacob, where Dr. Ward set the bone. Afterward, he made two attempts to start home, but was compelled to return on account of the pain resulting from the injury.—Ellenville Press.[26]

Berry money could make all the difference. A July 1871 *New York Times* noted earnings served "to replenish many pocketbooks, and pay many store bills, and often add to the already well begun account at the savings bank." A July 1886 *Times* said berry money earned during "an abundant season is a godsend." In July 1897, *The Times* estimated pickers would earn an average of 6¢ a quart, combining for a total of $4,800 a day or $100,000 for picking seven days a week for a three-week season. (Figures were for the counties of Ulster, Orange, Sullivan and Greene.)

The Times reminded its readers:

> Few people living at a distance from the huckleberry-growing country realize the real import of a big huckleberry harvest. It means shoes, stockings, and warm raiment for the coming Winter and a thousand little comforts to the myriads of people, young and old, who swarm the mountain sides and upland meadows culling the wild and delicious fruit. This year, fortunately for the berry pickers, the harvest will be an enormous one.

Fortunately, too, for the railroads, which ran special "huckleberry trains." In good years everyone benefited—pickers, peddlers, railroad owners, consumers. A September 1878 story from Ellenville noted the berry trains had stopped for the season, adding, "It is safe to say that 10,000 bushels have been marketed at home and abroad this year, of berries picked between Lake Mohonk and Wurtsboro. The profits of the crop probably exceed all the profits of agricultural operations on the land on both sides of the mountain in the same locality for a mile in width."[27]

People near and far relished Shawangunk huckleberries. The market extended from Buffalo to Boston and south to New York City and likely beyond. *The Ellenville Journal* observed in June 1879, "Shawangunk mountain berries always command the highest price in the market." An August 1884 *New York Times* reported, "… it is said the Buffalo people, having a tooth for delicacies, have decided that the berries picked from the Shawangunk and Blue Ridge Mountains are far superior to those found further west." The greatest demand was in New York City. Two-thirds of the crop destined there were "either produced in Ulster County or reach the market through it," noted the July 1897 *Times*. Berries picked from Sam's Point to Paltz Point, including in the Trapps area, ended up on kitchens and dining room tables from Lake Erie to the Atlantic Ocean.

Most years, crops ranged from good to great. Yet berries could fail because of drought or frosts at flowering time in May or picking time later in summer. In June 1880, Lewis Wynkoop said, "the huckleberries on the mountains are drying up on account of the drouth [drought], and that there will not be more than a third of a full crop." In late July, Mother Nature came to the rescue: "The huckleberry crop has been cut short by the drought, and shipments are not averaging as much as last year. The late rains, however, will give a good crop of late berries."[28]

It was fire that transformed berries into the mountain's blue-black gold. "The only cultivation the berry gets," reported the July 1886 *New York Times*, "comes from the great forest fires which sweep over the mountain slopes in seasons of drought, destroying dead branches and choking shrubs, and clearing the soil for new and thriftier growths." Picker and peddler alike had learned to fan the flames of their livelihood. An April 1871 story reported "an extensive fire southwest" of New Paltz. "The huckleberry pickers were as usual supposed to have started the fire for the benefit of their favorite fruit."[29]

Many of the big fires in the Shawangunks occurred around Ellenville and south. Pickers set the fires in the spring and fall, reported an August 1884 *New York Times*. The fires generally "were condemned by popular sentiment," the paper said, for they could destroy homes, farms, harvests, businesses, and good timber. Setting fires was considered a misdemeanor by law, subject to fines, jail time, or both. "But few of the offenders are punished," *The Times* stated, "although some efforts have been made to bring the guilty parties to justice, and, in the interest of forestry, it is a pity the crime cannot be stopped."

Coverage of Shawangunk mountain fires continued down in New York City. A July 1886 *New York Times* was only partly accurate when it reported, "The lands are owned by the State, confiscated for non-payment of taxes, or by non-residents, and the ominous rattle of the native reptile alone challenges the intrusion of the casual berry picker." Yet, berry stands also might be owned by local farmers, speculators, the summer hotels, or timber and tanning companies. A May 1889 *New York Times* made two other points: owners generally put no restrictions on pickers except to "reserve the right to the best pickings"; and mountain lands "long considered worthless" were becoming valuable.

So began the modern-day real estate boom in the Shawangunks.

Caring for the general welfare, though, had little meaning for those facing a grim winter. Pickers needed berries, and berries needed fire; it was the only way to guarantee annual yields. Plus, the public demanded the best berries. In areas not burned over by fire, berry bushes "all run to wood," producing berries full of seeds, noted a July 1897 *New York Times*, which also stated: "This burning over of the mountains where huckleberries grow is due to the fact that the very best and choicest fruit grows where the ground has been burned over the previous Autumn. … The best crop of berries the bushes bear is on their first year, which explains why the berry pickers start their fires in the woods."

A June 1885 *New York Times* went so far as to reassure metropolitan readers with this headline: THE BUSHES NOT INJURED BY THE FOREST FIRES.

In July 1897, *The New York Times* said it best: "During the season the huckleberry is a staple article of food with the farmers. They are eaten from big bowls holding a quart or more. Flooded with rich cream and with the tops sprinkled with powdered sugar and nutmeg, the farmer dines like a King."

The Trapps people perhaps dined as royalty at home but, in reality, harvests from farm or forest only went so far. For mountain families, survival also meant reaping wood and stone to fill a far hungrier appetite, that of an ever-expanding industrialized nation with an ever-swelling population.

• 9 •

Wood and Stone Cutting

Summer 1880, the Enderly sawmill on the Coxing Kill. Hiram Enderly sets up the old up-and-down saw to gnash a log. A driver hauls away a load of milled wood. Nearby, the blacksmith sharpens dull blades and forges new metal in red-hot fire.

Up on the knoll called Dickie Barre, men bust out massive slabs of Shawangunk grit with black powder and sledgehammers. The slabs are shaped, "wagoned down" to shops for finishing into millstones, then bound for markets across America.

Smells of sawdust, smoke, and manure prick the air. Several times a day, carriages filled with guests pass by the mill, crossing the Enderly property on the way between Lake Mohonk and Lake Minnewaska. Hiram looks up, recognizes one of the drivers, and hollers a hearty hallo to his young neighbor Oscar Harp.[1]

The Woodcutters

Life in the Trapps depended on wood and stone. The characteristics of the northern Shawangunks shaped the character of its people. In 1880, Trapps resident George Coddington simply called himself "woodsman" for the Gardiner census taker.[2]

During the nineteenth century the bark of oak and hemlock, two common trees growing on the mountain, was highly desired for the tannin within. The union of tannin with the gelatin of rawhide turns the latter into leather, once crucial for a society dependent on horse power. Oak tannin made heavier products of leather; hemlock tannin, lighter ones.

By the mid-1800s, New York City led the world in leather production. At first, tan bark was transported from the forests to city tanneries, but then it was found to be cheaper and more efficient to tan the raw hides, shiploads of them imported from South and Central America, right in the bark woods or in the vicinity of them. The Catskills and other heavily forested regions boasted virgin woods, clean water, and nearby waterways for shipping, such as the Hudson River and Delaware and Hudson Canal, noted Alf Evers. So, too, did the Shawangunks.[3]

A tannery operated first by Rowland DeGarmo and later by his son, William, ran full blast on the Shawangunks' eastern haunches at Butterville. There, "oak bark was used exclusively for tanning," wrote Ralph Le Fevre. "[DeGarmo] would send around his teams to the farmers at butchering time and gather up hides, which he would tan on shares." When John Stokes cut oak bark at Mohonk, he likely hauled it down to Will DeGarmo, the closest tannery. The McKinstry family of Gardiner ran a more sizable tannery in the Wallkill Valley, with an annual capacity of 7,000 sides, cited Hamilton Child in 1871. One local resident recalled "seeing heaps of bark piled as high as a house" at the McKinstry tannery. Tanneries such as McKinstry's were often targeted by thieves. William Enderly and Alonzo Dingee of the Trapps struck the McKinstry tannery in spring 1878, as they had the Coe and Abrams woodlot three years earlier.[4]

Andries Van Leuven peeled oak bark on the east side of the mountain below Mohonk in May 1882. Andries was born in the Trapps but had moved to Gardiner. McKinstry or Riley Addis might have peeled bark in the Palmaghatt Ravine at that time, noted Nan Knickerbocker. She related that Alfred Smiley, while bushwhacking one day, found many hemlocks debarked in the ravine. "His love of nature was outraged," she wrote, "and he was determined to save those trees by buying the ravine." Alfred bought the property from a man who said he was the owner, but then two more men claimed ownership. Alfred stated "that before he finished buying it he had really bought the Palmaghatt three times."[5]

Bark peelers swarmed over the mountains from mid-May to about July 4 when the sap runs. By early July a new layer of bark grows, the sap solidifies, and the bark is "glued" back to the tree. During that period a year's supply had to be cut, noted H. A. Haring, who explained the process: the peeler made a horizontal gash around the hemlock at its base and a second about four feet up the trunk, the desired length for tan bark. He next made vertical gashes around the tree about twelve to sixteen inches apart between the two horizontal ones. The bottom sections of bark were peeled off. The tree was cut down within the peeled section to protect the rest of the bark. Once down, more bark was peeled in four-foot lengths up to the first branch. Bark higher than the first branch was too "broken up" to be worth the effort and cost.

The bark spud was the peeler's tool of choice. The wooden handle had a balled top, which enabled the peeler to more easily grab the top and exert pressure on the base. The wrought-iron base had sharpened edges. The base was elongated like a chisel but rounded

Teams of men cut bark from hemlock and oak for tanning leather. A bark spud at lower left, propped up against the stump, separated the bark from the tree. Engraving by R. H. Pease, Sc. [Latin, sculpsit] *for P. T. Barnum's* Illustrated News, *May 7, 1853. Courtesy of the New York State Library.*

like a spoon, thus fitting under the natural curve of the bark to peel it away. After the tan bark was peeled off, it was stacked up until winter for easier sliding down snow via oxen or horse teams, then hauled down to tanneries nearby or ones set up in the bark woods. The tan bark was crushed, then "cured" in a series of tanning vats. The tanning process was relatively simple, but finishing a hide might take up to a year. A. W. Hoffman noted that it was once common to see fifty tan bark teams in a row along the main roads. We can imagine such a steady stream of wagons coming down from the Trapps and Mohonk. *The New York Walk Book* cast a chilling figure—the Zadock Pratt tannery in the Catskills used over 200,000 cords of bark to tan 2,000,000 hides. Haring calculated that 3 to 10 hemlocks were needed to make 1 cord of tan bark, which translates to cutting down some 600,000 to nearly 2,000,000 trees.[6]

Industry hazards were many, with fire the most dangerous. Bradley Snyder observed about the Shawangunks: "In many cases only the tannin-containing bark was taken: all the wood was left to rot. The fires that later swept through the dry debris are believed to have been the most intense in local history." Case in point was an 1864 fire at Lake Mohonk when John F. Stokes lost $1,000 worth of marketable tan bark.

There was no such thing as forestry management. The trade collapsed for one reason—the hemlocks were largely wiped out. William DeGarmo left Butterville in early 1868 to take up farming in Kerhonkson. McKinstry and Addis were in business as late as the 1880s, perhaps when hemlocks and oaks could still be found in and around the Shawangunks. When the bark peelers left, many Catskill tannery towns became ghost towns. Numerous farms that had fed hundreds of men, their families, and livestock also were abandoned, Haring wrote. Yet, things were different in the Shawangunks. Trapps and other mountain families continued to farm, log, cut stone, and pick berries, but they now had a new source of steadier income working for the Smiley families of Lakes Mohonk and Minnewaska.[7]

• • • • • •

They were shrouded in soot, they stank, they faced death. They were the charcoal burners, and their job "was not for any normal human being," wrote Eric Sloane. Yet, many a normal human being, including Trapps men, burned charcoal by necessity. "An important way of making a living on the farm in the 1800's was that of making and selling charcoal," noted August Wohltman.

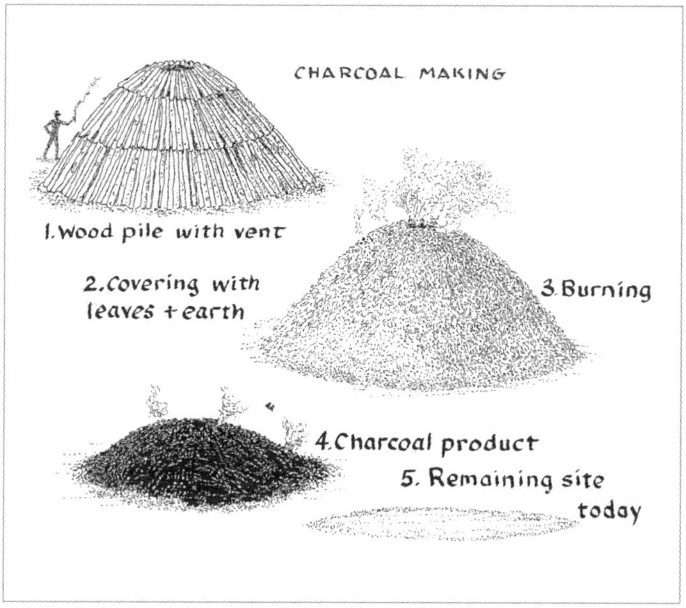

The Shawangunks, and the Trapps in particular, were renowned for charcoal. Burning a mound of "Shawangunk chark" might take weeks and was very dangerous work. Illustration by, and courtesy of, Jack Fagan.

Burning twice as hot as wood, charcoal was the fuel of choice until replaced by cheaper coal in the later 1800s. John and Richard Polhemus wrote that about 600 acres of wood in northwestern Connecticut were once turned to charcoal to keep a blast furnace running for one year; three furnaces swallowed up 36,000 acres of woodland in 20 years. Wood demand was once so high for tanning, charcoal, building, and fuel, it is no wonder many forests were clear-cut by the mid-1800s.[8]

A charcoal burner might have a signature style for building a good mound, but mounds were similar in design. About twenty-five to forty cords of logs were piled around a center stake, then covered with green or wet vegetation and topped with sod. Small openings around the base allowed for air. Daniel Smiley II likened the mound to "a flattened beehive." He added that any available tree was used, but "the heavier the wood the heavier the charcoal." One cord of four-foot-long wood typically yielded some thirty bushels of charcoal, noted Jared van Wagenen, Jr.

Building a mound was the easy part. The danger began when it was lit. Small mounds took about one to two weeks to char, but mounds "30 feet or more round" averaged a month, noted Sloane. Fair weather or foul, perhaps taking shelter within a hut, the charcoal man lived next to his mound as if tending a feverish wife. Mostly it was a lone man on twenty-four-hour watch. To turn his back was to court disaster. Too much air, and the

Detail of "The Charcoal-Seller," Appleton's Journal, *February 12, 1870. Courtesy of the University of Michigan Library, Making of America Digital Library.*

mound could break into flames. Charcoal burners breathed in noxious fumes and were enveloped in pitch-black soot. Sloane called it "the blackest job in America." When the charred coal was ready, the burner plugged up the vent holes and let the mound cool. Yet, peril still lurked deep within. "Knowing how and when to walk on the mound," Sloane wrote, "was an art in itself, and many a man fell through into the furnacelike heat."[9]

Shawangunk charcoal, or "chark," was carted locally to homes, iron forges, and businesses, including the Smiley resorts. It was also shipped down the Hudson River to New York City. Charcoal "was hauled to the river for market in large bins with flaring sides with an iron shoe blocking at least one hind wheel down the steep parts of the Minnewaska old stage road," noted Sanford A. Cross. When the charcoal men came through town, it was a big event; these men looked like creatures from another world.

Mountain charcoal was prized for broiling and baking well into the 1900s. A 1943 *Poughkeepsie New Yorker* (today's *Poughkeepsie Journal*) article noted that Mohonk "guests long have known that no steak in the world has the same flavor as one broiled over Shawangunk 'chark.'" One Mohonk baker walked up the path from the Clove at 3 AM, his way lit by lantern, to start the ovens with charcoal, which gave an even heat to bake the day's bread, recalled Daniel Smiley II, whose family at Mohonk bought charcoal "from individual burners in the Trapps." The Smileys of Minnewaska must have, too, for it was readily available and was made practically at their front gate. The Van Leuvens of the Trapps were particularly renowned for charcoal. Henry Osterhoudt "Osha" Van Leuven was the first of his family to burn it, Kenneth E. Hasbrouck wrote. Henry's son, Hiram, known as "Daddy Hype," became "one of the county's famous figures" for charcoal, noted the same Poughkeepsie article.

Trapps resident Edwin Davis sold chark to John June of Dutchess County. "For the past two years," a local 1892 article said, "Mr. June has purchased all the charcoal that he uses in his tin business of parties living in Ulster [County]. He knows good charcoal when he sees it." An 1896 article reported: "Smoke almost continuously rises from the wood lot purchased by Daniel Hedden [of Gardiner]. The lot lies near the precipitous rocks on the opposite side of which the popular summer resort Lake Minnewaska is situated.—The wood on the lot is being converted into cordwood and charcoal."[10]

• • • • • •

Wood fueled the nation, but also built it, and that made sawmills essential. When George Davis built a massive sawmill on a two-story stone foundation during the mid-1850s, it was the most substantial structure in the Trapps and gave work to every able-bodied man in the vicinity. Davis dammed a millpond above roaring Peters Kill Falls, then built a wooden sluiceway to channel water to the mill below the falls.

In September 1874, fire destroyed George's sawmill. George had no insurance and, as far as we know, did not rebuild. Yet, George won a contract in 1877 to furnish eight-foot ties for the Wallkill Valley Railroad. These were likely hewed ties, preferred by the industry, Walter Needham noted. Hand-hewn ties were much more labor intensive than milled ones, but also gave work to local men. In the early 1890s, M. H. Pike wrote about the Davis sawmill: "The remains of a substantial mill, half-way down the gorge, and near the road, tell that this wild creature [the Peters Kill] was once made the servant of man's need for lumber."[11]

The Enderly mill on the Coxing Kill operated much longer as an industrial center of the Trapps. The Enderly mill, less substantial than Davis's, made use of a dramatic geologic feature—a deep crevice in the rock through which the Coxing Kill runs. (Today, this crevice is a popular swimming hole known as Split Rock.) It was a perfect spot to hold an overshot waterwheel, the most efficient type of waterwheel for a "small creek [that] doesn't have that much flow," noted Jeff Muise. A dam was built on a rock ledge above the crevice, creating a millpond. A wooden sluiceway carried water to the top level of the wheel; from there it spilled over and turned the wheel. Hence the name "overshot wheel."

The Enderly mill used an up-and-down saw, the kind that "only cut on the down stroke," noted Muise. Needham said of the saw's slowness "that the lumbermen used to

The Enderly sawmill, located at the "split rock" on the Coxing Kill, was one of two sawmills in the Trapps. Nothing remains of the Enderly sawmill, and the site today is a popular swimming hole on Mohonk Preserve land. Courtesy of DSRC-MP.

say if they wanted to saw a big log they would put it on and start the mill—and go home to dinner [lunch today]. Pretty soon after they came back from dinner they could start the next cut on the same log." We can imagine the Enderly men starting the saw, then walking a few steps for the main meal of the day. An old local saying described the up-and-down saw as so slow that "it was said to only cut one log a day."[12]

Every sawmill scrap was sold, even sawdust. Sawdust, used for packaging and preserving, once filled America's icehouses, including those at Lakes Mohonk and Minnewaska. Sawdust was packed into the double outer walls, set about a foot apart, as well as in between layers of ice. (Charcoal also was in demand as an effective insulator, keeping ice from melting and meat from spoiling.) In January 1892, David McKinstry, of the Gardiner tannery family, advertised sawdust at "the mill near 'the traps.' double-box load one dollar." Trapps sawyers likely sold sawdust, too. No one could afford to miss a sale.[13]

• • • • • •

Hoop shaving came into its own on the heels of the tanning industry. With hemlocks gone, in many places hardwoods grew up in their place. White birch, mountain ash, chestnut, white oak, maple, and hickory—they were the next to fall to the hoop shaver's axe, noted Richard Lionel De Lisser. New hardwood forests became "hoop pole farms," second-growth, managed woodlots of sprouting saplings destined to secure staves, the curved pieces of wood forming a barrel.

Hoops were a major income provider for subsistence farmers. John Burroughs wrote: "No man had any money until he sold his hoops. When a farmer went to town to get some grain, or a pair of boots, or a dress for his wife, he took a load of hoops. People stole hoops and poached for hoops, and bought, and sold, and speculated in hoops." Hoops were used to buy or barter items families could not make themselves. Sam Reavin of Accord, a hamlet west of the Trapps, wrote: "This land was not that productive, for most farmers, as far as having a cash crop. The biggest problem was getting that couple hundred bucks you needed every year." Large farms down in the valleys could raise dairy herds for milk, but mountain farmers "did everything they could think of" to make do.[14]

The Shawangunks generated hoops, and hoops generated news. These two news items from the Trapps ran in February and November 1885, the first concerning centers of the cement industry a few miles north of the Trapps:

> Our leading hoop dealers are busy drawing them to Rosendale, Lawrenceville and Binnewater. Hoops are not got out in so large quantities as last winter owing to low prices, from $3.25 to $3.75 per thousand. I.C.U. [the correspondent's pen name]
>
> George Van Leuven, a boy of 18 living in the Traps, recently shaved 1200 hoops between 6 in the morning and 6 in the evening.

Young George shaved 100 hoops an hour, or 25 hoops every 15 minutes, during that 12-hour period. Another young Trapps man did not survive his work shift. John Vandemark "was found dead, under a load of 900 hoops" when he did not come home as expected around New Year's 1886, reported a New Paltz paper.

Two leading hoop dealers of the Trapps were Lewis Wynkoop and Abe Billingham. In May 1882, Lewis's hoop shop went up in smoke; arson was suspected. At the time the shop stored about 8,000 hoops and some poles ready for shaving. The loss was about $50.00, part of those "couple of hundred bucks" needed to carry the family through the rest of the year.

Abe shipped 100,000 hoops from the New Paltz train station to New York City in October 1893, the first time he had done so from there. Two months later he shipped another 80,000 out of New Paltz via rail, a much faster mode of transport than the D&H Canal. Faster delivery benefitted customers and, in turn, generated faster payments to the dealer and hoop shaver. Hoop shavers from the Trapps "leave most of the money which they receive with our village merchants, and that helps the place." As the village of New Paltz profited from hoops, so too did many other local communities.[15]

At the industry height in the 1880s, some 8,000 to 9,000 hoop shavers worked in the Catskills alone, noted H. A. Haring. In 1887 one hoop dealer predicted Ulster County shavers would produce 45,000,000 hoops a year, with about two-thirds coming from the Shawangunks—earning more money than that county's "whole grain crop."[16]

Hoops were cut in the fall, recalled Mrs. John L. Hoffman, when it was easier to sight poles about 1 ½ to 2 inches in diameter. Some stumps could yield up to twenty to thirty "lusty straight-grained saplings," a hoop dealer told *The New York Times* in the 1890s. He added that every sapling could yield between two and five poles, and every pole, two hoops. Hoop cutting rotated with bark cutting, the harvest of either "controlled by the available supply of labor," Haring wrote. Poles stacked at the woodlots were hauled in winter down the snow on sleds or stoneboats, De Lisser wrote. He noted that the hoop shop or hut, deserted in summer, came to life each winter, heated by a great stone fireplace. An old hoop shop in the Town of Rochester measured twelve by ten feet by about eight feet high, a fairly standard size.[17]

Haring described a hoop shaver at work:

> At the 'shaving hut' the saplings were split lengthwise with a draw-knife. The workman sat beside a specially constructed bench, which much resembled the bench used by a shoe cobbler. He worked with his feet and body buried in a heap of shavings. He bent over the narrow bench. He leaned forward, with arms stretched ahead, to get a long reach with his two hands on the draw-knife; then he drew the sharp knife towards his body, leaning backwards like an oarsman

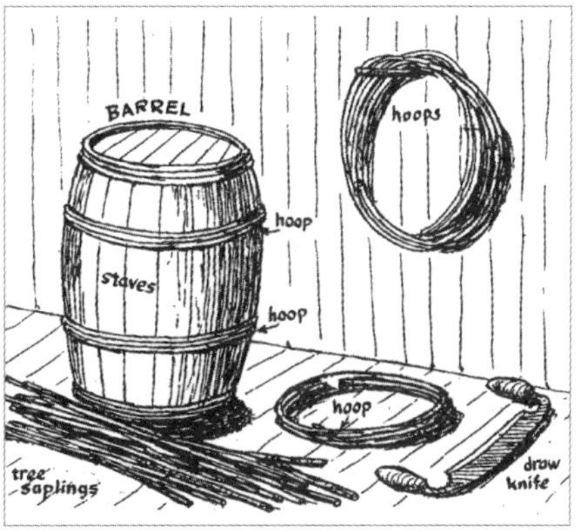

Left: Come fall and winter, hoop shavers moved into small outbuildings, called hoop shops, to work at the shaving horse. Detail of photograph in Richard Lionel De Lisser's Picturesque Ulster. *Courtesy of HHHC-EML. Right: A draw knife, lower right, was the hoop shaver's tool of choice. Illustration by, and courtesy of, Jack Fagan.*

rowing a boat in order to get a long pull. Next he shoved the sapling forward as it lay on the bench, and then repeated the long motion of the knife.

Hoops were then smoothed, tapered, and sometimes notched, depending on use. Expert shavers might finish more than 1,000 hoops in a day, Alf Evers noted. After hours bent over the bench, called a "shaving horse," the shaver straightened up inch by painful inch only to lash the hoops into mounds of 100 or more.

Wooden barrels held most everything, including apples, milk, crackers, fish, nails, and cement. Eggs were padded with wheat hulls or wood shavings, noted Michael Swift. Some containers required particular wood for special hoops: hickories for apple, pork, vinegar, and beef barrels, and for kegs; oaks for strapping boxes; ash for butter tubs, firkins (casks), and flour and sugar barrels. Each barrel typically needed four or six hoops.[18]

Local coopers or dealers bought finished hoops and shipped them across the country. Nearby Rosendale provided the largest market for cement hoops, with each barrel needing twelve to secure its hefty load. "From 1825 to 1900, Ulster County supplied the country with 50 percent of its natural cement needs," Jonathan D. King wrote. "At the height of the boom, Rosendale was choked with companies ... producing five million pounds a year." The cement was "durable as granite," Jeffrey Simpson noted. The jewels in Rosendale's crown include the Brooklyn Bridge, the Statue of Liberty pedestal, the wings of the U.S. Capitol, and the Washington Monument. Rosendale cement also was the standard for projects across the region, from construction jobs to patching dry walls.

The local hoop trade crashed when Rosendale cement was replaced by faster-setting Portland cement and when manufactured metal hoops, known as patent hoops, flooded the market for a fraction of the cost and labor of wooden ones. It was a one-two punch for Trapps families; yet, for a time they adjusted. Men shaved white birch bands for orange crates shipped south to citrus-growing regions, Nan Knickerbocker noted. Metal replaced birch, too, yet shaving wooden hoops remained a gainful profession locally, especially for Trapps men when the Smiley resorts closed in winter. "Nearly all the poles 'got out' in Ulster County are sent to markets in the West, where they are used by coopers," stated local papers early in 1887.[19]

• • • • • •

Many a Trapps man considered himself a farmer, but he, like many men of that time, was "half-farmer and half-woodsman, wise in the many ways of turning the trees of his town to use," wrote Alf Evers. As such, they became very attuned to market needs. If they heard a certain wood was in demand, they would fill the supply. Similar to picking berries, cutting wood was a big part of their ability to survive. Yet, logging was hazardous, and accidents might result in the loss of much-needed income.

In April 1890, Trapps resident Thomas Lyons "met with a painful accident on Monday of last week," it was reported. "He went to the woods on that day for the purpose of cutting timber, and while at work, the axe glanced and cut his foot between the first and second toe, severing it in two about half way to the instep. In this condition, he walked home, a distance of nearly a mile, when Dr. Fullerton [Fuller], of Kerhonkson, was sent for who arrived in the latter part of the day and dressed the wound. This will necessitate him laying up for several weeks."

Trapps men cut cordwood to stoke the brick industry of Kingston, but when coal replaced wood, they lost that income, too, noted Knickerbocker.[20] Trapps men most likely

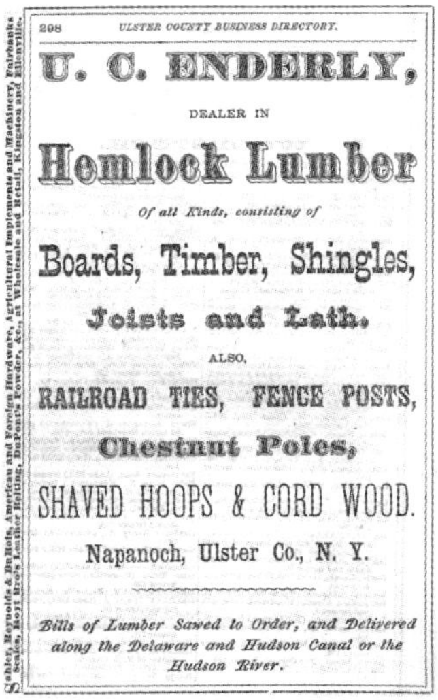

Local lumber and stone dealers, particularly those located along the Delaware & Hudson Canal, provided steady markets for products from the Trapps and its sister hamlets on the Shawangunk Mountains. Uriah C. Enderly of Napanoch, New York, placed this lumber ad in Hamilton Child's 1871–72 Ulster County Directory. *Courtesy of Robi Josephson.*

sold wood to the Smileys for various needs, but as the Smileys at both lakes expanded their properties on the mountain, they were able to harvest more of their own wood. The Smileys used native Shawangunk trees, such as white pine, hemlock, cedar, oak, maple, and chestnut, the last until it succumbed to blight in the early 1900s.[21]

• • • • • •

Like grain and saw millers, blacksmiths once were central figures in American society. Sara Bader presented a partial list of items crafted by these magicians of the red-hot forge: "knives, rakes, shovels, nails, hooks, latches, candlesticks, chains, hinges, scythes, metal rims for wagons, shoes for horses and oxen." Without the blacksmith, "doors couldn't be hung, fields couldn't be plowed, and nails simply wouldn't exist," she added. Blacksmith skills also were indispensable for stonecutting, to hone sharp the iron tools needed to work the great stone slabs. The ruins of a large hearth, likely a forge, remain at the site of the Enderly sawmill. The Coxing Kill and a spring provided the water so critical to the trade. The Mohonk and Minnewaska properties also needed full-time blacksmiths. Horse farrier William Coddington of St. Josen likely moved to the Trapps about 1880 for such work.

Walter Needham of Vermont said that in his grandfather's day the blacksmith was "a big man in the community" because he gave new uses to "precious metal." His grandfather

The blacksmith once played a key role across America, including in the Trapps. "The Country Blacksmith's Shop," illustration by Edwin Forbes, Appleton's Journal, *July 10, 1869. Courtesy of New York State Library.*

picked up "every bent nail or scrap of iron" to save for the blacksmith, who then crafted a new piece of iron into any desired form, including ax heads, a vital tool on the farm. Post axes cut out holes for fence posts. Other axes felled trees, others scored lines along the trunks, and broadaxes hewed the timber square, Needham explained. Powered saws would replace scoring and broad axes, but a man could still earn cash making fence posts by hand. John H. Voorhis, who sold fence posts among his many occupations, ran local ads in December 1875: "Fencing and grape vine posts for sale on the premises of the subscriber, near Geo. Davis', in the Trapps. The posts will be delivered if desired."[22]

The Stonecutters

They dragged it out of fields. They crushed it for roads. They busted it out of quarries and shaped it into great blocks. Shawangunk rock, both the underlying shale and the conglomerate caprock, had many uses in nineteenth-century Ulster County and beyond.

Men learned the skill of laying stone as boys working alongside their fathers and grandfathers. For farmers, a stone wall or fence was more than a row of rocks. John Burroughs wrote: "Every line of fence has a history; the mark of [the farmer's] plow or his crowbar is upon the stones; the sweat of his early manhood put them in place." After the spring sowing and fall harvest were the usual times for building fences, he noted. Three briefs appearing in spring 1890 stated that Simon Coddington of Rochester and Hiram Van Leuven, Jr. of Gardiner "split a large quantity of stone for bridges"; that "one hundred rods of stone wall [were] laid this season" by Hiram; and that Simon Coddington "spent a few days with friends in the Traps."

Mountain folks knew that Simon and Hiram, Jr. were closer than the paper indicated. When Simon, a stonecutter from Lawrence Hill, walked to the Trapps—and walk he did—he likely used either of two routes dating to colonial days—the Shawangunk footpath and King's Lane. Simon Coddington and Hiram Van Leuven, Jr. were not only stonecutting partners; the two became brothers-in-law in 1890 when each married a younger sister of the other.[23]

Shawangunk conglomerate, or "grit," was abundant for millstones and crushing stones, for bridge abutments, for foundations. Some thirty-two house foundations, or cellar holes, remain in the Trapps, nearly all built of stone dug from mountain quarries. Usually only one room of a multiroom house, the original room before others were added, had a full cellar. Foundation walls typically were built of smaller stones and dry-laid without mortar. Three surviving foundations of note are the Davis Mill on the Peters Kill and the farmhouse and barn at the Enderly mill site on the Coxing Kill.

The Enderly barn measured twenty-four by sixty-four feet, with a smaller addition. The barn was the largest in the hamlet, reflecting the high volume of activity at the farm and adjoining sawmill. Thus, its foundation stones were larger than ones used for farmhouses. The men who built the Enderly and Davis mill foundations hoisted the great stones using hand-operated gin poles (portable derricks) with the help of horses or oxen, then transported the stones using wagons or stoneboats.

The foundation of the Davis mill was even more massive, about thirty-eight by fifty-six feet. Great blocks of quarried stone were dry-laid to well over ten feet high. Two rectangular openings partway up either long wall and directly opposite each other held, we believe, the great axle for the immense wheel. By contrast the Enderly sawmill was built directly on bedrock, where little or no stone foundation was needed. If a stone foundation existed there at all, it is long gone.[24]

• • • • • •

When crushed, Shawangunk shale or slate made a good bed for unpaved roads, both for public ones and for carriage ones on the Mohonk and Minnewaska properties. Exposed shale banks dotting the mountain made removal of the softer rock more convenient than quarrying the harder conglomerate or the shale under it.

Before 1800 a public road across the Shawangunks joined the Wallkill and Rondout Valleys. The section from Gardiner in the Wallkill Valley through the old Trappen gap became known as "Trap Road" for its terminus. Trap Road approaching the gap was one of the steepest sections on the mountain. Maps of the period show the grueling grade characteristic of the times. Mountain roads, with wide switchbacks to turn horses and wagons, were purely functional to the point of straining horses and men to their limits. In August 1889 the Town of Gardiner put out for bid repair work for the uppermost section of the road. Gone were the days when landowners were required to maintain public roads along their properties.

George Enderly of the Trapps won the job repairing the road. George was a teamster and farmer but, like all Trapps men, took on other work as it came his way. It took George only two months to put the road "in first class shape," reported a New Paltz paper. An adjacent hogback (shale ridge) rising above the road bed provided stone for the road and its repair.[25]

Elsewhere on the mountain the Smileys at Lakes Mohonk and Minnewaska were well into one of the greatest periods of private construction in the region. These one-time educators were busily transforming the northern Shawangunks—marred by acres of stumps, rotting trees, and fire scars—into a mountaintop haven. Their mountain hotels, with requisite support buildings and vast network of new carriage roads and walking paths, needed stone and tons of it. Construction gave new work to men of the Trapps and its sister hamlets. Through the herculean efforts of the Smileys, the Shawangunks had joined the Catskills and Adirondacks as a "must-see" tourist destination. News articles gave Mohonk a ringing endorsement and priceless publicity. In August 1885, *The New Paltz Independent* noted that the Smiley "roads are kept in superb condition."

The superb roads of the Smileys were very much by design and very much reflected the times. "One of the most controversial details about road design was whether roads should be straight or curved," American artist and author Eric Sloane observed. City parks and cemeteries, such as New York's Central Park, Woodlawn Cemetery, and Green-Wood Cemetery, helped popularize natural settings, where curved lines added grace, tranquility, and the element of surprise. Albert K. Smiley told his guests: "Our roads, you will notice,

Left: Albert and Alfred Smiley's extensive network of carriage roads on the Mohonk and Minnewaska properties was built primarily by hand. Work crews included Trapps men. Horses, hand-operated derricks called gin poles (left), and a portable steam drill (below) provided extra power. Observing the construction was a popular activity for visitors. Photograph below by Albert K. Smiley II. Photographs courtesy of MMHA.

many of them, are curving. We like that. Some people want to see half a mile, but we like roads that curve all the while so that you are getting surprises."[26]

While forests were being cut at a prodigious rate, the Smileys were at the forefront of a national stewardship movement. The Shawangunks became, once more, a little frontier, this time not for pioneers but for those coming "for the season" from urban centers or for the day from nearby villages and cities. Visitors on a day excursion between the two lakes rode through a little frontier town called the Trapps, complete with frontier families—tourists and residents a curiosity to each other—where lived some of the very men who helped build the carriage roads and resorts.

The exceptional quality of the Smiley carriage roads was a marriage of the engineering skill of the Smileys and the stone-working skill of local men, Peter Manning observed. Road crews, Trapps men among them, dug, drilled, and blasted their way through tons of rock. Undercliff Road, completed in 1903, was considered a wonder of its day. "It traverses more than a mile of rugged talus slope beneath the Trapps cliffs," Bradley Snyder noted, "opening up splendid views and a near level route between Mohonk and Minnewaska. … The workmen earned room, board, and $1 per day, and the road was completed at a cost of about $1 per linear foot." Workers wielded picks, shovels, crowbars, and sledge hammers, aided by horse power, a hand-operated derrick called a gin pole, and a portable steam rock drill called Black Maria (pronounced Mariah).

The work fascinated others, and a popular day activity was to ride to a construction site and observe the goings-on.[27]

• • • • • •

America's millstone industry began during the Revolution, when the British blocked the import of stones from Europe, Steve Hirsch noted. As a result, New York and other states became big millstone producers, with New York's Ulster County supplying the bulk, noted D. H. Newland. Newland dated millstone quarrying in Ulster County to about 1800, possibly earlier. In particular, Ulster's Shawangunk conglomerate or grit was once prized for its hardness, purity, and grain.

Color varies from red to white. The whiter the rock, the finer the grain. The finer the grain, or particle size, of the stone, the better the quality. Stonecutters searched exposed outcroppings or just below the surface for "a ledge with a 'grain' suitable for hand shaping,"

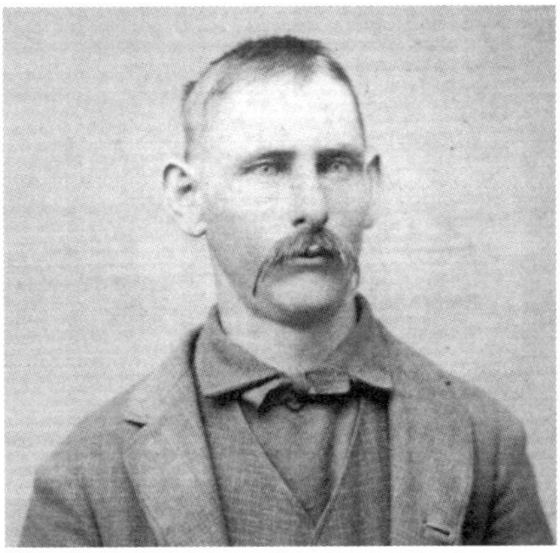

Top left: A millstone is born. The final stage of dressing the millstone with a grinding pattern was completed off site in a stonecutter's shed. The vertical lines in the stone were holes resulting from drilling or exploding black powder. Illustration by Karl Beard. Courtesy of DSRC-MP. Top right: Simon Coddington of Lawrence Hill quarried stone on Dickie Barre ridge. His stonecutting partner, not shown, was Hiram Van Leuven, Jr., of the Trapps. Courtesy of Simon's granddaughter, Marian Whitford. Above: Hiram Van Leuven, Sr., known as "Daddy Hype," and his wife Anna Maria, both at right, pose at their Trapps home with their son Eli and his first wife, Kathy. The Van Leuven men were skilled stonecutters and charcoal burners. The Van Leuven house is gone today but its cellar hole remains. Courtesy of DSRC-MP. Donated by Eli's grandson Roger Van Leuven.

said one cutter named Old John Lawrence. Mountain land "was worthless for farming or anything else," *The Accordian* noted, "so there were no questions of mineral rights or boundaries. Stonecutters worked where they wanted, and moved their shacks, forges and tools from one area to another at will."

The main quarries were not along the cliffs facing the Wallkill Valley; cutters could not get the stones down because of sizeable boulder rubble at the base. Yet some, such as the Schoonmaker brothers of Accord, opened quarries below the rubble "near the Traps" where the "stone stand perpendicular," reported the New Paltz papers in the 1870s. Most quarries were located on the west or Rondout Valley side of the mountain, with some up on the ridges just west of the Trapps. Like sawmills, stone quarries were once centers of local industry and a primary occupation for Trapps men.

The process started with a swing and a prayer. Each section of bedrock was tested twice, before it was quarried and again in block form, Daniel Smiley II said. A stonecutter swung a sledgehammer high, then whacked the bedrock hard, he described. If the hammer bounced and pinged, the rock was "live," or good, clean of fissures or internal cracks; if the hammer landed flat, the rock was "dead," or bad, with internal cracks that over time would weaken the stone.[28]

Cutters quarried the stone first into huge blocks. Black powder, drills, sledgehammers, and other specialized tools were the implements of choice for this part of the process. To cut stone along a seam or ledge, a number of holes were drilled at close intervals and stuffed with black powder, which when ignited could "bust out the stone," a phrase used by local cutter Simon Coddington, said his daughter Evelyn DeWitt. The block was chiseled on site into a rough circle called a blank, ridding extra weight and making it easier to transport.

Stonecutters often worked in teams. We can imagine Evelyn's father, Simon, walking east from his home at Lawrence Hill to meet up with Hiram Van Leuven, Jr. or another one of his partners from the Trapps. While stonecutters did the brunt of the work, one other teammate was sometimes overlooked—the blacksmith. *The Accordian* noted that the "blacksmith was an important member of the crew, constantly sharpening, straightening and tempering the tools."[29]

The stonecutting team watched each other's backs, for the trade, like logging, was dangerous. Stonecutting made for widows; a man could be crushed or develop weakened lungs from working with stone dust. Simon Coddington was one such victim of a lung disease called silicosis. His wife, Laura, a Van Leuven born in the Trapps and Hiram's sister, was left to run the farm and raise alone their youngest five of ten children.

The best "mill rock" in the Shawangunks could be found on Dickie Barre, Simon Davenport told his nephew Guy, "but it was too far back to get the big mill stones cut out with only a trail over the rocks most of the way." Yet, local stonecutters did get the huge blanks out of "far back" quarries such as Dickie Barre by taking advantage of a natural gully in the cliffs. First, they improved the gully slightly by clearing away any vegetation. After maneuvering a blank with gin poles and horse or oxen power, they could then move it down the steep stone slide, or dugway, that became scoured smooth over time. The dugways were just wide enough for one stone, perhaps slowed or stopped with a kind of block and tackle. In winter, snow and ice made sliding easier. (Remains of a former dugway are still visible today along the Mohonk Preserve's High Peters Kill Trail over Dickie Barre; the dugway, on the east side of Dickie Barre, enabled millstones to be taken down into the Trapps.)

Some stones, though, never made it off the mountain. An August 1941 *Mohonk Bulletin* explained, "An imperfection in the rock or a miscalculation in a blow made it a trying profession and accounts for the fact that so many unfinished stones may be found today." These stones were simply abandoned at a quarry; any found in the woods today might have been quarried up to 200 years ago. Ruth H. Smiley of Mohonk noted the frustration felt by one stonecutter when a stone went bad: "Often part way through a flaw would appear in the stone and it would crack. Then said he, 'We'd have to go off and get drunk before we could continue again.'"[30]

Left: Weeks of work—and income—were lost if a millstone broke as it was nearly finished. Shawangunk stonecutter Chris Schoonmaker was forced to abandon this millstone after it split in half. Photograph by Robi Josephson. Courtesy of Bob Larsen and Barbara Rubin. Right: Small jugs of alcohol, called "hooch," might keep a stonecutter company if weeks of work went awry. Photo by Ted Reiss. Courtesy of Bob Larsen and Barbara Rubin.

Most blanks, some weighing over fifteen tons, made it down to the stonecutter's shop or larger "manufactory" to be finished. In 1880, Alligerville, Ulster County, was "the only place in the state where millstones are manufactured," as noted in New Paltz. One such business in Alligerville engaged thirty-five workers. The stone was finished into crushing stones, called chasers, or grinding stones, called millstones. Chasers were larger stones set on edge and rolled, much like a wheel. Chasers were more roughly finished than millstones, as crushed products did not need to be ground so precisely.

Millstones were chiseled, or "dressed," on one flat side with grooves, which formed grinding edges and helped catch the product being ground. Each stonecutter had his favorite pattern wheels or templates. Patterns spiraled out from the center hole and echoed farm fields, with open areas called "land" and chiseled grooves called "furrows," noted Eric Sloane. Thus, a "land-and-furrow" pattern was a stonecutter's personal signature. It took about one to two weeks to finish a stone, noted Mohonk literature. The time depended on the stone's size and its ultimate use. The great stones worked in pairs at the gristmill. The lower one, called a bedder, was stationary; the upper one, called a runner, turned to create the grinding action.[31]

Millstones ground numerous products besides grain—chocolate, peanut butter, mustard, cider, spices, health items such as talcum powder and drugs, and building materials such as gypsum, paint, sand, and stone. Chasers crushed stone for building, glass, cement, and lime, tan bark for leather, and various plants for numerous products and dyes, grains for distilleries, and flax for linseed oil.

Millstones came in many diameters depending on use. Art Coddington was a younger cousin of Simon's and a third-generation stonecutter. Art sized a stone depending on use—twenty-six inchers for chocolate, forty-two for talcum powder, forty-eight for peanut butter. He said flour mills required different sizes than gypsum or mustard

Each stonecutter "dressed" millstones using templates such as these shown. Bob Larsen donated these metal templates, once belonging to stonecutter Chris Schoonmaker of the upper Clove, to the Mohonk Mountain House Barn Museum. Photograph by Robi Josephson.

factories; Gulden's and French's were the largest mustard companies to buy Shawangunk millstones. Pizza-sized millstones between eight and twenty-four inches ground such products as paint and spices, noted Guy Davenport.

Chasers ranged from about 4 ½ to 7 ½ feet. Fifteen-ton chasers, quarried in the Shawangunks, were used for the first time in 1888 to crush Shawangunk grit for an Ellenville glass company. That February, *The New York Times* reported the local stone, called granite or flinty granite, was found to be "an excellent substitute" for sand. The stones, cut from quarries above Kerhonkson near the Trapps, were shipped via the D&H Canal. Initial work was delayed after a wagon hauling one of the great stones, probably from the canal lock in Ellenville, caused a village bridge to collapse.[32]

"Kerhonkson and Kyserike were the main shipping points on the D&H Canal; from Rosendale stones were shipped on the Wallkill Valley Railroad," wrote Bradley Snyder. Charles E. Foote noted that the D&H provided a cheaper means of getting to market both farm products and "Esopus millstones," as Shawangunk stones also were called. The canal allowed the industry to expand, and vice versa. Foote added that the canal was enlarged in 1842 and 1851 to accommodate increased business and larger-scale products, such as millstones. When the D&H shut down in 1898, stones were shipped via the Ellenville and Kingston Railroad.

However shipped, stone was first "wagoned down" the mountain. Stone and wood cutters created and used "a network of small roads, now mostly overgrown," noted *The Accordian*.[33] Wagoning the stone was just as dangerous as other parts of the operation. One fatality made the front page of *The New York Times* on September 12, 1884. The headline broadcast a cruel reality of the industry: "CRUSHED TO DEATH."

> William Stevens, aged about 50, was driving down Clove Mountain, one of the steepest of the Shawangunk range, near Alligerville, yesterday, with five tons of millstone on his wagon. One of the wheels caught in a rut in the road and the wagon was overturned. The stone fell on top of Stevens, crushing him almost to a jelly and breaking every bone in his body. He leaves a wife and family.

Millstones shipped from ports along the D&H Canal between High Falls and Middleport (Kerhonkson) totaled about 350 tons in 1893, noted *The Accordian*. Ten years earlier, Ulster County produced grain millstones worth $120,000, representing 80 percent of total U.S. production of $150,000, Newland reported. Ulster stones were shipped all over the country and around the world. Art Coddington recalled that Shawangunk millstones were shipped as far as Sydney, Australia. Stones destined for that corner of the world to grind paint pigments were shipped as late as 1941, noted *The Mohonk Bulletin* that August.

Earnings rose and fell with the industry. In the 1890s, cutters earned as much as $300 for a forty-eight-inch stone, noted the same *Bulletin*. Knickerbocker confirmed that millstones from the west side of the mountain once garnered $150 to $300 apiece. When the boom days ended in the early 1900s, prices plummeted. In 1911, sixteen-inch stones sold for $3 to $4. Chasers fifty-four to seventy-two inches sold from $30 to $70 each and even lower, Newland noted. Steve Hirsch said a three-man team then might earn $50, which they would split. He added that farmers sometimes bartered stones for other goods.[34]

While the boom days held, wood and stone work sustained mountain families through the year, and that work was steadier than ever before. Up in the Trapps, life was fundamental. It revolved around family, neighbors, work, church, school, and—for the time being—life was good.

"Wagoning" stone down the mountain was as dangerous as quarrying. Working at a Rock Haven quarry west of the Trapps are, from left, quarry owner George T. Decker driving the wagon, Richard Jumes, and an unidentified member of the Croce family. Courtesy of DSRC-MP. Donated by George's grandson, Richard Decker.

• 10 •

Hunting, Guiding, and Working for the Tourist Trade

January 1903, the Trapps. The door of the Kimbark house flings open.

"It's my high and migh-ty hunter!" Amy declares, picking up baby Lina. "The paper says you're 'one of the most noted fox hunters' in these parts!"

"High and migh-ty!" chime Melvin, 7, and Wilson, 5.

"My father will surely be proud of you," declares Amy, kissing her husband hello.

"That old veteran surely will be."

"As am I."

"Daddy, show us them foxes you caught!" Wilson shouts.

"Those foxes," Levi corrects, as the boys wrestle him to the floor.[1]

Mountain Hunters

In September 1861, taverner Isaac Dingee shot an eagle measuring seven feet four inches "across its wings from tip to tip." That October, Isaac went raccoon hunting with four other Gardiner men. At the foot of the Shawangunks south of the Trapps, their dog "had a bear in the millbrook." The men shot the bear with a rifle and double-barreled shotgun, then "beat [it] over the head with an axe, until dead. The bear weighed 204 pounds and was 7 feet long. The party had quite a jubilee the rest of the night over their victim." In August 1864, Willet L. Munson was traveling through the Trapps from Wawarsing to New Paltz when he shot "two catamounts or wild-cats in the woods" near John C. Shaffer's Hotel.

Trapps men hunted and trapped by necessity for food, for income, and to rid their farms of "varmints." By the later 1800s, larger species such as bears, wolves, cougars (also called catamounts or painters), and deer were gone or largely so from the region. Their populations had been hunted out or their habitats cleared for farming and logging. Threats for farmers, though, still abounded, and opportunity didn't knock. It slinked in, it crept in, it swooped in. Livestock and crops were fair game for hungry wildlife. Outbuildings storing food, such as the corn crib, were elevated to keep harvests safe. Families were on perpetual alert, shot anything that moved, and skinned or ate whatever they could. Hardly a soul gave a thought to animal welfare.

Newspapers featured stories about hunting with regularity. City papers tantalized metropolitan readers with escapades from the country or from mountain wilds. Local papers spiced up the news with reports of mountain life for valley readers. Trapps hunters, such as Levi Kimbark, could still earn hard cash from hunting into the 1900s and get their names in the papers for it. Hunting and trapping in the Trapps, even as late as the 1920s, could be "fairly remunerative if one happened to have a lucky season," noted adventure writer Arthur O. Friel.[2]

As big game disappeared, raccoon, fox, and rabbit were the main animals trapped on the mountain. In fall 1884 and winter 1885, hunters in the Minnewaska region had "splendid luck," but luck ran out for one white rabbit they bagged. Another story sprang to life on New Year's Eve 1890: "Mr. Henry O. Harp lately set a trap under a tree only a few rods from his dwelling, in the neighborhood of the 'Traps,' and caught two large foxes in succession. He set it again, and on visiting it next morning found a different kind of game—a big wildcat, fiery-eyed and vicious. The animal showed fight, springing at its visitor to the length of the chain. It was big enough to make a good fight, too, for it weighed 35 pounds. A. A. Clyne, our local fur dealer, got the skins."

Trapps hunters also unloaded their catch in New Paltz with Elting Harp, a distant relative of the Trapps Harps. In January 1895, John D. Van Leuven, who dealt with Elting, was "a famous coon hunter" who caught seventeen over one season.[3] A hunter might be paid twice for his efforts, as was John McEntee. In January 1866 he was paid a bounty plus $13.50 for the hide of a cross fox, the hallmark figure on its back "an intermediate phase" of the North American red fox, noted H. E. Anthony. "Mr. John McEntee, of Gardiner, caught a cross fox in his trap one night last week," a news report stated. "The fur of this animal is valuable—the hide being worth from $6.00 to $7.00. The town of Gardiner, we understand, pays a bounty of $1.00 for all foxes killed."[4]

Trapps resident Levi Kimbark earned a wide reputation for hunting fox. Levi the boy had lived a short spell at the Ulster County Poorhouse. Levi the man had made the Trapps his home, marrying Amy Davenport, daughter of Civil War veteran John J. Davenport and Elmira Van Leuven. In January 1903 one New Paltz paper called Levi "one of the most noted fox hunters in this part of the country," adding he "spends a great portion of his time in winter in hunting foxes," each fox still bringing a bounty of $1.[5]

While some men hunted game, others hunted gold, silver, lead, and coal. Daniel D. Bell was a Town of Rochester farmer, carriage maker, and speculator who mined his way into the Shawangunks and the very heart of the Trapps. He was "a poor man, rich in inspiration," described Harold Harris, yet in 1870 owned real estate worth $4,000, no small sum then. Among other land deals he made a sweet one in April 1874 with Trapps sawyer George Davis and his wife, Sarah, to buy for the recorded figure of one dollar the mining and quarrying rights to between 1,500 and 2,000 acres of their land. Fossil, oil, and water rights were thrown in, too. Dan was not to damage the land other than what was necessary for his operations; if no mining was done within five years, any rights were null and void. Dan could also dam "Cocksen" Pond and build any required structures.

Sarah Davis was the document's primary signer, as she had been for George's earlier land deals. Up in Kingston, Sarah and George signed the papers, each marking an "x" witnessed by others. The deed stated that the judge interviewed Sarah separately, as before, to ensure "she signed the same without any fear or compulsion on part of her said husband." Five months after the ink dried, the Davis sawmill burned to the ground. Whether Dan ever found anything is unknown, and about a year later Alfred Smiley bought Coxing Pond, soon to be known as Lake Minnewaska.

The "classic" legend of Shawangunk treasure, according to George Heusser, is that of Old 99's Cave. Old 99 was a Wawarsing chief presumed to be the last of "the 99th tribe" of the Lenapes. Feeling he could trust his friend Benny DePuy, Old 99 blindfolded Benny, took him up the mountain to a cave, showed him the vast riches inside, and led his blindfolded friend back down again. Benny never sought out the cave while Old 99 was alive, fearing to betray his trust, but then one day Old 99 was gone, presumed to be dead. Again and again, Benny searched for Old 99's treasure but never found it.

Arthur O. Friel incorporated the cave into his two adventure novels set in the Trapps, *Cat O'Mountain* (1923) and its sequel, *Hard Wood* (1925). Steve Oaks, a main character of the latter novel, describes the cave's entrance:

> Everything was like Hampton said, 'ceptin' this: a piece o' rock bigger'n two houses had fell down right onto the little hole a-leadin' into the mine. It had fell down sometime into the winter while I was a-makin' fun o' "God Willin'." An' it jest kilt the onliest chance of ever gittin' into that mine.
>
> What's more, that rock warn't an overhang, li'ble to fall any time. 'Twas a great big wedge o' straight-up-an'-down rock that got pried out—some way—...to fall onto that one place. When I studied it all out I—wal, Hard, I got kind o' shivery. I ain't never been nigh the place sence. An' I don't make fun o' "God Willin'" no more.[6]

Many have hunted for treasures and veins, but all in vain. Kenneth E. Hasbrouck summed up their fruitless efforts: "Some gold was found in the rock of the Shawangunks, but not in sufficient quantities to pay for the work of mining. ... For generations a legend has existed concerning a lost gold mine. Many variations have been told and retold from father to son and many of the old residents have recollections of hunting for this mine in their boyhood."[7]

The New York Times told many a tale of fools' errands, including Dan Bell's claim of finding gold in the Shawangunks above Ellenville. Other than Dan, we have no records of miners operating in the Trapps or of the Trapps people doing so. Expensive mining ventures were unlikely for mountain families. In February 1868 a local story ran about coal and the interest it generated, no location or other identities given:

Our friend John McEntee, of Gardiner, has made a discovery which gives rise to many surmises and some think that the mountains west of our village cover immense beds of coal. While John was out gunning a few days ago near "The Traps," he came across a hole or fissure in the rocks from which hot air was continually escaping. On closer examination he found that the hot air had melted the snow in the vicinity, and the vapor-like fog had formed ice-cicles on the trees. He says the current of air from the fissure is very strong, blowing out a lighted pine knot. He also inhaled some of the air and felt no disagreeableness therefrom. Several persons have visited the locality since the discovery, and attest to the facts as we have recorded them.[8]

Mountain Guides

The earliest hotel keepers at nearby Mountain Rest likely hired guides or guided guests themselves up to the lake named Mohonk. Once John F. Stokes and then the Smiley brothers established the mountain hotel tradition there, Lake Minnewaska to the south became the new place for guided tours of the wilderness. When the Minnewaska hotels opened, Lake Awosting farther south became the next untamed destination. As each place became "settled," the search was on for new places to explore. Among the guides of this period, one emerged from the pack—a native son of the Trapps, Calvin Burger, known as the Old Rooster.[9]

Calvin, born ca. 1828, was the eldest of John and Ann Burger and a grandson of pioneers Ben and Lena Burger. He likely was born in his parents' log cabin, perhaps living there until marrying Rachel Smith, born in Orange County. This is all we know of her past, other than the fact that she and Calvin were literate. They wed on January 4, 1851, and by 1860 were raising a family in a log cabin probably built by Calvin.

Tragedy struck in 1862 when son Andrus, nine, drowned in a mountain stream fighting with an older boy in shallow water. By 1865 the Burgers had four surviving children, but their marriage had soured. Rachel took the children and moved into a log cabin a few hundred yards up the turnpike from their old place, where Calvin remained. Perhaps Calvin built the new cabin for Rachel and their children, but she was on her own financially. In 1867, Calvin placed a notice in the local paper stating that Rachel had left him and notifying local businesses that she was now responsible for debts of her own.

> Notice.—I, Calvin Burger, hereby forbid each and every person trusting my wife Rachel Amanda Burger, on my account, she having left my bed and board without any just provocation.

Calvin claimed Rachel left him, but did Rachel have cause? In 1880, Rachel, fifty, was raising their two middle boys, Ambrose, sixteen, and Sheridan, fifteen, who supported her as laborers. The oldest child, Rosetta, might have died or left the area. The youngest, Mary Edda, and her husband, Frederick Pratt, a laborer from Massachusetts, were raising a family near her parents' separate cabins. Calvin, Rachel, and Mary each lived on land once owned by Calvin's father, John, the long, narrow parcels carved from Great Lot 6 of the Grote Transport. With each generation the huge parcels of the old Grote Transport and Nineteen Partners Tract were becoming smaller and smaller, partitioned further among descendants of the original owners or sold off.[10]

Like many Trapps men, Calvin was a skilled Jack-of-all-trades and knew the mountain well. He watched traffic coming and going on the turnpike past his cabin door. He saw Ben Fowler, John Shaffer, and John McGibney do well with the hotel trade. He no doubt saw or heard about people going up to Coxen Pond (Lake Minnewaska), which gave him an idea. Put some rowboats on Coxen Pond and charge people to rent them, or take them hunting or whatever else they wanted to do. A man needed to make a living.

When the *Guide Book to Mohonk Lake* was published in 1875, Alfred had not yet bought Coxen Pond. The summit lake south of Mohonk was described as a "romantic sheet of water," a hidden destination for more adventuresome guests. "No house, or clearing, or road, or boat even, disturbs its primitive wilderness," the guide read. Yet there was a boat on the lake, and it belonged to Calvin, the first to put a boat on Coxen Pond before its purchase by Alfred H. Smiley. In October 1875, Alfred wrote to Albert in Providence: "We made a passable road from the turnpike to the N [north] end of Lake where CB's boat lay." Calvin closed his little operation for the winter and reopened in late spring. The following article appeared in June 1877; perhaps the stamps mentioned were for postcards or letters mailed home.

Calvin Burger of the Trapps was hired as a hunting guide. This illustration titled "The Guide" appeared in Appleton's Journal *on February 24, 1872. Courtesy of the New York State Library.*

> The same "old rooster," Calvin Burger, residing in the "log cabin" near Lake Minnewaski [*sic*] will be ready again this season to furnish boats for pleasure parties who visit said Lake. And don't forget the "stamps."

In 1875, Albert came to Mohonk from Providence and took a group of Mohonk guests to Long Pond (Lake Awosting, also called Big Pond). They rode by carriage to the inn near where his brother Alfred's new road left the turnpike. From there, "a man from The Trapps" guided the group on foot to Long Pond, wrote Nan Knickerbocker. They bushwhacked "through dense underbrush and over a stream and through a marsh," finally arriving at the pond, she added. Albert's guide was not named, but it likely was Calvin Burger. Alfred soon added Long Pond to his Minnewaska property.[11]

Calvin was a farmer, but thrived on his reputation as a guide and mountain man. The "Old Rooster," the man who told tales, relished playing the part of a mountain man. In July 1876, Ralph Le Fevre published "A Visit to Lake Minnewaski" by "D." in his New Paltz paper. By then, Cal was the sole operator of the boat business on the lake and had expanded his little concession. "D." wrote: "About 2 ½ miles this side of the Lake is the cabin of Mr. Cal. Burger, said to be the best mountain pioneer in the neighborhood, and so we found him. He is an important person to meet with as he has charge of the three boats which are on the Lake." At the north end of the lake, they found "a shingle nailed to a tree [that] bears the phonetic description, 'Lake Miniewasky!'" The party "at once entered a boat and pulled out into the Lake," D. wrote. The location of Cal's boats matched that in Alfred Smiley's letter a year earlier; the sign, too, likely was Cal's.[12]

Alfred Smiley bought out Cal's boat business for the 1879 opening season of Minnewaska, but kept Cal on as a path maker. Cal continued guiding guests, who ventured down from the lake to his log cabin, which added a wilderness tang to their adventure. Three briefs, among others about him, were printed in New Paltz, where townspeople must have gaped and gossiped at the sight of this mountain man on Main Street. Misspellings in the first article were intentional, attributed to a Minnewaska correspondent with "spicy" items to report. The third was excerpted from Cal's obituary.

> Calvin Burger Sole [sold] a Rattle Snake to a man in New Paltz for $3 cash. Bisiness is Bisiness.

> Cal. Burger who was one of the original settlers of the Minnewaska country and has quite a reputation as a guide was in our village on Friday.

> He had a wide reputation as a guide to some of Smiley's guests who when they wanted to go, at the opening of Minnewaska, through different parts of the mountain, would come to his home and he would go with them and show them through the woods.

In the 1850s a twenty-something Calvin left his mark, noted Ralph Le Fevre on his first visit to the Trapps as a young man of fourteen: "On the perpendicular cliffs all about this region, we saw inscribed in great letters 'C. B.' made by scraping off the lichen and thus revealing the rock beneath. ... We could not guess what those letters meant, but learning long afterwards that Cal. Burger by thus placing his initials in great letters on the rocks had hoped to hand down his memory to future generations. When we related this fact to Ed. Smiley [Alfred's son Edward] he made the practical comment, 'It gave us a lot of trouble when we came here to have those letters scraped off.'"[13]

The youngest mountain guide we have on record was Eli Benjamin from New Paltz. A September 1878 story featured Eli, thirteen, who lived on the mountain while his father, Robert, worked on the construction of Cliff House, Alfred Smiley's first hotel at

Lake Minnewaska. The mountain was a boy's dream, and it is not surprising that Eli led a party from New Paltz, which perhaps included newspaper owner Ralph Le Fevre. Eli told them the story of the Indian maiden, Minnewaska, who drowned in the lake. She jumped from a small cave at the "western end," Le Fevre's paper stated. Young Eli likely knew Cal and might have heard the story from the Old Rooster himself.[14]

Working for the Tourist Trade

By the mid-1800s, a new industry had arrived on the mountain to outlast all others —the tourist trade. Artists and writers depicted the first "advertisements" in their works of wilderness areas, piquing the curiosity of pleasure seekers and excursionists. The mountain hotels at Lakes Mohonk and Minnewaska offered the people of the Trapps and its sister hamlets additional income, and the Smiley families found a skilled work force built-in and readily available. While the resorts were only open during the summer season, here was a steady paycheck as well as some year-round work, work that helped free the Trapps people from a strictly subsistence lifestyle. If the crops were poor, if the berries failed, if anything else went awry, there was work.

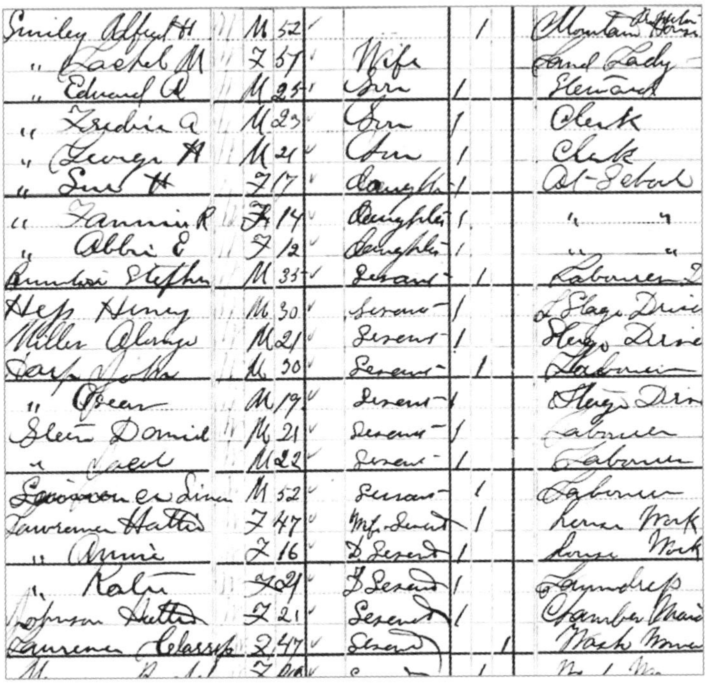

Smiley Alfred H.	M	52		Mountain Hotel
" Rachel M	F	57	Wife	Land Lady
" Edward A	M	25	Son	Steward
" Frederic A	M	23	Son	Clerk
" George H	M	21	Son	Clerk
" Sue H	F	17	Daughter	At School
" Fannie R	F	14	Daughter	" "
" Abbie E	F	12	Daughter	" "
[illegible] Stephen	M	35	Servant	Laborer
Hess Henry	M	30	Servant	Stage Driver
Miller Alonzo	M	21	Servant	Stage Driver
Harp John	M	30	Servant	Laborer
" Oscar	M	19	Servant	Stage Driver
[illegible] Daniel	M	21	Servant	Laborer
" Jacob	M	22	Servant	Laborer
[illegible]	M	52	Servant	Laborer
Lawrence Hattie	F	47	[illegible] Servant	House Work
" Annie	F	16	[illegible] Servant	House Work
" Katie	F	[illegible]	[illegible] Servant	Laundress
Johnson Hattie	F	21	Servant	Chamber Maid
Lawrence Clarissa	F	47	Servant	Wash Woman

Federal census records reveal work positions once available at Shawangunk Mountain hotels. Alfred H. Smiley of Minnewaska (at top) employed many Trapps residents, including Alonzo Miller and brothers John and Oscar Harp, listed on this 1880 census. Courtesy of the United States Government federal census records.

The year 1880 was only the second season at Minnewaska when the Town of Rochester census alone listed forty-one employees under Alfred H. Smiley, including Alfred, his wife Rachel, and their three sons; their three youngest, all daughters, did not yet work. Positions for men included stonecutter, carpenter, laborer, road crew, painter, repairman, blacksmith, stage or carriage driver, stable man, block man, harness maker, kitchen and dining room staff, office staff, porter, boatman, night watchman, farm or garden crew (the last primarily at Mohonk), and "operator of telegraph." By offering this scale of work, the mountain hotels at Lakes Mohonk and Minnewaska became, in a very real sense, unifying forces for the surrounding communities. People gravitated to the mountain hotels to find work, and thus to the mountain to live. New residents in the Trapps in 1880 were a horse farrier, carriage maker, and wagon maker (the last two were father and son). As well, the people of the Trapps and nearby hamlets began to have more contact with each other, with people from the region, and with those from the outside world.[15]

Stage drivers, hired in large numbers, were the literal "front men" of the Smileys, picking up guests at local railroad depots or Hudson River landings. Guests received their first impressions of the family and the business from these drivers. Alfred Smiley hired Trapps

Guests staying at the Smiley mountain hotels at Lakes Mohonk and Minnewaska could journey for the day to either lake via carriage roads. Carriages brought guests through the Trapps hamlet, visitors and mountain residents each curious of the other. Illustration by, and courtesy of, Jack Fagan.

"Laundry girls" working at Shawangunk Mountain hotels included women of the Trapps hamlet. Anna Enderly Hayes, third from right, worked at the Minnewaska laundry for many years. Courtesy of Anna's granddaughter Joan Hayes Wustrau.

men, such as Peter Enderly and Oscar Harp, to drive the Minnewaska stages, as well as Alonzo Miller from the Town of Shawangunk. Alonzo settled in the Trapps, where he later served as a school trustee. The continued success of the mountain resorts relied, in large part, on the caliber of transportation provided by the Smileys and by independent operators, such as Samuel Judkins of New Paltz. These two articles ran in New Paltz in 1879 and 1880:

> Now that the farmers are all so busy our streets would be dull enough were it not for the visitors going up to Lake Mohonk and the several stages from the same point. Two four horse stages one marked conspicuously "Lake Mohonk," and the other "Lake Minnewaska," occasionally awaken our citizens from their day-dreams, to recall the time, not so very distant, when honest John F. Stokes commenced business at Lake Mohonk.

> Quiet and gentlemanly men—the drivers of the Lake Mohonk and Minnewaski [*sic*] House stage lines.

An 1883 article noted Mohonk and Minnewaska stages "produce more travel on our streets than from farmers' wagons. The boarders from the mountain houses leaving by way of Judkins' stages, form likewise an important portion of the stream of vehicles."[16]

Positions for women included waitresses, maids, laundresses, and wash women, with staffing then divided more strictly according to sex. In 1880, Elizabeth "Libbie" Davis of the Trapps worked as a waitress at Minnewaska. A June 1880 article from New Paltz attests to the draw for young women from near and far:

> A number of girls—mostly laundresses, from Vassar College are employed at Lake Mohonk and Minnewaska each year; also a number of girls from Smiley's school, at Providence, R. I. The balance of the female help is composed of young ladies from this vicinity and the other side of the mountain.[17]

After the last guests and most staff had left Mohonk and Minnewaska for the winter, a clutch of trusted employees hustled about the buildings and grounds closing things up. Then came the day when they and the Smileys themselves were gone, too. The mountain, the lakes, the mountain hotels—all had entered their "winter rest," so-called by Minnewaska author William E. Doughty. "The few who spend all year on the mountain," he wrote, "often wish all the guests could see the vast stretches of white when deep snows come, leaving only the evergreens and the beetling crags to break the whiteness." Doughty's romanticism was far different from the stark reality for Trapps families who toughed out winters on the mountain and for the skeletal staff who stayed at the two lakes.

One article of January 1881 gives a glimpse of winter work, some unexpected, at Lake Minnewaska. This work supplemented hoop shaving, millstone cutting, and hunting and trapping for men of the Trapps. Alfred had just enlarged his first mountain hotel. The original section, opened two years earlier, was already being called the "old house." Alfred Smiley's three sons, George, Edward, and Frederick, "are spending the winter at Minnewaska, together with Simon Lawrence, and Mrs. Krom as housekeeper."

> The billiard room fell in a few days ago from the weight of snow on the roof. It was about 80 feet long and about 32 feet wide. It cost $900, and is injured to the extent of about $200.
>
> The addition to the Minnewaska House, built last fall, contains 83 rooms. This gives just about double the capacity of the old house.
>
> 100,000 laths are being drawn this winter to complete the finishing off of the rooms in the new portion of the building.

Shawagunk Mountains hotels hired Trapps men to cut ice for hotel use. This print shows the Knickerbocker Company cutting ice ca. 1846 at Rockland Lake, Rockland County, N.Y. Courtesy of the Library of Congress, LC-USZ62-2544.

Alfred opened his second mountain hotel, Wildmere, in 1887. About twenty-five men were hired to work over the winter of 1885–1886, and presumably the following one, to work on the new hotel.[18]

By far the biggest winter job was ice cutting. Enough ice had to be cut to last the coming year to keep food cold, to cool drinks, and to fill the icehouse. Even before John F. Stokes built his rustic mountain hotel at Mohonk, he cut lake ice there for picnickers, noted his son, Eli, in 1903. An icehouse, therefore, likely was the first structure that John built. Parties helped themselves to the ice, using an axe left for that purpose. John had to have ice and an icehouse for the summer trade at the tavern and inn. Once the Smileys established the Mohonk and Minnewaska properties, ice harvesting shifted into high gear. Beginning in 1870, from "two to four thousand tons of ice were harvested" from Mohonk Lake every winter, wrote Daniel Smiley II, whose great-uncle Alfred also had cut ice at Lake Minnewaska and Lake Awosting.

Until the development of power saws, ice cutting was done solely by man- and horsepower. Work began as soon as the ice was thick enough, and took about two to three weeks to finish. The first cakes were sawed by hand, the rest cut by horse-drawn ice saws. Tools were designed to handle the big cakes of ice, each weighing well over a hundred pounds. Men wielded pike poles (with hooks for pulling and curved tips for pushing), hooks, shovels, scoops, tongs, and ice-breakers; they raised the ice cakes off the lake surface via conveyor belts. Ice cutting, and with it winter work for thousands of men and boys, disappeared with modern refrigeration.[19]

Making Good

Some Trapps people worked at Mohonk or Minnewaska for decades; some Trapps families, such as the Van Leuvens, Enderlys, and Otises, worked for the Smileys at either lake for several generations. The obvious reasons for their longevity as employees were proximity, convenience, and steady work.

Yet, there was another reason. Nan Knickerbocker described Alfred Smiley as having the "happy knack of making and keeping friends," especially his employees who often became devoted to him. While a religious and moral man, Alfred never preached to others, she wrote. Knickerbocker demonstrated with the following story.

After Alfred acquired the Minnewaska property, he learned of "a man from a nearby village" who rented rowboats to visitors. The nearby village was the Trapps; the man, the Old Rooster himself, Calvin Burger.

Others warned Alfred that Calvin "had threatened to make trouble if he was interfered with." Untroubled by this, Alfred befriended Calvin and permitted him to carry on his enterprise until Alfred opened his first mountain hotel. Alfred then bought him out "for a good price."

"He gave the man a job at making paths," Knickerbocker wrote, "and the man made good."

The path maker made good for the Smileys because this Quaker family made good for him. The people of the Trapps and sister mountain hamlets came to depend on the Smileys, and the Smileys on them. Families gained steady work, some women and youngsters earning their first paychecks. Each spring the Smiley families and their employees returned to the mountain lakes as predictably as robins. Year after year, bonds of friendship and family were renewed. All in all the Smileys helped bring a period of good times, of halcyon days, to Trapps families. Yet, one still could, and very quickly, slide down the very slippery slope of life on the mountain.[20]

11

Living on the Edge

August 1900, 11:00 PM. The Wixon home in the Trapps.

A dream? A cry. Hush, baby Floyd. A smell? Fire!

"Georgie! Charlie! Maudie! WAKE UP!" Grab the broom! Bang the ceiling!

"GET DOWN!"

Little bare feet slip down the rungs. "Mamma?" "I'm scared!" "Me, too!"

"GET OUT! HELP! PLEASE, SOMEBODY HELP US!"

Out of the darkness, a voice shouts, "Minnie!"

Others storm in armed with buckets of water, tubs of dirt, shovels, brooms.

"You're too late. All burned down. Thank God, we're okay. Send word to George in Pennsylvania. We're all okay."[1]

Fire. Wild animals. Childbirth. Accidents. Illness. Life then was fraught with peril, especially for families living in isolated places such as the Trapps. With luck and a hardy constitution, people lived to tell the tale. Yet, tragedy, not survival, made news. Danger and hardship were standard fare in newspapers, journals, and books.

Philip H. Smith capitalized on the public's fascination with danger. The Putnam County native wrote and co-illustrated several books on American and regional history, including the now classic *Legends of the Shawangunk* published in 1887. Smith's lengthy subtitle included the phrase "Thrilling Border Incidents and Adventures." His intent was straightforward—to make the Shawangunks "better known to the outside world." Believing the range deserved "a Scott or a Cooper to immortalize the locality," Smith stepped in to fill the void. He spent a summer collecting stories up and down the Shawangunk ridges. His "rare and quaint" tales were for the most part violent; his main topics were the Indian wars, the Revolution, and wild animal attacks.[2]

Smith's half-dozen stories set in the Trapps added to its already rocky regard. Five concerned wild animals, the sixth the Cross-Gorsline tragedy. Smith began with Ben Burger's nighttime encounter with wolves. Wolves had hounded another young man named Hill, who late one night left "The Traps Tavern" located below the gap. Hill rode home west over the mountain, but as he gave …

> ... the reins to his horse, the frightened animal went galloping down the rugged mountain at a breakneck speed. The iron shoes of his horse sent the sparks flying at every step; and the clatter of hoofs, the shouts of the rider, and the sharp quick cries of the wolves in close pursuit, startled the night air and awoke the sleeping echoes among the mountains. A false step, or a failure to retain his seat, and all would have been over for young Hill. In this way the cavalcade went dashing down the defiles, and finally brought up before another hotel at the foot of the mountain. Here the pack turned off into the forest, and the panting horse and terrified rider sought the friendly shelter of the hostelry until morning.[3]

By the time Smith published his book, wolves were all but gone from New York State. Wildcats and dogs, though, remained a threat, shown in two articles of 1899 and 1903. The Williams children were likely Clara and Sadie, and their mother was Amelia.

> On Friday night last, Mr. Louis Denton, of Gardiner, captured a catamount weighing thirty pounds. The "varmint" has terrified the timid inhabitants of the "Traps" and vicinity for the past eighteen months, and it is with much rejoicing the mothers of small infants receive the above welcome intelligence.

> Two small children of Isaac Williams, who lives on the road to Minnewaska, a short distance beyond the Henry M. DuBois place, were bitten by a strange dog on Wednesday. The dog came to the house and Mrs. Williams let him in. After a little he flew on the children, biting one on the neck and one on the face. The dog was taken into the woods and killed.[4]

Smith told two snake stories about the Trapps. In the first, Rachel Burger alerted then-husband Calvin that a snake was living under their cabin floor. Over the winter Calvin "maintained at least an outward show of incredulity," Smith wrote, "knowing that any other course on his part would necessitate the taking up of the floor to search for the snake, or removing it from the cabin." One warm, spring morning, while Calvin was out, the snake emerged. Rachel killed it, saving her husband the task. Smith wrote that the "snake proved to be one of the largest of its species." We wonder if it grew with each telling by the Old Rooster.

Smith's other snake story ended in tragedy. An African-American boy named Jed, about nine or ten, worked for a farmer named Evans at the foot of the mountain near the Trapps. One day Evans sent Jed to "bring home the cows." When Jed did not return, Evans went looking for him and "was horrified to find him bound to a bar post in a standing position by a huge black snake, and stone dead," Smith wrote. "The snake had probably attached himself to the post, and, as the boy attempted to pass through, it had taken a turn around the lad and squeezed him to death."[5]

The snake that killed young Jed likely was a black rat snake, a constrictor that can grow up to eight feet. A black rat, or pilot, snake bit Deborah Williams, twelve, daughter of Isaac and Amelia, while she walked on the main road. "Restoratives were applied to counteract the poison, and she is now out of danger," an August 1892 report stated. Ten years later, George Voorhis, son of John and Sarah, killed a black snake six feet three inches long near his Trapps home.[6]

Philip Smith retold an old "man versus deer" legend about Ephraim Decker of

Shawangunk, "a famous hunter and famous man of strength," noted Ralph LeFevre. "In the vicinity of the Traps," Smith began, "are vast crevasses in the rocky ledges, some of them of unknown depth." One morning Decker chased three deer to the edge of a precipice called Table Rock, an overhang on Sky Top. Two of the deer jumped off the cliff and fell onto the rocks below. "The third, a huge buck," Smith continued, "took up a position on Table Rock, and facing about, boldly defied his pursuer." Decker only had his hunting knife, having tossed down his rifle during the chase. He grabbed the buck's antlers, but it tried to shake him off. At times, the buck lifted Decker off his feet and dangled him over the cliff edge, but he somehow regained his footing and the upper hand to win this "desperate conflict," Smith wrote. The legend gained such fame that the precipice was called Ephraim's Point until renamed Table Rock.[7]

• • • • • •

From the moment of birth, life in the Trapps was hard, uncertain, dangling. Women gave birth at home and often without medical help. Mortality rates were high for newborns, infants, and mothers. Sometimes parents did not name a newborn, often waiting for a minister to hold a baptism or perhaps to see if the infant survived. In 1870, 438 children in the United States were listed on the federal census as "Nameless"; 2,556 were listed as "No Name." Other nameless categories of past census records include "infant," "infant not named," "no named infant," and "not named." The 1870 census included No Name Wynkoop, the three-month-old son of Lewis and Sarah Wynkoop of the Trapps, who lived to be named Perry.

Ben and Laney Fowler had ten known children; three listed in their family bible as "Babe" did not live long enough to be named. Laney had her first child at aged twenty-one, her last at thirty-eight. Shortly after Ben's death in 1875, their seventh child, Susan Mary, known as "Mary," married George Enderly of the Trapps. George and Mary lost an infant daughter, Sylvia or Susan, to an unknown cause, then their four-year-old son, Jesse, "died when a wagon wheel slipped and crushed him," wrote descendant Joan Hayes Wustrau. George and Mary had lost their third and fifth children.[8]

• • • • • •

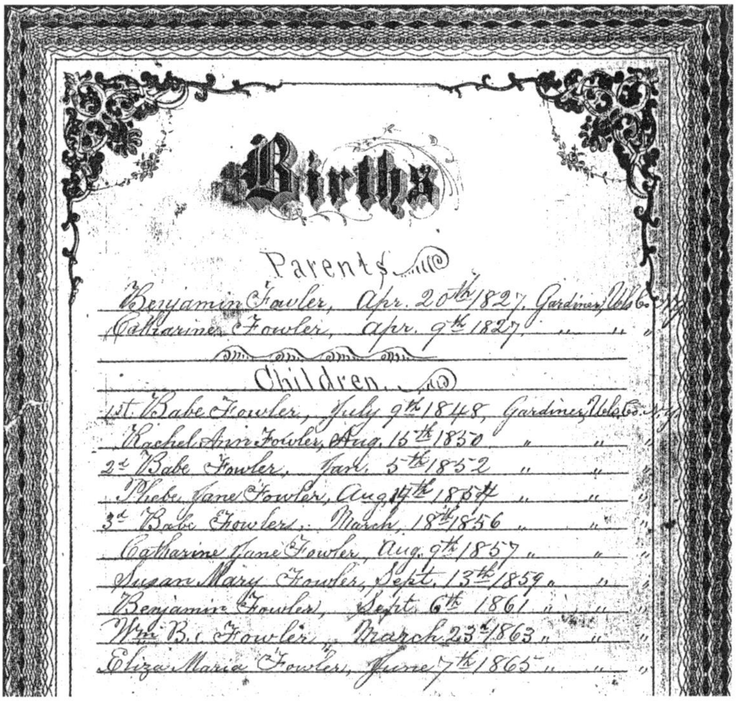
Births

Parents

Benjamin Fowler, Apr. 20th/1827, Gardiner, Ulst Co.
Catharine Fowler, Apr. 9th 1827, " "

Children

1st Babe Fowler, July 9th 1848, Gardiner, Ulst Co.
Rachel Ann Fowler, Aug. 15th 1850 " "
2d Babe Fowler, Jan. 5th 1852 " "
Phebe Jane Fowler, Aug 14th 1854 " "
3d Babe Fowler, March 18th 1856 " "
Catharine Jane Fowler, Aug. 9th 1857 " "
Susan Mary Fowler, Sept. 13th 1859 " "
Benjamin Fowler, Sept. 6th 1861 " "
Wm. B. Fowler, March 23d 1863 " "
Eliza Maria Fowler, June 7th 1865 " "

The Bible of Trapps couple Benjamin and Catharine Fowler lists their births and those of their ten children. Three infants listed as "Babe" did not live long enough to be named. Courtesy of descendant Joan Hayes Wustrau.

Accidents or worse could strike, snakelike, at any time. A young son of Calvin and Rachel Burger, named Andrus, about nine, drowned in 1862. What began as an accident quickly turned deadly. "He and another boy were in swimming, and had a clinch," a New Paltz paper stated, "in which the younger one was thrown into the water, which was not over two feet deep, and partially stunned. The older one made no effort to take him out, but left him to drown."

Thom Lyons had nearly cut his foot off in 1890 while cutting timber. Edward Bunton was working at Minnewaska in January 1892 when he was badly hurt "by a cake of ice which dropped on his shoulder, as it was being hoisted into the ice house."[9]

Lightning and fire also might strike at any moment. Lightning destroyed a barn of George Davis in 1860 and "badly damaged" six telegraph poles servicing the Smiley hotels in 1880. The Wixon family, profiled in the vignette, lost their home to fire in 1900. George Wixon (Wickson) likely grew up in the Trapps and remained there after his marriage. After the fire, George and his family moved into another house, but by 1902 the family left the area, for they disappeared from local records.

> A dwelling house belonging to George Wixon, situated a short distance west of Lake Minnewaska, caught fire and burned to the ground, on Tuesday night, last week. Mrs. Wixon was at home alone, her husband away for work at Seigfried, Pa. She had been ironing that day and retired early in the evening. About eleven her infant child became restless and she arose to attend to its wants, she discovered the house on fire. She alarmed the neighbors, which are near by but the flames made such rapid headway that nothing could be saved except a few light pieces of furniture. She sent word to her husband who immediately came home, bringing his tools along and will rebuild at once. The fire is supposed to have come from the chimney. It was a narrow escape for Mrs. Wixon and her children from a horrible death. There was $600 insurance on the house and furniture.[10]

In 1905 fire also destroyed the "old Fowler place," once owned by Trapps pioneer Peter Aldrich and later by Ben Fowler. Owned in 1905 by Edwin Davis, son of George and son-in-law of Ben, the house was "very old—at least part of it, which was nearly 100 years old." George Davis had lost a berry shanty and his sawmill to fire; arson was suspected in the first. Lewis "Huckleberry" Wynkoop lost his house to fire, too, but luckily he had insured it for about $300.[11]

• • • • • •

Illness, including numerous epidemics—influenza, cholera, scarlet and yellow fever, measles, typhoid fever, smallpox, tuberculosis, and diphtheria—claimed many lives. Stonecutters were at risk for silicosis. A stonecutter who died from the disease "would leave families of eight to 10 penniless and force mothers to take in laundry and send their children out to work," Priscilla Prutzman wrote. Death certificates, however, might list pneumonia or consumption (tuberculosis), the only clue to the real cause. No official records link stonecutting and illness. Yet, Arthur O. Friel did so in his two novels set in the Trapps. In the first novel, Friel wrote, "millstone cutters, despite their hardy appearance and muscular development, usually died fairly young because of the stone-dust entering the lungs caused tuberculosis." In the second, he repeated, "The stone-dust, lodging in the lungs, was likely in time to kill even the strongest worker."

Children and the elderly were at the greatest risk. Between 1873 and 1883, George Coddington and his wife, Rachel, a daughter of Ben and Laney Fowler, lost four of their five children to unknown causes and buried them one by one in the Fowler Burying Ground: Louisa, five months; Stanlie, four years; Levi, three months; and Luella, two and a half years. Rosetta Coddington was the only child of George and Rachel to survive to adulthood.

Winter could be especially cruel. In January 1885 scarlet fever killed three daughters, aged five months to thirteen years, of John D. and Mary Van Leuven.[12] In January 1892 the "grip," an earlier term for the flu, had a literal grip on the region. Of about forty houses in the Trapps, "there have been one or more persons sick of late" in all but about six, the local press reported. "La Grippe," often leading to fatal pneumonia, was responsible for "three funerals a day" in an area spanning from Newburgh to New Paltz. Catharine "Laney" Fowler, sixty-five, Ben's widow, became a victim later that month. It was reported locally that her "funeral on Monday was the largest ever held in the Traps."[13]

Life on the mountain was nothing short of rugged. In remote places such as the Trapps, neighboring families had their own built-in social service system. During winter, in particular, the elderly living alone might move in with a son, daughter, or neighbors on or off the mountain. Or receive the help of a young person, related or not, who might move in or work daily as a domestic or laborer.

In 1850, Eve Fowler, fifty-eight, lived with son, Ben, and his wife of two years, Laney, both then twenty-four. Fifty-eight is not elderly today, but the average life expectancy in mid-nineteenth-century America was only about forty. By this time Eve likely had separated from her husband, Sam, a War of 1812 veteran. Sam's whereabouts in 1850 are unknown; in 1855 he lived with another son down in Gardiner.

In 1885 widower John Enderly, sixty, returned to the Trapps after a fifteen-year absence. In 1900 he boarded with Hiram and Annie Van Leuven, both in their sixties. Simon Coddington brought Charlie Burger, Calvin's brother, over to his house at Lawrence Hill to spend the winter, remembered Simon's daughter Evelyn. This was probably in the early 1910s, when Charlie was in his sixties. Charlie's home was too cold for an older man to live in during the winter months. Simon lived in the Trapps during the early years of his marriage and thus knew the Burger family well.

John Enderly and Charlie Burger likely contributed by doing chores or paying board. Yet the human kindness shown to them reflected an inherent sense of responsibility towards others. In a very real sense this neighborly charity, wherein people depended on each other to survive, was part of a more literal social security system than exists now.[14]

Yet sometimes people had no place to go or no one to care for them. They then entered the only place they could get help, the state tax-supported county poorhouse. One had opened a few miles southwest of New Paltz in 1828. A broad range of people came to its door, including "the indigent, the destitute, the insane, the intemperate, transient farm workers, freed slaves, unemployed canal and aqueduct builders, 'debauched' women, unwed mothers, the friendless, the elderly, the disabled and the sick, [who] called this site their home," noted the Ulster County Poorhouse Project.

In 1853, Maria Cross was taken to the poorhouse after her horrific fall. We recognized two other names on the poorhouse's admissions and death lists. James Gorsline, a son of James and Martha Gorsline, was born in 1797 in the upper Clove. James, Jr. was an older brother of Sam and an uncle of Jacob. James was blind and apparently never married. For years James was able to live on his own on a small parcel of land, likely helped by his family and neighbors. By 1860, however, he could no longer take care of himself. On March 16, James, sixty-two, entered the poorhouse, where he died one month and four days later. James is believed to be among the many buried at the poorhouse cemetery, noted the Poorhouse Project.

Twenty years later, on January 3, three people were admitted to the same poorhouse—a New Paltz housewife named Cathy Maria Kimbark, forty-two, and her two children, Helena, eleven, and Levi, six. Cathy was admitted for two reasons—her womb had fallen, and her "husband in jail, ran away with his son's wife." Cathy was ill and in obvious despair. She was literate and had "good habits," but no abilities other than being a housewife, the admissions form reported. She had five children; Helena and Levi, both born in the Town of Rochester, were probably her youngest. A section on the form titled "Destiny" predicted a resident's future. Answers ranged from "will not remain" to "will always be a charge." In Cathy's case, it stated, "Will not recover."

It is difficult to piece together the family history without her husband's name. Levi left the poorhouse alone within three months of his arrival. Where he went is unknown, perhaps to another facility, such as the Alms House or Industrial Home of Kingston, or was "bound out" (indentured) to a family. Children were often placed without choice to other families. This was a reflection of the times, when many families scraped by with extra hands, not government help.

As a young man, Levi found his way to the Trapps. After likely spending most of his boyhood without his mother and siblings and perhaps never seeing them again, Levi married and had at long last a family to call his own.[15]

The Civil War

With a divided nation at war, a dozen men from the Trapps, and a few others with close ties there, put down their plows and axes and bid loved ones fare-thee-well. Many of those volunteered in August 1862 to serve three years following President Lincoln's call for more troops that summer. The war was well into a second year with no end in sight. Trapps volunteers were family and neighbors who often enlisted and mustered in together at nearby towns, including High Falls. They represented well the mountain regions of the Towns of Gardiner and Rochester. They joined their regiments at Kingston and departed from the Rondout, along with vast numbers of fellow New Yorkers surging south from other ports along the Hudson River. To a man, the Trapps volunteers served as privates, paid $13 a month. It was a goodly sum, when collected, to send back home.

In the 120th Regiment were John D. Van Leuven, Enoch Freer Dingee, Isaac Abrams Enderly, John James Davenport, and William H. Chase. Enoch and Isaac, as well as John and William, were brothers-in-law, brothers-in-arms. The 120th, known as the Ulster Regiment and Washington Guard, was part of the early defense of Washington, D.C., and fought in some of the war's bloodiest battles, including Gettysburg. All five enlisted in Company C, the regimental Color Guard. (The 120th stood on duty at Appomattox when Lee surrendered, but none of the above men was present.)

In the 20th were Lucas Schoonmaker, Jacob H. Markle (or Maracle), Abram Wynkoop, and likely Alex Bush, husband of Margarette of the Grass Widow Trial, which occurred during the first year of the war. Early reorganized as the 80th, the 20th (as the men preferred) was The Ulster Guard, which safeguarded men, munitions, and supplies. Battles of the 20th include Antietam and, with fellow volunteers of the 120th, at Gettysburg. Monuments to the 120th and 20th, two of 300 fighting regiments in the war, stand with many others honoring North and South in the fields around Gettysburg.

In the 156th were Hiram Van Leuven, Benjamin Horton, George Coddington, Eli Lockwood, Joshua Armstrong and son George, and Jacob and Alexander Wynkoop, likely brothers of Abram in the 20th. The 156th was the "Mountain Legion," its mascot an American eagle taken into battle "tied fast" to a board. In May 1864, *The Kingston Democratic Journal* wrote that the eagle "never flinches, but stands up to the rack like a true 'hero.' The roar of musketry and artillery is the signal for him to commence a loud 'yelling,' which he keeps up until the strife ceases. The 'rebs' have made him the mark for their rifles on more than one occasion, but like the Nation they have so traitorously assailed, he still lives, the proud emblem of our country." The 156th fought at the Siege of Port Hudson, Louisiana, and at Cedar Creek, Virginia, among other battles.

In the 8th U.S. Infantry was John H. Voorhis.

Little is known about some of these men, yet one thing is certain—going off to war was the greatest adventure of their lives. Most were farmers who had not yet seen a quarter-century. Some had never journeyed beyond Gardiner or New Paltz, let alone Kingston. One, Joshua Armstrong, in his forties, was a cooper from Gardiner who likely bought barrel hoops crafted by Trapps men. Several were wounded, took ill, or were captured. Trapps prisoners were sent to two of the war's most infamous Confederate prisons—Libby Prison in Richmond, Virginia, and Andersonville Prison in Georgia. All but two survived their bloody ordeals; whether the eagle did is unknown.

The war, as all wars do, took its terrible toll both on soldiers and on their families back home. Alf Evers painted a grim picture of life for some Catskill Mountain families after their men went off to war. It was a picture likely repeated in the neighboring Shawangunks: "The prices of food and clothing doubled. 'Crust coffee,' made from darkened crusts of bread, became common. The poor lit their homes with a lamp consisting of a saucer of fat with a bit of rag by way of a wick. With so many men gone off to war there was a labor shortage. Wearing pants like their men, some Woodstock women took over plowing and harvesting." So, too, did Trapps women.

Those at home mustered much-needed support to ease battlefront conditions for the regiments. Relief societies sprang up. People boxed up supplies, medicine, clothing, linens, food, and longed-for letters from home. The Town of Gardiner's Ladies of Guilford hosted a social in December 1862 to raise funds for the Soldier's Relief Association. Tickets, 35¢ each, helped purchase "hospital clothing for sick and disabled soldiers." Women and children became knitting and sewing battalions. Taking up needles like tiny bayonets, they charged at wool and cloth, churning out socks, mittens, sheets, and other stitched items by the thousands. In January 1862 the Ladies Sewing Society of New Paltz sent forty handmade items to the Sanitary Commission for dispersal to "sick and needy" soldiers. The New Paltz paper

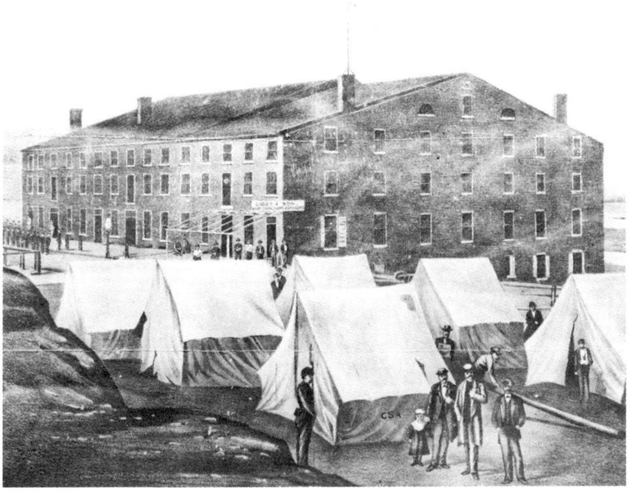

Some images of Confederate prisons belie the horrors experienced within by Union soldiers. Shown are Libby Prison, Virginia (left), and Andersonville Prison, Georgia (below). Courtesy of the Library of Congress, LC-USZ62-25518 (Libby) and LC-DIG-ppmsca-33768 (Andersonville).

stated one eighty-three-year-old member, Mrs. Maria DuBois of Springtown, "is willing to keep her needle at work as long as her eyesight is spared and a soldier in need."

Harry Rigby, Jr. recognized the efforts of those left behind during the Civil War: "In one sense, theirs is the greatest contribution, for they alone hold together the sober, workaday world to which militiamen must inevitably return." News then was often weeks or months old. While the war waged on, separations became unbearable. War mail was delivered locally only three times a week. Guilford helped rectify this in about 1861, as recalled by Rev. John N. Jansen of the Guilford Reformed Church:

John D. Van Leuven of the Trapps was held prisoner at both Libby and Andersonville prisons. He died of a common cold three decades after the Civil War and was laid to rest in a Trapps burying ground. Photograph by Robi Josephson.

> The whole community was in a state of unrest and anxiety. Many had friends and relatives who had gone to the front. Battles were constantly fought, and the suspense as to the result and the fate of friends became intolerable. ... This excited state of feeling originated what was styled a "Pony Express." Eighteen persons entered into an agreement, each to take his turn to go to New Paltz, which had a daily mail, and bring the papers to the store at Tuthill, distributing to those along the way who were subscribers. By this means at 6 or 7 o'clock in the evening of each day, we had the news which brought rejoicing or sorrow.[16]

The "Pony Express" was a success for nearly two years. Tuthill was the post office for many Trapps people, and we can imagine mountain families anxiously hurrying down at day's end for word of loved ones or neighbors. Did James and Phlinda Van Leuven hurry down to Tuthill to hear word of son John? Their gray-eyed boy, twenty-two, had enlisted in August 1862, but was captured in October 1864 in Virginia and sent to Richmond's Libby Prison. A New Paltz paper called the prison "famous," rather than the "infamous" it was.

John spent the last months of the war at Andersonville Prison, built in 1864 to ease overcrowding in Richmond's prisons. John survived that most hellish hell to be paroled and honorably discharged in 1865. He returned to the Trapps and married his neighbor,

Mary Elizabeth Slater, only to bury three daughters and in 1895 lose his own life at age fifty-six to a cold. Without modern antibiotics, a cold or cut could be deadlier than battle.

Enoch Freer Dingee, son of Isaac and Sarah, had married Mariah Enderly of the Trapps before the war. Enoch, twenty-three, enlisted with his cousin John Van Leuven and brother-in-law Isaac Enderly. He was injured in battle, sent to Convalescent Camp, near Alexandria, Virginia, and honorably discharged in July 1863. Enoch was five feet eleven inches tall with brown hair and grey eyes. His grey eyes, though, had been damaged; the left cornea was scarred and clouded, and the right eye was injured. Enoch also had developed rheumatism and "gastric derangement [related] to duty."

Enoch made it home to Mariah and returned to farming. His arrival was a great relief, for only months earlier Mariah and family had mourned the death of her younger brother, Isaac. Enoch and Isaac had mustered in together on August 22. Soon after marching south, Isaac took ill and was sent to Division Hospital in Falmouth, Virginia, where he died of disease on February 1, 1863. His widow, Anna Maria Harp of the Trapps, would never again set eyes on her husband, six feet one and a half inches tall with blue eyes and brown hair, who was buried in Washington, D.C.

John James Davenport had grown up in the upper Clove, just down the road from the Harps. John attended the Trapps School, then worked in the lime business. John, twenty-two, enlisted in 1862, but fell ill or was wounded. He was sent to Convalescent Camp, where Enoch Dingee was recovering. John was discharged in September 1863 and returned home, where the blue-eyed young veteran married Elmira Van Leuven of the Trapps.

William H. Chase lived in the upper Clove, a stone's throw from the Trapps and next to the Harp farm. He married neighbor Maria Elizabeth Davenport, sister of John. The brothers-in-law, twenty-one and twenty-two, served together until William was captured in battle in October 1863 at James City, Virginia. He was held prisoner for thirteen months and eleven days, part of that time at Andersonville. William was paroled and returned to active duty, but then was wounded in battle in March 1865 at Boydton Plank Road, Virginia. He lost the end of his right thumb and was still in the hospital when his company was mustered out later that year. He returned to his family and mountain farm, and during the mid-1870s served as a Trapps School trustee, where his children attended school.

Eli Lockwood was captured, date unknown, and held prisoner for 109 days in Richmond, Virginia, perhaps at Libby. He was not from the Trapps, but by 1900 had moved there with other members of his family.

Joshua Armstrong married Enoch Dingee's sister, Elisa or Eliza. Joshua and his son, George, served together, but only one came home. A letter made its way north and was published anonymously in New Paltz in October 1863. It read, in part, that George, "son of Joshua Armstrong of the Town of Gardiner, who was also with the regiment; died lately in a Hospital at New Orleans. He was wounded in the forheed [*sic*] on the 14th of June, in the battle before Port Hudson [Louisiana], from the effects of which he died. He was a good soldier, and adds another to the already long list of martyrs from this country." George was only about eighteen. Joshua, in his mid-forties, continued to serve until the following August, when he was discharged because of illness at Washington, D.C.

On March 3, 1863, Congress passed the Civil War Military Draft Act, which made "subject to military duty" male citizens between the ages of twenty and forty-five (as well as male immigrants who had applied to be citizens). The federal government established quotas for each congressional district, with the Trapps in the 13th New York district. Ben Fowler found himself the sole draftee for Gardiner on the list issued on June 30. The Trapps tavern owner was relieved of duty, though, for no records show he ever served.

The war ended and survivors returned home, but the nation was changing. Harry Rigby, Jr. wrote, "It was time to go back to work, to go back to the farm and relieve a younger brother of the chores so that he could get some schoolin'." Many men came back whole, but others were broken, as time went on becoming "fidgety, restless and withdrawn," Rigby, Jr. noted. Some moved into cities, while others moved west. Enoch Dingee moved to Michigan after 1890, likely to join other family; he died there in 1901.

Rigby, Jr. summed up the war's aftereffects in one New York mountain range: "Catskill clearings grew up again to brush, with only a cellar hole and a lilac bush to mark the abandoned marginal farm, the tracery of crude lumber roads leading to the tanneries reverted to stands of second-growth forest—but the settlements along the riverbanks grew lustily. The veteran had put down a rebellion but he had started a revolution." What men had once produced on their own farms for their family was being replaced by manufactured goods sold in stores.

While settlements were changing in the Catskills, Trapps families in the Shawangunks were able to remain there far longer thanks to the new hotels at Lakes Minnewaska and Mohonk. The Smiley hotels—the first opened at Mohonk only five years after war's end, followed by those at neighboring Minnewaska—provided the first steady paychecks for the Trapps people, thus helping to ease their hardscrabble, hand-to-mouth existence.[17]

Family Burying Grounds

Let us turn to a time some three score and ten years before the Civil War, when the Revolution was yet fresh in American memory. In 1792, in the dead of night in the Upper Clove, a wife groaned. George P. Frost, thirty-four, awoke to cradle his wife, Elizabeth, thirty, but she was gone, dead of a fit. Perhaps George sent out his oldest, William, eight, to tell Aunt Martha and Uncle James. James Gorsline was Elizabeth's brother; by 1790 he and his family were homesteading on a rocky parcel in the northern Shawangunks. George, who had grown up in Kittery, Maine, rose to the rank of captain during the Revolution. He and Elizabeth had lived in Dutchess County, New York, for eight years after the war, but then resettled near her brother only about a year before Elizabeth's sudden death.

Did Martha prepare her sister-in-law's body for burial? Did George, a carpenter by trade, make her coffin and a cross for her grave in his workshop? Perhaps James rode down the mountain to fetch the minister, and then Martha saw to the children, her two and George's four young ones. A place was needed to bury Elizabeth, and a small hill with good sandy soil was found within sight of James and Martha's place.

When family members passed away in isolated communities, such as the Trapps and Clove, they often were buried on the family farm rather than in town or church cemeteries. Several family burying grounds remain on the mountain. Two, the Frost and Davis Grounds, became community cemeteries for mountain people.

The Frost Ground is one of the earliest in the northern Shawangunks. Perhaps a minister from the Reformed Dutch Church of Rochester or Marbletown arrived to lead the little procession, with George guiding the horse and wagon bearing the coffin. James and Martha followed with the children, the youngest not understanding this sad adventure near home. Neighbors likely came, too, taking time away from their work and chores to pay their respects. After Elizabeth was buried, did George hammer a wooden cross, a proper headstone to come later? That night, George wondered how he would get by, yet he did with the help of James and Martha and their neighbors.

Yet tragedy returned a year later. George came back from a business trip only to learn that his youngest, Mary, about a year old, had died and was already buried. George went to her grave, where the dirt was fresh, the grave so small next to Elizabeth's. "I thought my troubles was to have no End," George wrote years later. He realized he had to board the children with his neighbors and went to each in turn asking for help, likely knocking on the door of the Roosas at the mill site. George's Dutch neighbors only spoke "broken English," but arrangements were made. He worked doubly hard to pay them, and with their help and that of his family, he got by. One year led to the next. George needed a wife and partner; his children needed a mother. One day George met such a woman, Agnes Green, and the two were wed in October 1796.

Elizabeth and Mary probably were the first of the family buried in the Frost Ground. Others soon joined them: George and Agnes Frost; Peter and Hannah Aldrich; Samuel and Tryphena Aldrich Jansen. They had carved out a life in the wilderness, family and neighbors in life as in death. The first generation of mountain pioneers was passing away. James and Martha Gorsline were likely buried here, too. Their son, Samuel, who raised his family farther up in the growing Trapps community, was buried there at the age of fifty-one only a few feet from where he was born. Amanda, a three-year-old daughter of his, died six months later and was buried beside him. Their inscriptions read, respectively:

Do Think As You Pass By
How Sudden Was My Death
And May You All Prepare To Die
Before You End Your Breath

Sleep Lovely Amanda And Take Thy Rest
God Willed Thee Home And Thought It Best

Sam's wife, Sarah, perhaps chose the inscriptions and arranged for a local stonecutter to make the headstones. Four unmarked field stones, or foot stones, near Samuel's grave may belong to Sarah and other children. Over the years about fifty people were buried in the Frost Ground.

Up in the Trapps at the mill site once owned by Teunis Roosa, Maria Calder Enderly died between 1840 and 1845. She probably was the first of that family to be buried at a site uphill from the homestead farm. Her headstone was broken off long ago, but her footstone, bearing the initials M.C.E., remains. Her husband, Johannes or John, died in 1846 and was placed in eternal rest beside her. Perhaps a dozen Enderly relatives are buried there, according to descendant Joan Hayes Wustrau.

The footstone of Trapps sawyer George Davis is all that remains at his grave. Nearby, the headstone of his wife Sarah leans against the one of their daughter Elizabeth, called Libbie. Photographs by Robi Josephson.

Up a steep hill above the Coxing Kill, also in the Trapps, is the Fowler Ground. Ben Fowler, forty-eight, was buried there in 1875, as were other relatives and grandchildren, including the four little ones that Ben's daughter Rachel Coddington had lost. An obelisk headstone, with the children's names on each side, was raised to mark their gravesite. Ben's mother, Eve Fowler, also might be buried in this family ground.

One difficulty in studying burying grounds is the one-time practice of removal, noted Richard L. Rider. Bodies were sometimes removed and reburied elsewhere, especially if the family moved, yet records might not follow the departed to the new resting place. Handmade depressions in the Fowler Ground, slumps in the soil about one and a half feet deep, suggest graves that were moved. Death certificates and church records also indicate many more people buried in burying grounds than graves found. Either the individuals were moved, or the headstones have not survived.

The Davis Ground began as a family plot, but then became the main burying ground for the Trapps people. Only a few headstones survive: those of Sarah Davis, wife of George; their daughter, Libbie; Albert Smith and his wife, Cornelia Ann Slater (also an obelisk); and Civil War veteran John D. Van Leuven. George's headstone is long gone, but a footstone bearing his initials remains. Since Libbie's burial, Sarah's headstone has leaned against her daughter's, a tender gesture, even in death, of maternal embrace.

A mock-orange tree once adorned the ground's entrance, recalled Joe Otis, who grew up in the Trapps some decades later. It was likely planted for its white blossoms of late spring and for the citrusy sweetness that welcomed and comforted one and all.[18]

Burying grounds remind us how fragile life was, especially in remote places like the Trapps. "There was little that could be done," Elwood Hitchcock wrote, "when illness struck, however. Sickness and death were common and expected, and accepted in early colonial days. They were a part of life."[19] For mountain families, life during the 1800s remained similar to earlier times. Illness and death were ever-present, like wolves in the darkness, waiting. After death, life for those surviving goes on as it must. In the Trapps hamlet, life went on for a time. The first generation passed away, and new generations were born. Yet, by the dawn of the twentieth century, the winds of change were sweeping across the mountain. For the first time since pioneering days, the population was diminishing. Work in the old trades turned steadily downward, and more people than not moved on, never to return. Epidemics also took their toll on an already dwindling population.

The final decline had begun.

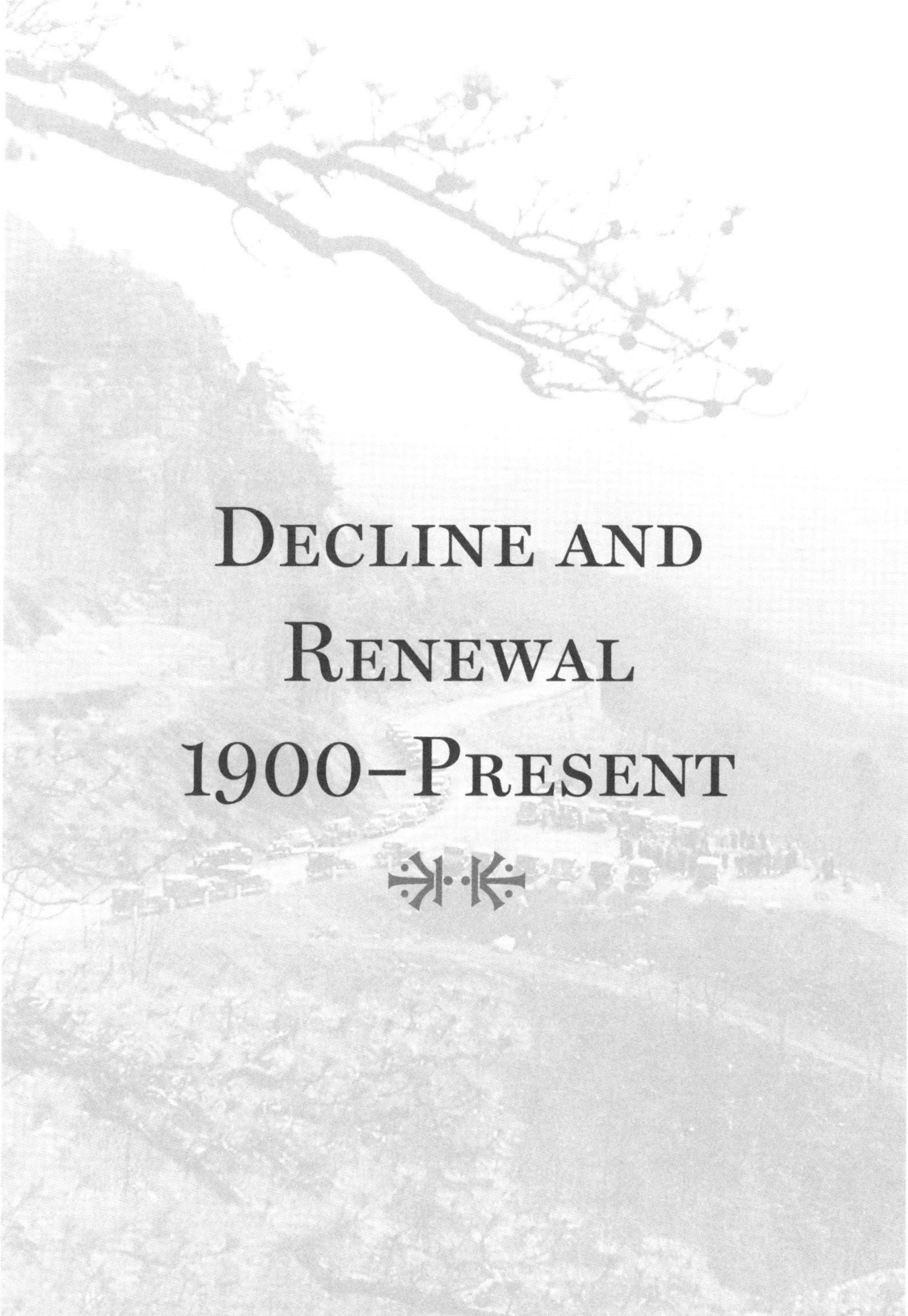

Decline and Renewal 1900–Present

• 12 •

Moving On

Fall 1910, the old Wynkoop place. Sisters Rachel Palmer, sixty-eight, and Martha McGibney, fifty-six, stare at their Trapps home, long abandoned. They link arms, hardly realizing it.

"I barely remember our Papa," Martha whispers.

"You were but a slip of a girl," Rachel says.

A breeze, tender as a mother's kiss, pulls them to the present.

"What time is supper tomorrow?" Martha asks. "I have something for Eddie."

"You spoil him."

"He's like my own." Martha drops her arm from her sister's. "It's late."

The two walk away from the house, from the past, from their sorrows. It was time to leave the Trapps and go home to the village of Walden.[1]

In 1900, Ralph Le Fevre journeyed through the Trapps on his way to Lake Minnewaska. "There are about forty families residing in the Traps neighborhood," he wrote in his New Paltz newspaper. "A great portion of the men work for Mr. A. H. [Alfred] Smiley while the mountain houses are open. In the winter getting out hoops is the principal industry. At this season of the year picking huckleberries is quite a business with the women and children." The number of households was holding from the heyday years. The slide, when it came, was as sudden as a spring melt. By 1915 the number dropped to some two dozen. By 1918, to seventeen. The downturn did not go unnoticed. Le Fevre ran a piece titled "Rural Population Rapidly Decreasing," excerpted below, in 1921:

> It is a matter of general remark that the rural population in Ulster County and elsewhere in the state is rapidly decreasing. This is perhaps nowhere more noticeable than in the Traps settlement. ... Sixteen unoccupied houses were lately counted in the Traps neighborhood. There are only about a half dozen children in school and the same number in Sunday School.[2]

The Trapps people were moving to more urban areas, such as Walden and Poughkeepsie, where industry jobs beckoned. As in the previous century, a few also left to find their fortunes in the West. The old trades no longer could support families. Advances in technology shut down mountain industries one by one, and marginal, rocky farmland had become exhausted.

Norman Van Valkenburgh wrote that "this knowledge faded as the old folks died, and their offspring sought escape from the hardscrabble life by moving to the hopeful bounty waiting in the surrounding villages." As Trapps homes were vacated, over sixty new ones sprang up in New Paltz from 1920 to 1924, stated that town's paper. For those wanting to move on, selling out to the Smileys helped them start new lives elsewhere. For those staying on, the Smileys gave them a place to work.

For the few who remained, the little Trapps Chapel remained open, too. A photograph from this period shows, displayed inside on a scripture board, Bible passages selected for one Sunday service—passages from Exodus.[3]

Nearby villages and cities became magnets for Trapps families. Down in Walden, Rachel Palmer's husband, Sydney, and son, Edson, both worked in a village knife shop.

Rachel's sister, Martha, and her husband, John McGibney, resided near the Palmers, living off savings from a lifetime of business ventures, including the hotel in the Trapps, an oyster saloon and a variety store, the last two in New Paltz.[4]

In 1893, Edison's invention lit up New Paltz for the first time and electricity provided by generator arrived at Mohonk Mountain House. During that decade, Trapps resident Alonzo Miller had moved earlier to New Paltz, where he painted houses and his son worked for the Wallkill Valley Railroad. Three other Trapps men found jobs on the latest type of passenger service serving the local area. Isaac H. Enderly, left fatherless during the Civil War, John B. Wynkoop, nephew of Rachel and Martha, and George Van Leuven, son of Charles—all worked the New Paltz–Highland line of the new electric trolley.

By 1910, John B.'s half-brother, Oscar, and his family also had moved from the Trapps down to New Paltz. Oscar Lyons might very well have ridden John's trolley run while commuting to work at the Hudson River Psychiatric Center in Poughkeepsie, where he was employed as a "fireman" and "stationery engineer" running the furnace. Their younger brother Charlie worked as a cook at the same hospital, but lived on the grounds. By 1920, Oscar had moved to Poughkeepsie, likely to be closer to his job.[5]

• • • • • •

One by one the mountain industries were falling away. Men could still make a living at the old trades, but less work supported fewer men. By 1900 the heavy burden of millstones was all but over. Today we find many of these great stones put out to pasture, as it were, as decorations. The reason was simple. "Steel equipment replaced stone," *The Accordian* reported, "and milling was done on a larger scale by a few mills, rather than many small mills. Most of the water-powered gristmills were abandoned." Production of millstones in Ulster County sank like a figurative stone, as it did across New York State and the nation. In 1883 the value of stones produced in the county was $120,000, state records noted. Five years later, it was half that. In 1883 total national value was $150,000; by 1910 it was just over $6,600.

The New York Times reported in 1928 that millstone cutting "was once a flourishing trade, but now has fallen upon evil days. The grit from Shawangunk Mountains is peculiarly well adapted to the manufacture of millstones. To this spot a few dozen men repair for a few months each year, and there engage in their business to no very great profit." Until mid-century, local stonecutters chipped away, their craftsmanship desired by home and overseas markets. A few industries, primarily gypsum, mustard, and paint, still needed the grinding power of the great stones. In 1972, Art Coddington, one of the last local stonecutters, said American gypsum companies once took "the most millstones ... about 600 pairs a year. That would be over a thousand stones that were made right in this valley [the Rondout]—for one year."

Yet a 1902 article stated stonecutting near the Trapps was still "quite an industry ... Already enough orders have been received to keep eight men quarrying for over two months. Two large stones will soon be sent to Germany, and orders have been filled from all parts of the United States." The Mohonk and Minnewaska properties also continued as places of employment for local stonecutters.

In 1905, Hiram "Daddy Hype" Van Leuven of the Trapps—turnpike gatekeeper, farmer, stonecutter, charcoal burner, wood cutter, laborer, and Civil War veteran—died at

Stonecutter Simon Coddington died from lungs weakened by stone dust, leaving his wife Laura, born in the Trapps, a widow with several children to support, including daughter Evelyn. Their Lawrence Hill farmhouse, west of the Trapps, is gone but its foundation and a root cellar, at left rear, remain intact. Photograph of Simon and Laura courtesy of their granddaughter, Grace C. Smith. Photograph of Evelyn courtesy of her daughter, Marian Whitford.

aged sixty-seven. Hiram had served the Town of Gardiner as commissioner of highways and overseer of the poor, the latter a position to which he was "repeatedly elected." Hiram passed his vast knowledge down to his sons and grandsons, but one son, Hiram, Jr., and his wife, Sarah, elected to move to Kerhonkson while retaining their jobs at Minnewaska, he as a superintendent and she as a servant. Two of Hiram, Sr.'s other sons, William H. and Eli, and some of their children stayed in the Trapps, where they farmed, worked the old trades, and worked for the Smileys. In April 1902, two Van Leuven brothers were "actively engaged in the manufacture of millstones," said one local paper.[6]

Orders for charcoal still came to Trapps men. In 1912 the Smiley family of Minnewaska paid $188.84 for charcoal to William H. Van Leuven. There came a day, though, when mountain charcoal couldn't keep up with cheaper prices from the coal regions. The days were ending, William's son Irv later said, when people "were accustomed to the sight of the high-sided charcoal wagons making their way to Poughkeepsie in the early dawn and returning at night, empty of their loads of charcoal, but now carrying provisions and articles that could not be bought at the little village stores." Besides local markets, charcoal continued to be marketed in Poughkeepsie and loaded onto barges destined for New York City.

Hoops, too, had declined with the downswing of Rosendale cement and upsurge of metal hoops, but Trapps men still found local markets. News stories on the industry, themselves once a winter staple, also dwindled. Knickerbocker observed that "again the Trapps dwellers lost their occupation." But they hadn't lost it entirely. Local markets could still be found. George Millham, a cooper by trade, had opened a cooperage in New Paltz in 1911. Two years later he "sold about 23,500 apple barrels," twice his usual output, reported a town paper. George's son Irving, born in 1905, told his own son, also George, about "the old timers" of the 1920s who "went back into the woods up on the mountain, where they cut the saplings then shaved the hoops at their shops." Cooper George used both wood and steel hoops, grandson George said.

Yet, families were resilient. They continued working for the Smileys. They continued the mountain trades of shaving hoops, of cutting wood and stone, of burning charcoal. And they continued the summertime ritual of picking huckleberries.

In 1900, Alfred Smiley constructed a carriage road connecting Lake Awosting and the village of Ellenville for Minnewaska guests arriving at that village via the Ellenville & Kingston Railroad, a new line being constructed north through the Rondout Valley along part of the old D&H Canal. Alfred's son Edward was the first to drive on the new road to

Awosting. In 1903 it was reported that pickers "heretofore have had to climb straight up the mountain side in stony paths or beds of dry mountain brooks. The Smiley roadway is of such easy grade it is a pleasure to walk upon it. Berry dealers can go up with their horses and wagons and bring down the berries of the pickers."

A tent was pitched or shanty cobbled together, and soon camps sprang up, like mushrooms after a rain, on the wide plateau above Ellenville and at mile intervals along the new Smiley road. Thus began the berry camps of the 1900s, still fondly remembered by the last generation to spend summers at them. Others lived near enough to trek to the berries each day, as stated in 1903: "Near Lake Awosting the picking is good and people residing in the Traps neighborhood come on foot or by wagon to gather the berries for market. They sell at 9 cents a quart, and some of them find their way to our village."[7]

The Smileys at both lakes continued to keep up construction with the same impetus as before. Before moving to New Paltz, George Van Leuven of the Trapps was injured in October 1897 while drilling at Minnewaska. George "had one of his eyes greatly injured by a fragment of a rock. Mr. [Alfred] Smiley gave him a sum of money to take him to the hospital in New York, where good medical attendance will be procured." During the 1890s, Alfred built a new road from Lake Minnewaska to Lake Awosting and put Trapps Civil War veteran Abram Wynkoop in charge of a small boardinghouse and boat rides at Awosting.

By November 1897, Alfred hired as many as sixty men to finish the new road and had surveyed another around Lake Awosting. He had planned to erect cottages and a deer park at Awosting, but changed course in favor of a summer camp for boys, which opened in 1900 and was privately run. In February 1901, Abram Wynkoop, seventy, passed away at the Trapps "home of his adopted son," George Wixon. Eight months later Abram's wife, Catharine, ten years his senior, also died at their home in the Trapps.[8]

• • • • • •

As one century became another, the law kept its eye on the Trapps as it did with other communities. In 1900, repeat offender James Schryver was arrested in Newburgh. "Schryver is a bad man," *The New Paltz Times* reported on Christmas Day, "and has committed a number of crimes in [Ulster] county recently. It is said that he has recently been making his home in the Shawangunk Mountains, in the vicinity of 'The Traps.'"

In 1902 a *New Paltz Independent* article denied two thieves had any relationship with the hamlet: "The statement that George Turner and David Williams, who were recently lodged in Goshen jail, charged with stealing chickens from Isaac Jenkins of New Hurley, are residents of the Traps neighborhood is not true. Neither Williams or Turner lives now, or ever has lived in the Traps country." New Hurley was a community in southern Ulster County.

Were Turner and Williams members of the notorious "Monroe-Turner-Williams crowd" of that period, as identified in *The Kingston Daily Freeman*? These were the extended families of "Shawangunk mountain desperado[s]" headed by William "Bad Bill" Monroe. Bad Bill, born in Alligerville, and his gang terrorized the citizens of Orange and Ulster Counties during the century's first decade. Bad Bill's "picturesque career" made fresh fodder for the newspapers, which duly reported his hide-and-seek games with the law from New York and New Jersey to California, his daring escapes from a moving train and from prison, and his being wounded while being chased by the law.

In 1907 two guards from Napanoch reformatory searched for three escaped prisoners who were last seen in the Trapps. Escapee Tom Monroe, brother of Bad Bill, was captured in the hamlet, where the two were known to have relatives. Was Willard F. Monroe of the Trapps another relative of the gang? Willard, a Civil War veteran from Cayuga County, New York, had moved there after that war.

Bill's taunts became legendary. This one made the front page of *The New York Times* in early September 1908. "I will be at the fair Thursday. There is a reward of $500 for my capture. Come and get me." Bill had signed and pinned this note on a tree during the Orange County Fair. He had posed as a beskirted widow begging for her seven children before revealing himself and "slipping away." That same week *The New Paltz Independent* reported that a Trapps resident denied Bill had been living there, "altho, when a boy, he lived at Dingee's hotel on the road from the Traps to Tuthill." Dingee had not owned the hotel for several decades, but it was still known as his. Bill, in his mid- to later twenties, was apprehended in California in November 1909 and sent to prison. It was not the first time, nor would it be his last.[9]

• • • • • •

On Thursday, July 24, 1902, a deluge unlike any other in living memory struck the Trapps and the Town of Gardiner. Ralph Le Fevre ran side-by-side stories on Friday, August 1 in his newspaper, *The New Paltz Independent.* He put the Trapps storm story at left, the Gardiner one at right. The Trapps headline thundered "Terrific Storm on the Mountain," then began:

> The worst storm that has ever visited any community in this section of Ulster county hovered over the Traps, between Lakes Mohonk and Minnewaska last Thursday. It began about twelve o'clock and lasted till nearly dark. Peals of thunder and flash after flash of lightning came from the clouds over head, together with a constant down pour of rain all the time. There was no let up to the rain, thunder and lightning. No storm of such severity has been known to have been there in over 70 years, at least that is what the oldest person there living says.

The "Gardiner News" feature began:

> Thursday a cyclonic storm accompanied by hail, rain and terrific electrical phenomena hung over the Shawangunk mountains on the west border of Gardiner for several hours.

Back to the Trapps: "Small streams rose until they were like rivers, taking trees, logs and all kinds of rubbish, together with boulders weighing hundreds of pounds were carried down the stream."

Readers, jumping back and forth, must have felt like they, too, were being tossed by the storm. The following was compiled from the two reports:

The Coxing Kill rose twelve feet, the Shawangunk Kill five feet.

Two-inch hailstones fell for over an hour to a depth of two inches.

Bridges and abutments along the Coxing Kill were swept away. One, an iron bridge two to three years old, was twisted "up in a heap," carried about an eighth of a mile, and lodged in a dooryard. The turnpike bridge across the Peters Kill was also damaged. Forty-two bridges were swept away in the Town of Gardiner. One "was washed away, the foundations were also taken and rock and debris, the accumulation of ages, were torn lose [*sic*] and carried down the embankment until the former bridge site was transformed into a deep gorge thirty feet broad by as many feet deep."

Carriage roads and drives around Lake Minnewaska and toward Lake Awosting were washed out in places. Alfred Smiley's property sustained some $2,000 in damages.

"The lightning," the paper said, "was constantly flashing, houses, barns, cattle, people and trees were struck promiscuously." A dozen people died, including a Rutsonville wife and mother of four struck by lightning while picking berries near Lake Awosting. Merritt Van Leuven, about forty, "was rendered unconscious but survived." A berry peddler, Merritt was caught in the storm while returning from Ellenville to his Trapps home via Awosting on Alfred's new carriage road. He "had to drive through water that came into his wagon box and when he reached Minnewaska was so near exhausted that Mr. [Alfred] Smiley took him and his team and cared for him nearly all day Friday when he returned to his home in the Traps." Merritt was the oldest of four known sons of Charles Van Leuven, Sr. and Rebecca Bunton.

Hiram Enderly's sawmill was "taken down Coxing creek two or three miles." (Whether he rebuilt the mill is uncertain.) Hiram also lost "his garden, earth and all." Gardens along the Coxing were "all swept away, potato patches all gone. A small island containing about an acre on which Oscar Harp had planted seven bushels of potatoes is wiped out." Only one thing saved Hiram Van Leuven's granary—"the great rocks that had washed around it." That granary helped Hiram's family and perhaps others in the Trapps survive the coming winter.[10]

• • • • • •

In January 1903, Alfred H. Smiley, two months shy of seventy-five, died in California. The founder of Lake Minnewaska was buried in Poughkeepsie Rural Cemetery. Sons George and Edward inherited the family business, and life in the Trapps went on much as it had.[11] That November, Sarah Davis, seventy-nine, widow of sawmill owner George Davis, died and was buried beside her husband. Neighbor Ben Walker died the same month.

For years Alfred had tried to buy the Laurel Inn, the tavern Ben had operated just west of the Minnewaska entrance. Ben finally sold out "on his deathbed" to George and Edward for $4,000, wrote Nan Knickerbocker. She added that the tavern was remodeled into a boardinghouse, where a sign was to be erected with the name "Minnewaska": "In those days some uneducated inhabitants of places in the vicinity used to call Minnewaska,

Minnewasky. One can imagine the surprise of the owners of Minnewaska when the sign was put up with the name Minnewhiskey. The workman who made the sign did not intentionally misspell the name. Needless to say, the sign did not last long."

Minnewaska employee Calvin Burger remained one of the most colorful figures in the Trapps. Knickerbocker wrote about the later life of the Old Rooster: "When he was too old to work, he asked Mr. Smiley [Alfred or one of his sons] to give him the left-over bread and rolls, and he especially asked for stale corn muffins which in some unrevealed way he used to dry out and keep for his food during the months when the hotel was closed." Calvin, seventy-six, died in 1905 in Wallkill at the home of his son, Sheridan. Ralph Le Fevre gave Calvin a thirty-eight-line farewell in his New Paltz paper, calling him "a genial man" who would be "greatly missed by many people, not only here, but on both sides of the mountain."

Late that same year, Calvin's ex-wife, Rachel, seventy-five, died in the Trapps at her log cabin, the one built next to Calvin's. Her three-line obituary ran in January 1906. Gone, too, was Elizabeth "Libbie" Davis, sixty-seven. Libbie, who never married, burned to death in 1911 when her clothes caught fire while "taking burning shavings from the oven" of the boardinghouse she operated out of her own home. William Enderly also passed away. Troubled in his early life, William became a property owner in Ellenville, where he was a laborer and vegetable peddler, likely among other occupations. It is unknown if he ever returned while alive to his birthplace, but in 1916 William at last came home to the Trapps. He was buried in the Enderly Ground near his grandparents and

Mary Edda Pratt, right, daughter of Calvin and Rachel Burger, grew up in the Trapps and returned for a later day visit. She stands on the porch of the Trapps Chapel, by then abandoned, with her daughters and granddaughters. At far left is her daughter, Madge Kidney. Courtesy of DSRC-MP. Donated by Madge's son, Clarence Kidney.

Left: This boy is likely Herbert Bunting, born about 1911. Herbert was the son of James William "Jim Will" Bunting and the former Cornelia Otis, sister of Jake Otis. The Bunting family left the Trapps for Detroit, Michigan, during the 1920s. Right: Ray Otis (left), son of Jake, and his cousin, another member of the extended Bunting family tree, are off to school ca. 1919. Both families lived in the Trapps. Photographs courtesy of Ray's son, Don Otis.

other family members. William's executor erected a dignified "Barre granite monument" made by Henry A. Swart to mark his final resting place. William's headstone is the most distinguished one in the Trapps today.

One of the few families to arrive in the Trapps during this period was the Otis family. Don Otis related how his grandfather, Jake Otis, had prospered as an owner of canal boats until the D&H Canal closed. Jake, his wife Minnie Purcell, and their young son Eugene moved up the mountain shortly after the new century to find work at Lake Minnewaska. The family, joined by Jake's parents, Johan and Mary Ann, settled a short distance west of Minnewaska along the old turnpike.

By 1910, Jake and Minnie were raising four children, and Johan had died. That year, daughter Helen, about seven, died during a diphtheria epidemic. In January 1916 tragedy struck again when Eugene, fifteen, took lunch to Jake, who was logging near their home. While crossing the Peters Kill, Gene fell through the ice and was swept over Awosting Falls and Peters Kill Falls. His body was found downstream near the Minnewaska powerhouse. Gene's friend, Tracy Van Wagenen, identified as a dwarf in an Orange County newspaper, also drowned in the same tragedy.

A few years after Gene had died, the Smiley family of Minnewaska bought the former Billingham house and made a proposal to the Otises. The two families had become close, working and socializing together. As well, Jake had helped his employers acquire property in the Trapps. The Smileys offered the Billingham house rent-free to Jake and Minnie, as trusted employees, for the rest of their lives and those of their descendants in exchange for their house near the resort's entrance. Jake, Minnie, and their surviving children, Myrtle and Ray, left their old place with its sad memories and settled into their new home. There, the family kept a large garden, Jake cut saplings and shaved barrel hoops for local sale, and he leased his work horses out to Minnewaska for carriage drives and hauling.[12]

Trapps men worked as laborers on the construction of the Catskill and Delaware aqueducts built during the first half of the twentieth century. Railroad cars, shown here, removed tons of rock from the tunnel bored under Bonticou Crag in the northern Shawangunks for the Catskill Aqueduct. Courtesy of Vivian Yess Wadlin.

• • • • • •

In January 1907, Eli Van Leuven had a narrow escape while working at a Smiley shale or slate bank in the Trapps. Neighbor John Voorhis reported to his old workplace, *The New Paltz Independent*, that Eli "was picking in the bank on a bunch of loose slate when it fell on him. It cut his head quite bad and tore the flesh under his eye loose and broke two bones in one leg and otherwise hurt him." Of about twenty-five men working there, one brought him to Kerhonkson, where a doctor put eleven stitches under his eye. Several hundred pounds of rock had given way when the frost went "out of the earth and slate," but Eli had sense enough to crawl out of the rubble. "It was a miracle that he was not killed outright," John wrote.

Trapps men also worked on one of America's first megastructures of the new century—the Catskill Aqueduct, built to slake the ever-thirsty metropolis of New York City. Thousands of men from near and far, many of them immigrants and African-Americans, worked from 1907 to 1917 to complete this massive project as it snaked its way from the new Ashokan Reservoir south. The aqueduct tunneled under the Shawangunks at Bonticou Crag, paralleled the foot of the mountain below Sky Top, then veered east under the Wallkill River. It just skirted the Trapps, but Trapps men worked on the project, including: foreman Hiram Van Leuven, Jr., son of Daddy Hype; teamster Cyrus Schoonmaker, nephew of Lucas; and laborers Edward Hotaling and brothers Theodore and Emil Gierisch. The Gierisches were newcomers from New Jersey, likely attracted by work and good wages. By 1910, Edward Hotaling and wife Mary had lived in the Trapps for at least ten years.

In 1912, George Davis, Jr. was the only enrolled Prohibitionist in the Trapps. That December, Albert K. Smiley, eighty-four, died in Redlands, California; his wife, Eliza, died the following year. They were buried next to Alfred in Poughkeepsie. Albert had bequeathed Mohonk to their half-brother Daniel.[13]

• • • • • •

The Trapps people continued working for the Smiley families of Lakes Mohonk and Minnewaska and continued to sell them fresh produce. During the 1910s, Minnewaska paid Ira Decker $8.96 for 64 quarts of berries, Jacob Slater $15.36 for 6 bushels, and Peter Brought $1.40 for 14 quarts. Lewis "Huckleberry" Wynkoop, who had moved to

Kerhonkson, continued working the berry trade with his oldest son, Melvin, and that landed Lewis in a jam.[14]

In October 1910, Lewis was on trial "for setting fire to the woods east of Napanoch" the previous August. Young witnesses John Slover, Arthur Rose, and Oscar Markle said "that while picking huckleberries on the mountain they saw him light a match and start a fire in the brush. They hastened to the spot, told Wynkoop there was a fire, but he went straight on up the hill, and they buckled in and put out the fire, though with much difficulty. On the evidence the jury found Wynkoop guilty."

Whether Lewis served time is unknown, but prosecutions of "firebugs" under new state laws landed a Trapps trio in hot water. In October 1911, Irving Van Leuven, twenty-two, Delford Van Leuven, eighteen; and Charles B. Gardner, twenty-two, all cousins and fourth-generation Van Leuvens of the hamlet, were arrested for setting fires "for a little excitement" on the old Minnewaska carriage road heading up toward Mohonk Lake. This time, though, telephone wires of the Smileys also had been cut by Charles, thus preventing calls for help. The fires burned for three days, but were mostly out by the next morning—the excitement, too.

An earlier complaint had prompted the state forest, fish, and game commission to place two undercover detectives in the Trapps. They posed as photographers employed by a Chicago company making a new form of entertainment sweeping the nation—motion pictures. The outsiders boarded with Irv's brother, Hiram (both sons of William), but as his house was full, they stayed in a shed until it became too cold. The men hunted and fished with the Trapps people and invited them to see the pictures being developed, thus getting "on the soft side" of the Van Leuvens. They softened one person in particular, Charles's wife, Maud Brought, a Trapps woman who promptly disappeared. Charles claimed to be happily married until the detectives showed up. "What had become of her was a mystery to her husband," stated *The Kingston Daily Freeman*—but not to the Ulster County district attorney, who had squirreled her away as a prosecution witness.

Maud dazzled the Kingston courtroom draped in a red dress, large green hat, pink neck ribbon, and wearing eyeglasses. The star witness stated that she lived apart from Charles and had been living in Jamaica, which was never explained. She swore she had sat on the same porch with them when the three hatched their plan. She and Charles differed on two points. She said he got "a pair of nippers" to cut the phone wire; he said he used a hammer and stone. She said the fire was set in August; he said in May.

Other Trapps residents testified, including Levi Kimbark, who said "there was practically a fire every day during the summer in the mountains." The courtroom was packed the week before Thanksgiving, and spectators included Edward Smiley and others from Lake Minnewaska. The trial fired up the presses from Ellenville to Kingston.

Del's confession stated that one of them had said, "Let's go out and set the woods on fire for a little excitement." The others replied, "All right, we don't care." Delford also claimed under oath that his confession was forced.

Irv and Del were found guilty; any decision regarding Charles is unknown. At sentencing Del's lawyer argued mitigating circumstances. Scarlet fever at nine years old had left Del afflicted with the St. Vitus' dance (Syndenham's chorea), in his case "a nervous twitching of both eye lids," and perhaps affected him mentally, causing him to be influenced by "a stronger mind."

With several indictments against Irv for violating fish and game laws, he was remanded with bail set at $500, then sentenced to three months in jail, where he spent two of those months waiting for a new trial "for snaring a partridge," reported *The Ellenville Journal.* The judge gave Del a suspended one-year sentence on parole and required him to report monthly. He exhorted Del "to keep in the path of honesty and rectitude" or he would otherwise be sent away.[15]

With summer fires almost a daily occurrence, plans got underway for a better method of spotting them. The Smileys erected a wooden fire tower and cabin on Dickie Barre in 1913, noted Daniel Smiley II. New York State had earlier set up a system of mountaintop fire towers, but the Smileys privately owned the "Dicka-Bar" one. Mohonk hired Trapps widower Sam Schoonmaker as fire warden "because he knew how the picker-firebugs operated," wrote Smiley, who thought the fire tower was a "joint project" of the two resorts. There might have been another reason Sam got the job. "Scuttlebutt has it," Keith LaBudde noted, "that the person hired as the warden was the person they thought might be starting the fires. Whether he was or not, the fire problem in that area disappeared." By 1920, Sam moved his family to the village of Highland in the Town of Lloyd but continued working as a fire forester, likely on Dickie Barre, and a few years later at the new stone tower on Sky Top, the fourth such observation tower built there.

By this time, berry pickers' summer camps were in full swing atop the Shawangunks above Ellenville, from Sam's Point near Cragsmoor north to Lake Awosting. In 1916 a

New York Times reporter dubbed these camps "tar-paper villages," a sight that "Alice in Wonderland had no better cause to rub her eyes and stare."

> Every Summer, when the first purple shows on the huckleberry bushes, the procession of pickers begins to arrive. Up the hill trudges a weary horse hauling furniture, family, and frying pan. From the heap a stovepipe sticks up, behind the wagon pants the family dog. The tar-paper village has waited, deserted, all Winter. Now it wakes to life.

Of the tar-paper shacks, the reporter wrote, "What more summery? Surely the breezes of the mountain must play through a tar paper house, bracing the dwellers therein."

Of the people, he wrote they were as exclusive "as any country club you might name." The berry picker might be "a queer nomad, but give him a chance and he's not an unfriendly one." One summer guest at Sam's Point tripped over a berry trap [wagon] and wrenched her ankle. The "burly young picker" later dropped off a get-well bouquet for her of pitcher plants, orchid blooms, and Indian pipes.[16]

Arthur Van Leuven, with a white horse, poses about 1920 with berry pickers at Billy's Camp along the Smiley carriage road from Ellenville to Lake Awosting. Arthur was a son of Eli Van Leuven of the Trapps. Courtesy of DSRC-MP. Donated by Arthur's son, Harold Van Leuven.

• • • • • •

Berry money and work for the Smileys helped one Trapps family survive when tragedy struck. After Laney Fowler's death in 1892, daughter Mary Enderly and her family had moved to a farm below the cliffs. By 1899, Mary had had fourteen children, with most surviving. On July 17, 1901, Mary's husband George hauled a load of stone from Accord to Mohonk and was on his way home. All was well until, within minutes, all went wrong. Grandson Arthur Hayes wrote, "something spooked the horses and he lost the harness from the team of horses. He tried to retrieve it by walking out on the poles next to the horses. The horses being spooked took off in a wild gallop. My grandfather got the harnesses tangled around his legs and was dragged all the way home. He had a huge hole in the back of his skull. Nothing could be done for him." George, cared for at home, lived but a few days.[17]

Mary, widowed at forty-two, was left to raise the children; her youngest, Rubin, was still a toddler. The older children pitched in. During the summer, daughter Anna worked as a laundress at Lake Minnewaska and the rest of the year at the Merrill–Van Laer boarding and day school for girls in Manhattan. In May 1909, younger sister Edna mailed a postcard of Niagara Falls to Anna at the school: "I have not got no dress yet. I guess I will have to wait till huckberries gets ripe. from Edna." In March 1912, Edward Smiley of Minnewaska sent a postcard to Anna in Manhattan: "Can you come again this summer and do some work and wait table/ black for morning and noon and white for supper, rubber heels." Black and white were the uniform colors, supper the evening meal.[18]

Death spared Mary for a time, but then embraced her once more in 1917. Her eldest, Benjamin, a thirty-eight-year-old laborer living in Manhattan, died of pneumonia in early February. Mary brought him home to be buried. One year later, a far more "dread spectre of the hills" crept in.

On September 12, 1918, three of Mary's younger sons—Samuel, Edwin, and Rufus—went down to Napanoch to register. Reuben also signed up, presumably that day with his brothers. It was World War I, and they were ready to fight overseas.

Rufus, called Ruly, was the first to fight for his life, but not on the battlefield. Mary's youngest, a tall lad of medium build with blue eyes and brown hair, was a fireman, or boiler man, at Minnewaska. Ruly took ill, and on October 9 slipped away, his life stolen by one of the Earth's most ruthless killers. Area doctors were summoned once more, to tend five more of Mary's adult children, but the killer was callous, cruel, unrelenting. Influenza,

Mary Enderly, second from left, is flanked ca. 1911 by her two youngest children, Arthur and Reuben. At far right is daughter Mildred with her first husband, Frank Terwilliger, and their son Hector. In October 1918, Mary buried six of her fourteen adult children during the Spanish flu epidemic, including Arthur and Mildred. Courtesy of Mary's great-granddaughter Joan Hayes Wustrau.

dubbed the Spanish flu, had swung into the Trapps, grabbing partners in its deathly do-si-do, sparing no one, claiming all ages, as it whirled around the globe.

Three local doctors cared for the Enderly family—Dr. William Fuller, Dr. Fred Coddington, and Dr. Myron Stephens, who himself succumbed to the flu after tending 400 patients "night and day" over a three-week period, reported *The New Paltz Times*.

October 16, 1918, dawned mostly cloudy with southeasterly winds. Temperatures reached 60° Fahrenheit. The mountain sported its prettiest colors; maples and oaks twirled their leaves, as flirtatious as girls at a barn dance. No one, though, called any dances that day. Instead, sorrow paid four calls to Mary Enderly over one heartbreaking, twenty-four-hour period.

3:30 AM, gone was Arthur, twenty, a single laborer.
2:30 PM, gone was Samuel, thirty-five, a married farmer with two children.
9:30 PM, gone was Edna, twenty-five, a single housewife, who died on her birthday.
Gone was Mildred, twenty-seven, a wife and mother known by her middle name Eliza.

Eliza, with a son from her first marriage, had remarried neighbor Emil Gierisch. On February 29, 1918, the two had celebrated their second anniversary. Three days after Eliza was gone, Emil also was dead from the flu.

On October 23, Mary lost Edwin, forty—tall, slender, black-eyed, dark-haired Edwin. Edwin, a mechanic, worked as a wagon repairman at Minnewaska and lived at the lake with his wife, Isabel or Isabella, called Bell, whom he had just married that January.

Seven family members lost in a span of two weeks.

Two brothers, Reuben and Ernest, called Ernie or Emzy, and a daughter, Anna, were the only children left to Mary. Yet, sorrow was not theirs alone. In October 1918, American deaths spiked to 195,000, one of the highest monthly totals in the country's history, if not the highest. Some 675,000 Americans died between 1917 and 1918. While worldwide figures vary, up to 100 million might have succumbed to one of the Earth's most deadly pandemics.

Schools across the country shut down in an effort to check the flu's spread among children. On October 16, the same day that Mary Enderly lost four of her own, Miss Martha Brooks, twenty-one, locked the door of the one-room Trapps School. The children (fourteen were registered students) did not trudge to school again until five weeks later on November 18, ten days before Thanksgiving.[19]

Yet, death still had unfinished business with the woman who had buried so many.

On Sunday, July 24, 1921, Reuben, twenty-six, died—tall, stout, gray-eyed, black-haired Reuben. Reuben had served in France while the Spanish flu had ravaged his family. A teamster and blacksmith by trade, he had worked at Minnewaska and at Coddington's Blacksmith Shop in Accord. One summer day he was out on the lake in a canoe when a struggle began. The New Paltz paper reported that Reuben's "cries for help were heard by James Fitzpatrick, who went to his aid, but after being dragged down by the unfortunate, drowning man, was obliged to leave him to his fate. The water is about 40 feet in depth at the spot where Enderly was drowned and there was much trouble in finding the body. Finally a professional diver was brot [*sic*] from New York who, after long effort, succeeded in bringing the body up to the surface."

Questions remain concerning his death. Arthur Hayes wrote: "There must have been an argument and another man hit him across the face with the oar. They found Uncle Rubin's body six or seven days later with a big mark across the face similar to the oar. Rubin did not know how to swim. No one ever was accused or arrested for his death." The Kingston paper reported there was "a question of 'horse play'," the paddles of his canoe

After Libbie Davis's death in 1911, her home became a summer boardinghouse called Hill Crest operated by Connie Mack. This house is one of the few remaining today in the former Trapps hamlet. Courtesy of DSRC-MP. Donated by Harold Bittner.

"taken in play by some girls." His insurance company rejected a cash settlement for Mary, but the Department of Labor awarded her either about a dollar a day for life or $3,600 plus funeral and counsel costs.

The paper noted that Reuben was "the last survivor of his family," but was incorrect on that point—Reuben was not the last. Two yet lived.

Emzy lived in New Jersey with his family. His sister, Anna, also called Elizabeth, also had married and had lived in the states of Washington and Pennsylvania before Mary called her home to help care for her ill siblings. In a horrible *déjà vu*, Anna soon became her mother's companion in widowed grief.[20]

• • • • • •

After Libbie Davis died in 1911, James and Carrie Mack bought the boardinghouse from Libbie's brothers. James, born in Rosendale, married Carrie Enderly, whose father, Hiram, had inherited the family mill site on the Coxing Kill. The Macks first lived in Rosendale, where James worked at a powder mill, then moved to the Trapps, where James farmed and taught school and Carrie ran the boardinghouse. They also took in the overflow from Laurel Inn and later worked there, recalled neighbor Roger Van Leuven.

One day in 1912, three women in Brooklyn—Rose Bittner, her cousin Bertha Veres, and Mabel Veres, wife of Bertha's brother Frank—saw an ad for Hill Crest, Carrie Mack's boardinghouse. The Vereses were an Austro-Hungarian immigrant family. Frank was born in Europe; his baby sister Bertha was born in America. Rose, Bertha, and Mabel came up to stay with the Macks, said Rose's son, Harold "Hal" Bittner. This was the beginning of the Veres family's time on the mountain, one of the first city families to spend summers in the Trapps hamlet.

In April 1920, Mabel bought land with a house in the Trapps. That same year, Rose and her husband John brought their family up from Brooklyn for the first time and boarded with the Macks. The next year, Bertha wed Trapps native Winslow Blackwood Van Leuven, the only child of Eli Van Leuven and his first wife, Kathy Lawrence.

Decades later, Raymond M. Veres, known as Ray, wrote about the family's arrival in the Trapps for the New Paltz paper, the *Huguenot Herald*. Ray's mother, Mabel, wanted to get the family out of the city. The "horseless buggies" were "a menace to children"; the city streets were "filthy"; the air was "too smoky"; and "dreadful little brats" lived next door.

Ray's father, Frank, said "he'd be damned" if all that "were going to run him out of his house."

Guess who did not have the final word.

"So it came to pass," Ray wrote, "that they paid six hundred dollars for the clapboard and shingled house, all rambling and termite-ridden, utterly isolated from the niceties and conveniences of a big city ... or even of a small town, for that matter."

The house was located in "a picturesque, remote mountain setting" in a hamlet called The Trapps near his Aunt Bertha, "her mountaineer husband" Winslow, and other relatives. Their new place "was set off cozily in a small stand of hemlock," Ray described, "not to mention worm-eaten apple trees, stumps and blackberry briars."

Mabel did not get her way completely. The family returned to the city during the winter, where Frank worked as a printer. He recently had a pay increase to $15 a week and was in line for a superintendent's job with a "juicier" $10 a week more.[21]

• • • • • •

The 1920s roared everywhere in America, except, it seemed, in places like the Trapps. That decade was the hamlet's last hurrah. The population was down by more than half, yet Trapps men continued to work on important building projects. A South American explorer gave the hamlet a starring role in two novels, following the earlier example of a local preacher.

Edward Hayes and his wife, the former Anna Enderly of the Trapps, pose with their firstborn child, Arthur, in Washington state in late 1912. Arthur, eleven or twelve pounds at birth, was a few months old when the photo was taken. Courtesy of Arthur's daughter, Joan Hayes Wustrau.

In 1921, Ralph Le Fevre ran a front-page story in his New Paltz paper from the *Highland Post*. "Fred" might have been Frank Van Leuven, son of William. George Muller owned a drugstore in the village of Highland.

> George Muller has one of the members of the cat family on exhibition at his drug store, known as a Bob cat. The animal was shot Monday morning near Lake Minnewaska by Fred VanLeuven. VanLeuven was on his way to work and saw the bob cat coming steadily towards him. When it was about 25 feet away he started shooting at it with a 32 caliber revolver, firing five times and hitting the mark four out of the five times. George Muller had purchased the body of the bob cat and intends mounting it. This specimen weighs about thirty pounds.—Highland Post.

In March 1921, Lucretia Decker Enderly, her children, Carrie Mack and Lester Enderly, and Lester's wife, Grace, sold much of the farm and mill to Mohonk Mountain House. The Enderly family was gone from the mill site six score years after John Enderly had acquired from Teunis Roosa the land at the great split rock on the Coxing Kill.

During the 1920s, the Enderlys held an auction at which John and Rose Bittner bought a pony for their young son Hal. Hal Bittner, Evelyn Coddington DeWitt, and Don Otis recalled other details about mountain life then. An open-air dance pavilion attracted people from all over to square dances and church meetings during summer. Families put a lit candle or lantern in the window to alert neighbors they needed help. With so much cleared land then on the mountain, any such light was visible for some distance. A neighbor could quickly race up to Lake Minnewaska to fetch the house doctor when one was available. A steep path off Lyons Road called The Crevices was one shortcut, used every day or for emergencies, from the Trapps to that lake.

St. Patrick's Day 1923, birthday of the late Albert and Alfred Smiley, dawned a normal work day for Edward Hayes, husband of Anna Enderly. Their son Arthur described their story. The son of Irish immigrants, Edward was baptized in Gardiner in 1884. "When he was still a teen," Arthur wrote, "he stowed away on a boat, landed in China and got involved in the Boxer Rebellion," the anti-imperialist movement at the turn of the twentieth century. Edward, along with his brother Jim, next worked on the Panama Canal for a short time. Meanwhile Anna worked at Minnewaska and in Manhattan, where she met Edward. He moved to the state of Washington to work as a coppersmith and copper miner. Anna followed him, and the two were married in 1912.

There, Edward joined the Industrial Workers of the World, or "Wobblies," an organization of the radical union movement. Edward went to prison for his participation in the Everett Massacre of 1916, also known as Bloody Sunday, but with Anna's intervention was released. The couple and their two children moved back East, first to Pennsylvania, where the children survived the Spanish flu, and then home to Gardiner to help her widowed mother during the family's ordeal with the epidemic.

Edward found work logging and a job with the Foundation Company, under contract to help construct the Rifton Dam for a new hydroelectric project on the Wallkill River north of New Paltz. His son, Arthur, ten years old at the time, years later described what happened: while his father was working on a scaffold, others were dynamiting ice below him when a blast "blew up the scaffold." His father fell and tried swimming to shore, but sank when "a big slab of ice hit him in the head." They never found his body. His mother collected $52 a month in compensation, and his sister and he each received $11 a month. At eighteen, they lost their monthly check. Their mother's was reduced to $42.62, and she returned to work at the Minnewaska laundry for many years.

The Kingston paper was the first to break the news on the front page of its March 17, 1923, evening edition. The Gardiner paper reported, "In one spot he came up partly out of the water and seized the edge of a mass of ice in an attempt to rescue himself but was quickly drawn under it."

The impossible had happened again to Anna and her mother, Mary.[22]

That same year, Edward Smiley retired after forty-four years at Minnewaska. He sold out to his brother George, and later moved to New Jersey where he continued working in the hotel business.

One year later a local news brief reflected the sorry shape of roads in the Trapps and an even sorrier turn of events for the hamlet's schoolteacher: "Raymond Everts who teaches the Trapps School near New Paltz drives a car to and from school, parking it some distance from the building because of the very bad roads. One evening last week, when he went to his car to drive home, he found it stripped of tires and rims." Raymond, about twenty-five at the time, had grown up on a farm in Gardiner and had been a factory worker prior to teaching at the Trapps School.

Art Lapp grew up in Alligerville in the 1920s and recalled many mountain families of that period:

> The mountain men in the Clove and the Trapps lived off the land, names like Ayers, Bush, Yeaple, Brooks, Shears, Lawrence. Each scratch farm would have a cow and chickens. They picked huckleberries, trapped, hunted, and fished. They had high boots up to the knee; would walk far distances to work. Many worked at Mohonk. They cut hoop poles, and made their own whiskey in stills that the Revenuers would find sometimes.
>
> When I was a kid, stonecutters were still cutting millstones and stone for fireplaces all through the woods, cutting by hand. Everyone had work to do. It is unbelievable how they survived.[23]

• • • • • •

During the 1920s, itinerant Methodist minister Charles Edward Hewitt and wife Margaret bought land near Mohonk, noted Michael F. McAllister. There, the couple ran an early type of Fresh Air Fund camp until 1950, Mabel DePuy wrote. Hewitt's published works include three religious romances set in New Paltz and vicinity: *Hazel* (1906); *The Outcast* (1913); and *The Hand in the Dark* (1947).

Johnathan Du Boise is the main protagonist in the *The Outcast*, set in post–Civil War New Paltz. He is a Quaker and pacifist, but one who wants to preach a nondenominational faith, much to the dismay of his parents and love interest, Rachel Doolittle, daughter of a prominent local Quaker preacher. Their relationship is fraught with misunderstandings, especially when Johnathan heads up to the Trapps to start his new church.

The main Trapps characters are: Joe Van Leuven, who can hold an hour's worth of tobacco before spitting it; Joe's sister, Liz, "a tall shapely girl with bright snapping eyes and lustrous red hair"; and her jealous mountain suitor, Jim Cornell. Liz falls for Johnathan, riling up Jim to hatch a deadly plan for his rival.

Hewitt portrayed Trapps people as "simple-hearted," "unlettered," and "neglected." A church was not erected in the Trapps until the 1880s, but at the time of the novel mountain families were members of Reformed or Methodist congregations. Likewise some Trapps people could not read or write during the book's time period, but others could; the hamlet's school had been open for over ten years. "Neglected" was the only adjective to approach the hamlet's isolation. Local readers would have instantly recognized Hewitt's description of the break in the wall of cliffs in this scene of Johnathan's initial journey up the mountain:

> The progress now led parallel to the summit, and quite near it, for about three miles. On the one side a mighty wall of ragged ledges and huge boulders extended, and upon the other the Hudson's Highlands could be seen thirty miles away, spread in beauteous panorama, the road finally passing through a gate-like formation of the gray-white Shawangunk Grit. It was from this reach of cliffs and boulders from which The Trapps village had taken its name.

The novel's climax takes place in the Trapps woods, where Jim and two cohorts attempt to tar and feather Johnathan. A real mountain man, no name or region given, was reportedly given the same punishment only five years before Hewitt's novel came out. The man's neighbors shamed him for leaving his family and "giving his company too much to a woman in the neighborhood."[24]

• • • • • •

Americans turned the pages of two more novels set in the Trapps, courtesy of a daring adventurer. In 1920, Arthur Olney Friel lived in Brooklyn and worked as a newspaper editor. Two years later the Yale graduate embarked on a series of expeditions to South and Central America, which became the basis for his pulp fiction adventure novels. Arthur also found his way to what he called "wild cat country" in upstate New York—the Shawangunk Mountains. He set two novels, *Cat O' Mountain* (1923) and *Hard Wood* (1925), in the Trapps, both serialized in *Adventure* magazine.

While researching his novels, Arthur boarded with Frank and Mabel Veres for "a couple of summers," their son Ray told a local reporter years later. Arthur also stayed with Jake and Minnie Otis. Minnie told her grandson, Don Otis, that Arthur "liked to sleep late and couldn't get up in the morning." Arthur dedicated *Cat O' Mountain* to another Trapps couple, James and Caroline Mack, "whose unfailing kindliness to the mysterious prowler of The Traps will long be remembered by 'The Detective.'"

Friel's foreword to *Cat O' Mountain* roused readers with its synopsis of early mountain history. A writer of his time, Friel described wild animals as brutes and the Dutch as saviors for ridding the region of native peoples. The "traps and countertraps" set during their conflict resulted in the name "Traps" for the "death-strewn hole in the hills." Years later, white and Indian settlers returned to the Trapps to farm. They "intermarried and interbred … followed their own inclinations … [and] achieved a reputation somewhat dubious. The tongue-wagging folk outside declared the Trapsmen were 'wife-swappers' and 'moonshiners' and other things. And perhaps they were."

How did the Trapps people react to reading this about themselves? Did Friel meet Trapps people of blended heritage or create his characters based on local stories? To date, no definitive evidence has been found of racially blended families.

While Friel's foreword was sensational and incorrect, his knowledge of the mountain was true to life. The break in the cliffs is "The Jaws of the Traps ... a long mountain wall with only one way into it." The Coxing Kill is "gentle" and "gurgling." The Peters Kill is "ramping, brawling."

Still, Friel came to understand a deeper truth—the old ways were all but gone.

"Yet there are," Friel wrote, "in the region round about, two or three old men—taciturn, abrupt, whole-souled old fellows—who were born in the Traps and who will die not far

Trapps men helped construct and operate a powerhouse built on the Peters Kill during the 1920s. A generator inside provided electric power for the mountain hotels at Lake Minnewaska. The powerhouse is long abandoned today. Photograph by Robi Josephson.

from the Traps. From them, and from the whispering ghosts, which by day and by dark, have drifted along beside me on the silent trails and talked to me in weird crevasse and uncanny old house, I have learned the tale which is here set down. It is a tale of Yesterday, in a land of Yesterday, chronicled by one who was there—yesterday."

All of Friel's yesterdays became wrapped up in the Trapps and its people. In *Cat O' Mountain*, the stranger in a strange land was one Douglas "Hamp" Hampton, a former New York City reporter. Attracted to the roughness of the Shawangunks, he blunders into the Trapps just in time to save damsel-in-distress Marion "Marry" Oaks, a young woman about to be attacked by a panther. Hamp suspects that Marry does not belong, for she spends hours sketching the mountain scenery, considered a wasteful occupation, nor does she look like her parents, her father part African-American and her mother part Native American. Hamp ponders modern genetics amid racial stereotypes as deeply ingrained as the Shawangunk grit beneath his feet.

Good and evil clash, with a classic cat-and-mouse game and romantic bungles. It's even more fun to guess the basis for some characters' names. Snake Sanders's surname was taken from the stream west of the Peters Kill; Sam Codd and Jonah Hay were named for the Coddington and Hayes families; Andy Mack, for the Macks; and the Van Hoovens, for the Van Leuvens. David McCafferty, with a gift of gab, likely was Oscar Lyons.

This overlook is as popular with visitors today as it was when U.S. Route 44/55 opened in 1930. The first paved road over the Shawangunks and through the Trapps hamlet replaced a much steeper unpaved road from Gardiner; its outline is visible below the parked cars. Photograph by Daniel Smiley II. Courtesy of MMHA.

Uncle Eb Wilham was based on Abe Billingham: "A big-framed, eagle-nosed, long-jawed old man he was, with white mustache drooping around his mouth and ragged wisps of white hair sticking out from under his nondescript felt hat. His shoulders ... were wide and bulky; and the gnarled hands of those reins were corded with sinew. Seventy if a day, he still looked powerful enough to handle many a man of half his years; and the direct gaze of his steely blue eyes betokened fearlessness of heart, simplicity of nature, and honesty of soul."

Suffice to say that good and love win out. Along the way, Friel also made sure that Hamp was asked not to judge the Trapps people by its few bad apples. Coupled with that, though, one character tells Hamp, "Them that's borned into the Traps lives into the Traps an' dies into the Traps. Ther' ain't no place for us outside."

Two years after *Cat O' Mountain*, Friel continued the story of the Trapps people in the novel *Hard Wood*. Romeo and Juliet meet the Hatfields and McCoys in this tale of Harry "Hard" Wood, son of upstanding, johnnys-come-lately to the Trapps, versus Jerry "Copperhead" Cooper, a member of a "vicious breed" of old-timers. Hard's love interest is, of course, Jerry's sister, Jane Cooper, a female St. Francis of the Trapps whom Jerry blinded at the age of two. The Trapps is a place where revenge truly is sweet and where the Biblical eye-for-an-eye comes in handy when no judge is. Yet, as hard as the Trapps people are, most, again, are hard-working, good-hearted folks.

On September 29, 1924, readers across America opened *The Saturday Evening Post* to find a story titled "Thirty Days After Death" by Arthur O. Friel. Once again, Friel set his tale, this time of classic swindle, in the Trapps. Character Mercy Davis was named for the Davises and character Abe Billings for Abe Billingham. An unnamed "party of quiet-spoken men" buys much of Abe's land to build a summer hotel. Abe is a childless widower, again with "bushy white hair and walrus moustache," living in a "little yellow house." In 1920 the real Abe, seventy-two, was a widower living in the Trapps. Did he actually have such white hair and a moustache, and live in a little yellow house? That November, Abe Billingham sold his general store building and land to George Smiley and left the Trapps for Esopus, a town on the Hudson River.[25]

• • • • • •

The 1920s also brought two substantial projects to the hamlet, among the last worked on by Trapps men. The Peters Kill was harnessed to bring electricity to the Minnewaska hotels, much as the stream had been over fifty years earlier to power the Davis sawmill. All available men would have hired on to help build a stone powerhouse and accompanying milldam, a veritable "Hoover Dam" of the Trapps. Trapps resident Jake Otis cut stone for the project, said grandson Don Otis. After the powerhouse was running, John Bittner, a summer Trapps resident from Brooklyn, landed a full-time job working at the site as a general mechanic, recalled son Harold. John brought his family up to the Trapps, but the Bittners returned to Brooklyn by 1930.

The new cement dam blocked "the full width of the Peters Kill," wrote Russell Dunn. The metal pipeline concentrated a far greater force of water than the old milldam's wooden sluiceway open along the top. The electricity sterilized drinking water from Lake Minnewaska, but was not yet used for refrigeration; Trapps men continued to cut ice for the latter purpose, reported the local papers.

After the 1923 completion of the Lincoln Highway, the nation's first, New York State joined the federal mission of paving America with a network of modern macadam highways. Three were to cross the Shawangunks. U.S. Route 44/55, the most northerly through the Trapps, was constructed during the late 1920s and opened in 1930. Ray Otis, in his early twenties, worked as a shovel operator for a company called Harp & Conway, said son Don, who noted that many local men worked on the new road. Winslow Van Leuven, thirty, who hired on as a laborer, lived with his family on the old dirt turnpike soon made obsolete by the new road.

"Construction of this road is one of the greatest improvements of the kind ever made in this part of the country," *The Wallkill Valley World* stated. "It will not only open up a short way over the mountains ... but will be worth a great deal in developing a beautiful mountain resort section. Its value to Gardiner is great."

The value of the new scenic highway, christened the Minnewaska Trail, *was* great as New Yorkers and those from surrounding states took to the highways and the Shawangunks became a relaxing day trip. The old Trap Road from Gardiner was transformed into a modern thoroughfare, its rough places smoothed, its steep switchbacks evened out to one longer (though at times still nerve-wracking) hairpin turn. Yet, its long-term benefits were less so for the Trapps. The outside world came to the hamlet as never before, but passed through faster than ever before. The new highway, as well, became the latest means by which more Trapps families left to join the larger world beyond. "Little by little," observed Knickerbocker, "the people moved away until only a few families were left."[26]

• 13 •

Fading from Memory

Evening, late August 1930. The Otis home in the Trapps. Jake and Minnie Otis cane chairs with George and Lillian Smiley, their employers at Lake Minnewaska.

"May I use your phone a minute to call Cliff House?" George asks, putting down his needle, pegs, and the seat he is weaving.

"Of course. Let's all take a break," Minnie states.

While Minnie puts up the kettle, Jake turns up the wick on the kerosene lamp.

"One day we'll get electricity," Minnie adds, "and running water, too."

"That'd be something," Jake says.

"Wouldn't it now?" Minnie agrees.[1]

In April 1930, Town of Gardiner enumerator Stanley A. Gray did not need long to record his census for the Trapps. He drove on the new paved highway, and then along the hamlet's still unpaved roads—the old turnpike spur, Clove Road, and Lyons Road, a section of the old Van Leuven Road.

Gray stopped at about a dozen households, down from a high of about forty to fifty.

Most of the Trapps people rented for $5 or $6 a month from the Smiley families of Lake Minnewaska or Lake Mohonk on land their parents or grandparents had once owned. In those cases each Trapps family had sold out to or traded with the Smileys, who extended to these trusted employees the courtesy of remaining in their homes. One exception was Agnes Bouton, the new schoolteacher, who paid $12 a month to Trapps resident Mabel Veres Warren.[2]

Over that year and the next, tragedy unfolded for a young man born in the Trapps. Abner Schoonmaker was a son of fire observer Sam Schoonmaker and May Della Davenport, and a grandson of Trapps Civil War veterans Lucas Schoonmaker and John J. Davenport. In 1930, Abner was arrested for third-degree burglary, a felony. At the time, Abner lived in the Trapps or Poughkeepsie. He was sentenced to ten years in Sing Sing Prison in Ossining, New York.

Abner arrived in the grip of "bloody convict riots," stated the *Ossining Citizen-Sentinel*. Four inmates had been stabbed to death between December 1929 and July 1931. Abner, who worked in the prison brush shop, was the fifth. His murder early on November 12, 1931, made the front page of that afternoon's *Citizen-Sentinel*. Abner was stabbed through the heart two weeks shy of his twenty-fourth birthday. He had served less than two years of his ten-year sentence and was buried in nearby Valhalla.[3]

Late spring 1930 found a far happier scenario unfolding in the Trapps. Agnes Bouton, twenty-two, born near Corning, New York, had left a teaching position in Livingston Manor for one at the tiny mountain school. Agnes and her neighbor Ray Otis, also in his early twenties, began courting and the two tied the knot in mid-June shortly after Agnes closed school for the summer.

On August 1 of that year, Dr. J. Wilson Poucher and Byron J. Terwilliger were visiting Ulster County cemeteries for a compilation of burial listings. They came to the Trapps, where they met newlywed Ray Otis, who took them to burying grounds in the hamlet and received a nod for his help in the credit lines: "Copied August 1, 1930, by Dr. J. Wilson Poucher, Byron J. Terwilliger and Raymond Otis." The now classic and indispensable *Old*

Minnie Otis holds her grandson, Don Otis, about six months old, in 1931. Don was the first of three grandchildren born at the Otis home in the Trapps. The maple trees across the road were tapped to make syrup. Courtesy of Don Otis.

Gravestones of Ulster County, published in 1931, boasts 22,000 inscriptions collected by the pair.

Ray and Agnes, also known as Hope, settled into the store half of the old Billingham place across the center hall from Jake and Minnie's half. Agnes retired from teaching to raise a family, and the couple soon welcomed their first-born, a son, Don, followed by daughter June and another son, Joe. The Otis place had a barn, large garden, granary, wood and equipment sheds, two-seat outhouse, two wells, and a watering hole Jake built for the cows. The main well was covered by a well house about twelve by fourteen feet, which contained a wood stove and dry sink for doing laundry every Monday. That well always had cold water even on the hottest summer days, Don recalled, with a tin cup kept handy for the family as well as for travelers or visitors to help themselves.

All were welcomed by Minnie. Oscar and Charlie Lyons visited frequently, as did Irv Van Leuven. They did not come for dinner but "would drop in and sit for a while," Don said. When Minnie Smith or Anna Lyons called, the women sat in Minnie's living room and smoked pipes. Minnie filled her small wooden one with Mechanic's Delight, the same tobacco that Jake chewed.

The Otis house was a gathering place for family, neighbors, and travelers, including Charlie the Peddler. Minnie made lunch, usually hot, for Charlie. "He would show up, always

Three generations of the Otis family lived at this Trapps home beginning in the later 1910s. "King of the Trapps" Abe Billingham had operated a general store, small hotel, and the hamlet's post office there during the later 1800s. Courtesy of Don Otis.

on foot, once a year carrying a pack," Don said. He sold all kinds of items, such as clover bloom salve. "He never stayed overnight," Don added. "We never knew where he stayed."

Despite a house filled with guests, Minnie would get "blue" at times over the loss of her eldest child. She kept some of Gene's things, a few items of clothing and his strap-on skates, in a trunk in a room on the second floor. The three Otis grandchildren were forbidden to enter the room or touch the trunk.

Jake and Minnie were the first to have a telephone in the Trapps. The main line between Lakes Mohonk and Minnewaska ran along the base of the low cliffs behind the house. A side line ran up to the Dickie Barre fire tower and another to the Otis house. The fire observer needed a means of calling out, as well as a backup line at the nearest house in case the line at the tower was not working, Don explained.

The Otises also owned a radio, one of four Trapps families to do so in 1930. It was the first year census takers asked if a family owned a set. This usually indicated electricity, but that utility would not come to the mountain for some time. Jake and Minnie used a car battery and several C batteries to power a set about two feet high and wide, said Don.

The Otis family gathered 'round the radio to hear *The Shadow*, *Jack Benny*, and *Amos & Andy*. Minnie enjoyed Burl Ives; Ray, Will Rogers; and the kids, *The Lone Ranger*, *Little Orphan Annie*, and other children's programs aired after school.

"When I was growing up the hamlet was changing," Don told a *Kingston Freeman* reporter years later. "A lot of people from New York City used it for summer houses and there were still some of the old timers—the Smiths, Van Leuvens and Davis." When Don grew up during "the heart of the Depression," the Mohonk and Minnewaska properties provided the little work available. In earlier times, mountain people supported themselves by charcoal burning, hoop shaving, and cutting wood, but "[t]he need for all of those things vanished."[4]

• • • • • •

Several Trapps men were living out their "sunset years," so-called by Peter H. Harp, off the mountain. In 1930 three Civil War veterans from the Trapps, all widowers, lived in the village of Walden: John H. Voorhis, ninety-three, with his son; Eli Lockwood, eighty-nine, in a boardinghouse; and John J. Davenport, ninety-three, with a grandson. Lewis "Huckleberry" Wynkoop made his final home with a son in Walden and died in 1936 aged eighty-nine. He was buried in Bruynswick Cemetery in the shadow of the mountain where he once had peddled berries with son Melvin and brother-in-law Abe Billingham.

Abe, eighty-three, lived in St. Remy, south of Kingston, where he was the victim of a classic swindle. Abe, who on occasion gave out small loans, agreed to lend $100 for two months but because of his bad eyesight had the borrower pen the note. "But the borrower," Peter H. Harp wrote, "substituted the word death for date, and the note read '2 months after death I promise to pay.' Abe did not discover the change until 2 months later when he requested payment and then was informed that the note was not due." Did Abe remember a story in *The Saturday Evening Post* a few years earlier? A story in which he had appeared as a character? What was the name again? Ah yes, the story by that adventure writer, Friel. "Thirty Days After Death" it was called.[5]

By 1930 brothers Oscar and Charlie Lyons had returned to the Trapps, a rare turn of events given the general exodus from the mountain. The two lived next door to each other on parcels left to them by their Irish immigrant father, Thomas. Hal Bittner described Oscar as "a very talkative and knowledgeable man. He read [every] book and newspaper he could lay his hands on." Oscar worked at the psychiatric hospital in Poughkeepsie with his wife, Matilda, a nurse. Hal added, "Oscar was buying up as much land as he could afford" to resell to the Smileys. Hal described Charlie as "a very quiet man [who] kept to himself."[6]

During the 1930s, news stories about the Trapps were, like its population, dwindling. In November 1932, *The Highland Post* reported that Walter Elmendorf of Clintondale shot a 9-point, 170-pound buck in the Trapps. In late 1937, the New Paltz paper reported: "Joe Speers [*sic*] who rolls down from the mountain fastnesses near Minnewaska every Wednesday to print the Main Street Market throw-outs in the Independent office, says two inches of snow fell in the Trapps Sunday night and it's all still there just as it fell, except that some of it has turned to ice." The next week, the same paper reported a fire at Joe's house, which was owned by George Smiley. Likely caused by "an over-heated chimney," the fire was "put out by Albert Smith, a neighbor, who saw the smoke pouring from the roof." Insurance covered what little loss Joe had.[7]

On June 25, 1937, *The Wallkill Valley World* ran a page-one story about a once-in-a-lifetime event at the Trapps School—a fiftieth reunion of the 1887 class to be held on July 3. The one-room mountain school had changed little since Edwin Baldwin, then seventeen or eighteen years old, taught there from September 1887 to April 1888. Edwin continued teaching at New York City schools and perhaps had just retired that June.

About forty-five pupils had attended the Trapps School in 1887. Twenty-seven were still alive and had been notified of the reunion, the *World* reported. The article served as an invitation to other surviving students. Days later, the New Paltz and Kingston papers reported the event was "the happy fruition" of "untiring efforts" by Edwin Baldwin and former student Melvin Wynkoop. More than eighty attended, including: twenty students of the class of '87; Edwin Baldwin and the other teacher of that year, Florence Schoonmaker; current students and their families; current teacher Leonard George and wife, Ethel May, also a teacher; and George and Lillian Smiley of Lake Minnewaska.

Families attending that day received a commemorative certificate. Today, Doris Van Leuven Hall and Don Otis own the only two known to still exist. The certificate listed the three school trustees and two teachers of 1887–88, the twenty-eight surviving students, and seventeen deceased ones. Of the trustees, William Dingee had died in 1888, and Abe Billingham, about ninety, had died in February, missing the reunion by six months. Edward Smiley was the only surviving trustee, but did not attend.

Mr. Baldwin inscribed the Otis family copy as follows:

> This token of a half century of pleasant friendships
> and happy memories is printed and presented to his
> former pupils, and to the friends attending the
> semi-centennial Reunion by their friend
> July 3, 1937 Edwin F. Baldwin

Don Otis, six, had just started at the school that spring with younger sister, June, four. The day lingers in Don's memory: eating hamburgers and hot dogs; a ceremony held by the flagpole at the schoolyard's "little bit of lawn"; speeches by Mr. Baldwin and Mr. Smiley; firecrackers snapping in the afternoon air, the staccato popping disturbing to Phil Dilschman, Tillie's son, a nervous, shell-shocked veteran of World War I.

The New Paltz paper stated that reunion-goers enjoyed "[s]inging, photographing, and a social time over boxed lunches" and "Mr. George's art work and delphiniums."

Leonard George taught not even a dozen children most of the year. A few times "it was just the three of us," Doris Van Leuven Hall said of herself, her older brother Roger Van Leuven, and their younger sister Ethel. Don and June Otis attended the Trapps school until it closed in 1943. Don remembers upwards of thirty students some years between Labor Day, when Cliff House at Lake Minnewaska closed, and Columbus Day, when its sister hotel Wildmere closed. During that four-to-five-week period, children of Minnewaska employees attended the Trapps school until the resort shut down for the winter.

A 1932 photo (next page) shows eight pupils: William Bayer (Bayard); Roger Van Leuven; Walter Lyons, son of Charlie Lyons and his wife, the former Anna M. Ost; Sylvester Smith; Ethel; Doris; Anita Randon, standing; and Freddie Evans. Walter, Sylvester, Ethel, and Doris balance on the schoolyard seesaw. Freddie is perched on the swing at right hanging from one of three trees next to the school. The students made the swing and seesaw with Mr. George, Roger recalled.

Nothing more is known of Anita Randon or Freddie Evans, but Sylvester Smith perhaps was a grandson of Albert O. Smith, Jr. Billy Bayer, ten, had moved with his family from Brooklyn to the Trapps. His father, Frank, had been a chauffeur and perhaps had found the same work at either Smiley resort. Within a few years the Bayers left the mountain for Gardiner.[8]

By 1930, Ambrose and Sheridan Burger, sons of Calvin and Rachel, had moved to Walden. Their grandfather, John Burger, had bought a long, narrow parcel of 150 acres

Left: Eight pupils attended the Trapps School in 1932. From left are Walter Lyons, Roger Van Leuven, Sylvester Smith, William Bayer (Bayard), Roger's sisters Ethel and Doris Van Leuven, Anita Randon (standing), and Freddie Evans on the swing. The one-room school was closed in 1943 when only one student remained. Courtesy of DSRC-MP. Donated by Doris Van Leuven Hall. Right: The graded readers and lesson books of educators James Baldwin and Ida C. Bender were once popular across America, including at the Trapps School. Courtesy of Don Otis, Trapps School pupil from 1937 to 1943.

---

from the old Great Lot 6 of the Grote Transport. Over the years the family sold off smaller parcels until 1905, when Calvin sold just about "his last sixty acres" to Albert K. Smiley, noted Larsen and O'Neill. All that was left of the original family holdings was a one-acre parcel that Rachel had deeded to Sheridan in 1896.

After the paved highway came through the Trapps, Sheridan saw a way to make a living during the Depression—open a gas station to sell gas and refreshments to the new tourists coming over the mountain. Nothing fancy, a little soda pop and candy. Sheridan erected a two-story clapboard building and painted on the front "CLIFF INN," named for a short row of cliffs that ran up the back, and "S.P. Burger, prop." He contracted with Gulf Oil Company to install two gas pumps out front. Each hand-cranked pump had a decal reading "Gulf No Nox," a premium grade of leaded gasoline sold then. There was plenty of room for several cars to pull off the highway.

Opening day occurred about 1932 and was quite an occasion, complete with Sheridan, Ambrose, their family, and neighbors posing for the camera at the pumps and along the new highway. The Cliff Inn was a perfect rest stop, the only station for miles around. Enter Carl Carmer, who was exploring the area for his now classic book, *The Hudson*, first published in 1939. Late one autumn afternoon, Carmer was driving west of New Paltz on a "steep and winding" road when he found himself on the mountain, where he met one of the "B'ys from the Traps."

Carmer had pulled up to Cliff Inn, where "the slanting highway" still does today. He entered a cabin of two rooms furnished with a small iron stove and oil lamp that glittered fitfully. Inside he found an elderly man wearing spectacles and a "big white mustache," contrasting an otherwise gray appearance in the soft lamplight.

"It's all I got left, jest one acre," said the man about his mountain land, "and they're a-tryin' to git *that*. Uncle Charlie owned most of this here mountain once. A lawyer come along an' says, 'Charlie, jest you sign these here papers an' you get this bran'-new five-dollar bill.' So Uncle Charlie signed an' lost all his propputy. Danged if I don't think that wife o' his didn't witch him into it."

Carmer had met Sheridan Burger, sporter of said handlebar moustache. Charlie was Sheridan's uncle and Calvin's brother. Much like the general store or tavern of old, Cliff Inn had become a big gathering place for the Trapps people. Neighbors met there evenings, Don Otis said, when part of the local entertainment was Sheridan's talking crow, a wild bird he had tamed and named Jim. "Sheridan would have Jim perched on his arm," Don recalled. "He'd talk to it and the bird would answer back." The crow could say "Hello! Hello! Hello!" and "Sheridan!," the second with a little imagination on the part of the listener.

Sheridan, who had owned a carpentry shop in Walden, and his wife Lucinda, likely his second, ran the place all year. He was "pretty handy," Don said, describing a unique set of rustic furniture Sheridan made and kept inside in the back—a table, chairs, and bench

Sheridan Burger poses at his gas station, Cliff Inn, on U.S. Route 44/55, the main road over the mountain through the Trapps hamlet. The woman at right is likely his second wife, Lucinda. Courtesy of DSRC-MP. Donated by Joan Hayes Wustrau.

Neighbors and friends pose ca. 1932 at the grand opening of Cliff Inn gas station on Route 44/55. Kneeling at left is Trapps resident Ray Otis holding his son Don. Standing from right are proprietor Sheridan Burger and his brother, Ambrose Burger. Courtesy of Don Otis.

Brothers Sheridan and Ambrose Burger, both wearing suspenders and hats, help customers at the grand opening of Cliff Inn gas station. Courtesy of Joan Hayes Wustrau.

of sassafras poles and saplings. He sawed flat the surfaces of the table, chair seats, and benches, but left the wrinkled bark on all the furniture legs as well as the rungs, arms, and spindles of each chair.

Sheridan offered his family's well of fresh mountain water to all who came by. Water was free and helped attract customers, Don said. A sign pointed to a few stone steps leading to the well up behind the station. Don also recalled the Cliff Inn as a center for berry pickers and buyers, who gathered around a long, wooden bench out front.

Sheridan ran the gas station less than a decade, selling it and the family's last acre to the Smiley family of Lake Minnewaska about 1940. Ambrose and Sheridan both died in 1943 and were buried in Walden. The last acre of the Burger family, then the last of the Burgers themselves, was gone from the mountain after more than a century.

The Smiley family kept the station going during the summer season. They switched to Mobil gas, pulled out one pump, painted over Sheridan's name, and hired others to operate the gas station. From 1946 to 1953, the Black family spent summers at Minnewaska, living the first two summers at Cliff Inn and then at the old Otis place near Laurel Inn. The Blacks, from Florida, learned about work at Minnewaska through their friendship with the Smiley family of that lake, some of whom were the Blacks' winter neighbors in the Sunshine State. George Black, Sr. worked at Minnewaska first as a lifeguard then switched to the maintenance department, working under Ray Otis. While the family lived at Cliff Inn, his wife Marion, called "Mickey," operated the gas station.

Mickey Black and her children (from left), George, Jr. "G.G.", Betty, and Jean, pose in 1947 at Cliff Inn for Francis J. Prand, a Minnewaska guest and professional photographer from New Jersey. Cliff Inn, with its bench out front, was a main gathering place in the Trapps hamlet for residents and locals, including berry pickers. Courtesy of DSRC-MP. Donated by George Black, Jr.

The mountain was a playground for the four Black children—George, Jr., called G.G., daughters Jean and Betty, and little Bobby. If the family stayed after Labor Day, the children attended school in New Paltz. "My 'memories' of those idyllic summers at Cliff Inn," G.G. wrote years later, "are mostly just feelings of being content, safe and carefree. … Any child would have dreamed of nothing better and it was mine! And those summers at Cliff Inn were spent with only an outside hand water pump and an outhouse!"

Many "idyllic" memories linger in G.G.'s memory over sixty-six years later—making "little rings and necklaces from dandelion stems" with one sister; visiting his neighbors, Oscar Lyons and the Otis family; picking berries; seeing his first snow on the mountain. One summer, Mickey and the children were picking berries around the cliffs above the gas station. She told them to hold her ankles as she inched out on her stomach over a cliff to reach a berry bush. It was a straight drop down to the carriage road below.

"Be careful!" she called out to them.

"Don't worry," they called back. "If you start to fall, we'll let go!"

Mickey enjoyed telling that story to guests at Minnewaska.

One October morning in 1953 at the old Otis house, the Black family awoke to a winter wonderland. The children dashed out to play in the snow, only to have to board the school bus (a station wagon provided by Minnewaska) a few minutes later for New Paltz, where not a flake had fallen. By afternoon all the snow on the mountain had melted. The family soon returned to Florida, those few moments the only time the Black children saw snow at Minnewaska. Summer 1953 was the last one that the Black family came to Minnewaska. G.G. was twelve years old going on thirteen. Sometime after 1947 when the Blacks had left Cliff Inn, the Smiley family of Lake Minnewaska closed the gas station, the pumps were pulled out, and the Trapps people lost a gathering spot.[9]

• • • • • •

On December 7, 1941, Don Otis, then ten, had just taken the family horse, Old Harry, and the cows down to the Coxing Kill for a drink. He came back to find his mother, Agnes, very upset—Pearl Harbor had just been attacked.

Three former Trapps residents served in World War II, cousins Arnold Van Leuven and Roger Van Leuven, both born in the mountain hamlet, and Hal Bittner. The job of air-raid warden Agnes Otis was to ensure that all homes in the Trapps were blackened at night and to hold air-raid drills. Everyone came to the evening drills she held about once a month at the school, son Don said. Agnes served as warden for some time and may have been the only one in the hamlet to do so during the war.

Don and his family, hungry as all Americans were for war news, listened to the broadcasts of commentators Gabriel Heatter and Walter Winchell. They also heard Mayor Fiorello LaGuardia's public service announcements when he read the Sunday comics in July 1945 during the newspaper strike in New York City.

Gasoline and sugar were two big commodities rationed during the war. Ray Otis rigged his truck to start on gasoline then switch to kerosene, a trick perhaps learned from neighbor George Speer. The truck "smoked like crazy" whenever it ran on kerosene, Don said. Don recalled rolling cigarettes, also rationed, by hand for his father. It was then that the Otis family first used Caro syrup for canning fruit. During that time the Smiley family of Lake Minnewaska stored sugar and flour at the Laurel Inn for hotel use, as by then the inn was no longer used for guests, Don recalled.

The few remaining Trapps families continued working for the Smileys and hired on for other work that came their way. Don Otis's first job at Minnewaska was working in the garden; his second was hauling ice. One winter day in 1943, he was asked to fill in as a teamster and found himself staring at the back end of a horse team. Don was too young to cut or load the ice, but not too young to haul it. He was twelve years old.

Once the men loaded up the sled—"and they loaded it pretty heavily," Don said—the driver climbed onto the ice, sat on a horse blanket, and took the reins. It was a hard haul up the steep hill, so drivers stopped to rest the horses about halfway up. A pointed rod attached to the back of the sled was stuck into the ground to act as a brake and take the pressure off the team.

While taking his first load up the hill, Don rested the horses, but they were frisky and started backing up instead of moving forward. The rod, with that massive weight bearing down on it, started to lift up the back of the sled. Don—twelve years old, new on the job, and about to be crushed by tons of ice and horse—jumped off and grabbed the first thing he saw, a broken branch about seven to eight feet long. He jumped back on and smacked the horses' rumps to prevent fatal injury to him or to them. "We went up the hill like a shot," Don said. "I didn't have any trouble for the rest of the day."[10]

• • • • • •

The last megastructure built near the Trapps was the Delaware Aqueduct, officially begun in 1937 to assist the Catskill Aqueduct in quenching New York City's thirst. Mary Enderly's grandson, Arthur Hayes, worked in 1943 on this aqueduct, which originates at the Rondout Reservoir on the border of Ulster and Sullivan Counties. Arthur and his family had moved into Mary's house below the gap after she died. His daughter, Joan Hayes Wustrau, wrote that he had been making $2 a day on a nearby farm, but then this better-paying job came his way.

"When Arthur was thirty-one years old," Wustrau noted, "he got a job on the Delaware Aqueduct. He started out as a laborer and worked underground. He made $7 a day. If you worked above ground, you made a dollar less—$6 a day. By the time he left this job, he was a driller in the tunnel and was making $10 a day. The engineers made $13.20 a day."

The new aqueduct ran southeast from near Grahamsville through the Shawangunks under the southern end of Minnewaska property near Lake Awosting, ultimately joining its sister pipeline from the Ashokan Reservoir to the north.

During the 1930s and 1940s, Trapps men cut cordwood for the Smileys, who still needed vast quantities of wood to run the resorts at Mohonk and Minnewaska. The Smileys of Mohonk had a barn behind Irv Van Leuven's place in the Trapps, which they used as a central stacking place for cordwood, remembered Don Otis. They dragged the wood out of the forest to this barn by horse and sleigh, he said, then hauled it away by truck. The Smileys likely had similar stacking places elsewhere on their land.

Winslow Van Leuven cut wood and ice for Minnewaska. He and son Roger cut cordwood in four-foot lengths, the length used to heat water. Minnewaska paid $1.25 a day in wages to cut wood during the Depression. A man was lucky to make $6 to $7 a week then, Roger said. Winslow also shaved hoops, and Roger tapped a stand of maples then boiled the sap down to syrup, selling it for 50¢ a quart. (Oscar and Thirzy Harp also ran a sugar bush operation in the Trapps during the 1920s, but not later, said Don Otis.)

Don recalled that his grandfather, Jake Otis, harvested hoop poles into the 1930s, and his father, Ray, cut wood for Mohonk and Minnewaska into the early '40s. As Ray only had work at times, Jake helped Ray financially. Ray, with $100 in hand (perhaps given by Jake), bought a Model A Ford in 1932 up in Kingston. Thereafter once a year, Jake had Ray drive him to Kingston to the wholesale fish place to buy a barrel of salted mackerel. "Pop had to strap the thing on the back of the car, and he hated that," Don recalled.

The Otises were fairly self-sufficient during the lean years of the Depression and rationing during World War II, recalled Don and his brother Joe. They had a couple of dozen chickens and two or three cows. Don raised a registered Holstein calf given to him by the 4-H program. Minnie and Agnes churned butter and made cheese. The kids sold milk and cheese to the neighbors and sometimes had more money than their parents. The family grew "everything" in their garden—tomatoes, potatoes, carrots, beets, Swiss chard, corn, and "some crazy thing called New Jerusalum artichokes," which looked like potatoes and were very spicy. They harvested fruit from previously planted peach, apple, pear, and plum trees. They grew gooseberries and currants. Flowering hop vines, curling over the lilac bushes, provided yeast for bread, but never beer, as Grandma Minnie was a strict teetotaler. They picked wild and domesticated grapes for jelly, wild strawberries "as big as cultivated" ones for jam and jelly, and huckleberries once or twice a summer for family use. Fruit was canned or dried on a sheet on the front porch tin roof.

The Otises also tapped a big row of maple sugar trees across the road for syrup. Those maples were some of the few trees left standing in the Trapps. Over time, so much forest had been logged out that the stone tower on Sky Top was in plain view from parts of the hamlet. It was in that stony basin—nestled between the ridges of Lake Mohonk to the north, of Lake Minnewaska to the south, of Dickie Barre to the west, and of the cliff escarpment to the east—where the Trapps people had endured for over a century.

In the not-too-distant future, much of the land would revert back to forest and be preserved, but in the 1930s the Otis place was typical of Trapps farms of earlier times. The well house and cellar were cool enough to store food in summer. Cured ham was hung in the well house during the winter. Apples and potatoes were kept there, too, but moved into the cellar during the coldest part of the year to prevent freezing. Milk was stored in pans on a swing shelf hanging in the middle of the cellar and suspended from the floor joists. After a day or so the cream would rise to the top and be taken off with a skimming ladle. Jake pickled "all kinds of things" in huge crocks in the cellar.

It was during this period that the Trapps people began to receive public recognition for their work. In his 1946 history of Minnewaska, William E. Doughty recognized the "loyalty and labor" of employees in a section titled "Unsung Heroes," singling out the Lyons and Otis families of the Trapps. Charles Lyons and his wife, Anna, had been employed for several decades each. Jake Otis, his son Ray, Ray's wife Agnes, and their son

Joe—three generations of the same family—were acknowledged, too. Jake's wife, Minnie, and Joe's brother Don also worked at Minnewaska.

Decades later, Larry Burgess published a similar tribute to Mohonk employees in his history of that lake. Burgess stated, it "was nothing out of the ordinary" for employees to have between fifteen and thirty years of service. Eli Van Leuven, son of Hiram and Anna Maria, worked for four decades at Mohonk. The Van Leuven and Enderly families share a record not likely to be broken again—four generations who worked at either mountain lake. These families became like extended members of the Smiley family and later the Phillips family of Lake Minnewaska, who could not have built up or operated the mountain hotels without them.[11]

Berry picking remained a summer mainstay, with Lake Awosting and Sam's Point among the prime picking spots. Four- and Five-Mile Posts along the Smiley Road from Ellenville to Lake Awosting were "the most popular campgrounds" for pickers, Alice Schoonmaker noted. Wendy McCormick quoted an unidentified camper as saying, "Each campground had at least 30 to 40 or 50 families. Now these families came from all over the United States." Some came from the local area, including the Van Leuvens of the Trapps, whose one daughter was a champion picker able to fill a quart in sixty seconds flat.[12]

Winslow Van Leuven worked as a laborer in general construction and also cut cordwood for the Smileys. His half-brother Arthur had earned money in berries, son Harold later told Carleton Mabee. Winslow followed suit. When school closed for the summer in June 1934, he and wife, Bertha, pregnant with their fifth child, packed up the children in their 1930 Chevy truck, bought used by Winslow for $20, and headed to Smiley Road via Ellenville. The Van Leuven kids were Roger, twelve, Doris, ten, Ethel, almost nine, Alice, three, and Richard on the way. (Doris was named for a daughter of Fletcher and Ruth Smiley of Lake Minnewaska.) Picking berries was the only cash crop left and the most important one, for berry money paid the year's rent. The only other work was for the Smileys. It was the Depression, and good jobs were hard to find. The family moved back down to the Trapps when school started.

Bertha, the entrepreneur of the family, set up a store at Four-Mile Post. Winslow found work not far away running a bulldozer for $1.75 an hour on the construction of Route 52 over the mountain near Cragsmoor. The job had an unexpected benefit—empty dynamite boxes with which he could construct a cabin. Roger helped his father put up short box boards as siding and "narrow tree trunks" as studs, Roger told Marc B. Fried. Doris added

Frank Veres painted the berry camp of his sister and brother-in-law, Bertha and Winslow Van Leuven, at Four-Mile Post along the Smiley carriage road connecting Ellenville and Lake Awosting. Starting in 1934 the Van Leuven family, with five children, left their Trapps home to spend summers at this two-room cabin. The family lived in one room and Bertha operated a store for berry pickers in the other. Photograph by Robi Josephson. Courtesy of the Van Leuvens' daughter, Doris Hall.

that their mother ran the store out of one room, and the family slept in the other rooms. Winslow and Roger even put down a wood floor, which not every berry cabin had.

Winslow, Bertha, and the children left the Trapps in December 1937, the same year the family car had been stolen, Roger and Doris said. (The car was found later in Newburgh.) Doris explained her parents' decision to leave: the family had had a lot of good times in the Trapps and that was where Winslow had grown up, but times had changed and it was time to go. The family moved first to Walker Valley in the Town of Shawangunk, and then west over the mountain to Ellenville. Bertha told Fried that they spent summers at Four-Mile Post until about 1947 or 1948 when they moved to the Sam's

Point camp along the public road. Winslow passed away in May 1961, and two years later Bertha sold the Sam's Point cabin.

And the family's champion berry picker? Daughter Ethel, born in 1925. Fried wrote that the Van Leuvens were "hard-working people and the berrypicking exploits of [Ethel] were legendary." Fellow picker Hype Addis told Fried that she "could sometimes be discovered in the bushes literally surrounded by pails full of huckleberries, which her father, Winslow, would carry out for her, replacing them with more empties." Ray Quick told Fried that "he once timed her by the second hand of his watch as she picked a quart of berries in one minute." Her exploits were even more remarkable given that the berries were not the larger, domesticated ones. For every "tame" berry, Ethel had to pick two or three wild ones. Fellow picker and older brother Roger said he could only "pick 10 quarts an hour when the picking was good."

Commercial berry picking declined after World War II. By the 1960s it was over, the causes many. The Smiley carriage road from Ellenville to Awosting washed out, and the Smiley family of Lake Minnewaska discouraged "the squatters' camps and commercial traffic over their vast landholdings," Fried wrote. Domesticated berry growing came into its own as an industry in the Pine Barrens region of New Jersey, which has soil similar to that found atop the Shawangunks. As well, Alice Schoonmaker noted, "Yearly forest fires disappeared in the 60's." John Davenport told Priscilla Prutzman that "fires on the mountain were prohibitive [prohibited]." Recurring fires, which had kept the berries productive and competitive vegetation down, were no more.

Perhaps the most critical reason was economic. In a 2006 documentary on the industry, Fried said that during World War II new "war-related factories" opened across the country, and with men away to war, women entered the work force in large numbers. By war's end, the country's economics had changed completely. Money in berries was gone, and so, too, were the pickers and their colorful camps.[13]

• • • • • •

The economic trickle down was felt in the Trapps. Two institutions closed there, spelling a further shift from a community of local families to one of summer people. For decades the school and chapel had tied mountain families together, but by the war years few were left to attend or sustain either. Don Otis attended school with "the big boys"—Roger Van

Leuven, Walter Lyons, and Hal Bittner. Also, the Van Leuven girls, the Grey sisters from a hamlet to the west called Granite, the Taylor kids from New York City, and Betty and Joseph Burger. Never more than a dozen during much of any given school year. When Don and his younger sister, June, graduated sixth grade in 1943 bound for middle school in New Paltz, that left only one youngster, the third and youngest Otis child, Joe. Joe never attended the Trapps school and instead was sent to New Paltz, effectively closing the one-room mountain school established in 1850. "The school has never been officially discontinued and stands unoccupied," wrote Kenneth E. Hasbrouck in 1953. "Possibly a renewed population may require its use again some day."

A renewed population never came. A movement to consolidate rural schools into the New Paltz Central School District had begun in 1929, but the Trapps School board voted not to merge. Instead, the mountain school paid tuition for its children to attend school in New Paltz, Don Otis explained, which was a lot less expensive than paying New Paltz taxes. The Trapps School remained an independent district for several decades. Records at the New York State Education Department suggest that with the passage of New York Education Law Section 1501, School No. 8 at Minnewaska, Gardiner, was officially incorporated into the New Paltz district on July 1, 1969.

The Trapps Chapel also faded away. Along either wall were stained-glass windows "like a regular church," Roger Van Leuven said. He and his sister Doris Hall recalled that wagons once pulled onto a large flagstone to park or parked inside a shed behind the chapel. Don remembered the chapel bell that "mysteriously disappeared."

Churchgoers dwindled until only Oscar Lyons was left. Oscar had a "big full" beard and "always sported a well-worn floppy hat," described George Black years later. George knew Oscar from boyhood summers spent in the Trapps, and summer evenings found the Black family sitting on Oscar's front porch on Lyons Road with other Trapps neighbors. There, Oscar spun tales of mountain life in his "gravelly" voice, said George, known then as G.G. His parents instructed their children to address Oscar as "Mr. Lyons," but George considered him "the Old Mountain Man."

One summer evening on a family visit to Oscar's house, George and his younger sister, Betty, did some poking around as kids are wont to do. They went around Oscar's house and peered under his back porch. There, in a little room hanging from a hook was a dead fox. "We never told mom and dad because they would have scolded us for sneaking around like that," wrote George years later.

Sitting on the porch of Oscar Lyons's house in the Trapps are, from left, Bobby Black, Tillie Dilschmann, Oscar Lyons, and Bobby's father, George Black, Sr. Courtesy of DSRC-MP. Donated from George Black, Jr. to Alfred B. Smiley to DSRC-MP.

As kids will protect their own escapades, Oscar Lyons became very protective of the Trapps Chapel. By the 1940s the chapel was no longer used, but Deacon Lyons tried to save it. To no avail. In his 1953 history of Gardiner, Kenneth Hasbrouck noted the intersection was called Chapel Corners and published a photograph of the chapel with a board nailed diagonally across the front window. This photo likely was the last one taken of it. The day came when the chapel and Oscar Lyons were gone from the mountain. Oscar, aged about eighty-seven, died in 1955 in New Paltz. Yet, in one small way, the chapel survived as Oscar might have wished. The name of Chapel Corners is still sometimes spoken today.[14]

• • • • • •

Late in 1947, Thirzy Harp of the Trapps passed away. Thirzy La Forge had grown up on a farm in the Town of Shawangunk, but made the Trapps her home upon her marriage to Oscar Harp, son of Henry O. Oscar continued the boardinghouse that his father had operated in the 1800s, but Thirzy took over the business, establishing the Clove Valley Dude Ranch. Oscar lived a few years after his wife, who had bequeathed the ranch not to him or their son but to their daughters and a daughter-in-law. The women sold the property, which later became popular as Smitty's Dude Ranch.

By 1950 only a handful of old-time residents were left in the Trapps. Kenneth E. Hasbrouck noted, "Much of the section is now owned by the Smiley family and some of the old houses are rented out to summer residents. The remainder of the area is still owned by various families and lots have been sold to people for summer homes."

In 1955 another mountain family sold out. Fletcher Smiley sold Lake Minnewaska, established by his grandfather Alfred, to long-time employees Kenneth and Lucille Phillips. Some Trapps people—Minnie Otis, Ray and Agnes Otis, Charlie and Anna Lyons—continued to work at Minnewaska. By then, Ray and Agnes spent winters at the resort, returning each spring to their Trapps home. The Phillipses added a nine-hole golf course near the lake and a downhill ski area called Ski Minne on Route 44/55. They opened the resort year-round. Yet while the resort bloomed anew, the Trapps was fading, like a withering plant, from time and from memory.[15]

• • • • • •

One of the few left in the Trapps was Irv Van Leuven. Born in 1888 and bred in the hamlet (and youthful offender thereof), Irv met and married Catherine "Cassie" Mead. Cassie was a former showgirl, according to local lore. The couple set up house on the "Traps Turnpike" in 1920 and left the mountain a few times, but always came back, reported the *Poughkeepsie New Yorker*. The couple eventually settled in the Trapps in his parents' two-story plank house; from the road, what looked like a clapboard addition at the front was actually an insulating facelift. After his grandfather died, Irv's father, William, and his mother, Anna, had sold out; the 242-acre homestead farm, dating back to Irv's great-grandfather, Henry O., became part of Mohonk. The Smiley family at Mohonk made the house available to Irv for the remainder of his life, as well as the use of the five-acre parcel on which it stood.

At one time, Irv ran a boardinghouse, likely at his home, "for hunters and fishermen," a New Paltz paper reported in 1957. A good host knows how to tell a good story, and Irv could "tell many a good tale concerning legends of the area," noted Hasbrouck. Most game had been hunted out, but Irv still lived the life of a trapper and relished telling tales of the hunt. As such, stories about Irv made the news.

Irv Van Leuven shaves a barrel hoop (below) at his Trapps home, then holds up a finished one (right). The dirt road at left was the public road called Clove Road. By demonstrating the old trades such as hoop shaving, Irv helped to preserve a way of life that was quickly passing. Courtesy of DSRC-MP.

A front-page Valentine's Day article of 1929, also from New Paltz, reported that Irv recently had trapped two large bobcats. The second one was "the largest seen in this vicinity for years. Its skin from the tip of its nose to the end of its hind feet measured a good five feet." At least three times earlier Irv had trapped the second cat, but each time it had been caught by a front paw and managed to escape, losing either toes or the paw itself. The cat finally was caught by a hind paw and was unable to free itself.

Irv said bobcats live mainly on small game such as mice, partridge, and rabbits, not "on a diet of calves and deer with an occasional tender member of the Outing Club or tough old hunter to vary the menu." He clarified that bobcats would take a fawn or "quickly clean up the lambs in a flock of sheep if anyone was foolish enough to bring sheep into the mountains."

"The cats, if caught in a trap and unable to pull away, are very quiet if anyone approaches," he continued. "Their most formidable weapons are their rear feet, the long claws of which are as sharp as needles. ... They are formidable when at bay and can dispatch the best dog in short order, but if left alone are as timid as a fox and no more feared by those who know their habits."

Irv Van Leuven's house in the Trapps never had running water or electricity during his lifetime. The house was later torn down and replaced. Courtesy of DSRC-MP.

Mildred Bendix, granddaughter of Irv's sister Mildred, took summer trips with her family to the Trapps, where they picnicked, picked berries, and visited Irv and Cassie. One childhood memory is permanently tucked in Mildred's mind—the sight of animal skins tacked up on the walls of the summer kitchen, a separate building across the road next to the Coxing Kill. Irv and Cassie slept there in summer, with the stream cooling food, kitchen, and cooks. A trip to Irv's was an exciting adventure for children, akin to meeting Daniel Boone of another time.

Don Otis has vivid childhood memories of Irv bringing venison to his family during the winter. "Irv always shared the venison with us," Don said of Irv's out-of-season generosity during the Depression.[16]

Like his parents and grandparents before him, Irv made a living by farming, charcoal burning, hoop shaving, woodcutting, trapping, and berry picking. He also found work where available, including as a maintenance worker on the new highway over the mountain and for the Smileys. Because of his vast knowledge, Irv was greatly admired and respected by the Smiley family. At Mohonk, brothers Daniel Smiley II and Albert K. Smiley III, known by his middle name "Keith," had an ardent interest in the mountain's cultural history and were intrigued by Irv. Keith had worked "as a young man under the guidance" of his boss Alton Quick and of Irv Van Leuven, whom the brothers had visited in the Trapps.

Irv and Cassie worked the last charcoal burns on the mountain during the 1940s. Daniel had documented one such burn, and an article in the *Poughkeepsie New Yorker* credited the couple for reviving "one of [the Shawangunks'] oldest industries." Cassie had contracted with Mohonk Mountain House to make charcoal for broiling in its kitchen. For many years guests there could attest "that no steak in the world has the same flavor as one broiled over Shawangunk 'chark,'" the article said.

Irv told the reporter that he had learned "all the tricks of the trade" from his grandfather, Hiram "Daddy Hype" Van Leuven, "one of the county's famous figures" in his day. After building and lighting their mound, Irv and Cassie hovered over it for days, watching for "danger signs"—flames breaking through, loam loosening up, or smoke seeping out. When the wood was completely burned, they raked out the charcoal and loaded it into burlap bags for hauling to Mohonk. Daniel Smiley II recorded 13 cords of gray birch and red maple, at $4 per cord, produced 522 bushels of charcoal, which sold for 44 1/4 cents per bushel. Profit was $169. Daniel wrote, "I hope that sometime we can start burning charcoal again and compare it with what we buy."

The revival, though, did not last, and those who saw the charcoal burns, including Don Otis, a young neighbor of the Van Leuvens, actually bore witness to the last days of this now-lost art.

Daniel Smiley II noted that he "first met Irve [*sic*] in the mid 1920s" while camping at the Coxing. For many years Irv worked for Mohonk cutting wood and helping survey boundary lines in the Trapps. As a result the two became good friends. Daniel recorded that Irv planted peas and potatoes in late fall, which Irv explained "was the way it was done in nature, and they were ready when soil moisture and temperature conditions were right for germination in the spring." Don Otis recalled that Irv had "the nicest vegetable garden in the Trapps," and remembered taking the Otis family's white horse, named Old Harry, down to plow Irv's garden in the spring.[17]

The years passed, and Cassie died in 1949. She and Irv had been married twenty-nine years. They had no known children. Irv must have spent what savings he had to give her a proper burial. He chose a casket lined with eggshell-colored satin and ordered a flower car, noted funeral records. Cassie was buried in the Highland Cemetery in the Town of Lloyd, where she was from. His showgirl was gone, and he was alone.

In the early 1950s a young Bob Larsen met Irv, then in his early sixties. At the time Bob was climbing in the Gunks and camping at "Wickie Wackie," an informal campground on an open ledge along Clove Road across the Coxing Kill from Irv's house. ("Wickie Wackie" was named after an African-American country club operating farther down in the Clove.) Bob recalls Irv walking over one night to investigate their bonfire and remembers that they had a friendly exchange, with climbers likely reassuring Irv that they were being careful with the fire. Irv was curious about new happenings in the Trapps and the new types of people who were coming there.[18]

Irv developed lung cancer and, by early March 1956, had entered Golden Hill Hospital south of Kingston (also known as the Ulster County Infirmary). A reporter found him there for a feature published on Sunday, March 11, 1956, in the *Poughkeepsie New Yorker*. The article included a photo of Irv with his deeply etched face and wearing a plaid bathrobe. His drooping eyes are framed above by thick, dark, arched brows and below by heavy, half-moon bags. The dying laborer told his last tale, about his life in the Trapps, about its people, his family, his beloved garden.

Irv's care at Golden Hill was "fine," but the Trapps tugged at his heart. "I have one garden where I grow nothing but glads [gladioluses]," he told the *New Yorker* reporter.

"Some are as tall as your head." What had started as a dozen bulbs had grown to 140 prize-winners. "I want to get home in time to pick my strawberries," Irv added. "It's the first crop."

Perhaps he made it home that spring, his last.

By December, Irv was admitted to the county infirmary in New Paltz, where the county poorhouse used to be. Irv, sixty-eight years old, died there soon after. *The New Paltz News* ran a front-page obituary twenty-nine lines long. The Ulster County Welfare people paid for Irv's funeral. He was buried in a pine casket in the New Paltz Rural Cemetery near his parents, a brother, four sisters, and other relatives, including distant ancestors Andries Van Leuven and wife Phebe Aldrich, daughter of Peter, the first Yankee homesteader in the Trapps. Irv's gravesite is covered only with grass, empty of any headstone.

Irv Van Leuven was born during the hamlet's heyday and had witnessed its decline. He was a typical man of the Trapps, a stonecutter, charcoal burner, wood cutter, and hoop shaver—versatile skills once necessary for survival in a harsh environment. While the world around him was changing, Irv was the last full-time Trapps resident to live a nineteenth-century way of life, to live the old ways of his parents and grandparents. Once Irv was gone, barely a handful remained, and one day soon they, too, would be gone.[19]

• 14 •

Restoring the Past

October 12, 2000. The Trapps. Evelyn DeWitt, ninety-five, and Thelma Atkins, eighty-three, sit shoulder to shoulder on a boulder next to the Van Leuven Cabin. On a calm, clear autumn day, the two are girls once again at a place they had known as children.

They've joined a hundred others to celebrate the cabin's restoration and opening by the Mohonk Preserve. Nearby are Evelyn and Thelma's cousins who once lived in the Trapps: Roger Van Leuven; his sister, Doris Hall; and Harold Van Leuven. Also present are brothers and former residents Don and Joe Otis and Trapps descendant Joan Wustrau.

After the speeches and ribbon cutting, people step into the cabin. An exposed wall shows worn, fading insulation—wallpaper, newspaper, green gingham. They peer at the spartan, rustic place that families once called home … and wonder.[1]

The death of Irv Van Leuven in 1956 marked the final act for the Trapps hamlet. No applause sounded for the 150-year run. The theater was all but empty, and the curtain came down on another old-timer. Ice cutting bowed to modern refrigeration after a century of continual harvests begun by John F. Stokes at Mohonk Lake about 1860. Trapps resident Ray Otis cut ice at Minnewaska until the last harvest there about 1960; at Mohonk the last blocks were hauled from the lake to the ice house in 1964.

While the times they were a changin', some things were slower to do so. As late as the 1960s some roads in the Trapps remained unpaved. A 1962 newspaper ad in New Paltz for the Wickie Wackie Country Club directed revelers to take Route 44/55 to the "Top of the Mountain," turn right onto the "First Dirt Road," then proceed three miles to find food, drink, and dancing on Friday and Saturday nights. The "Top of the Mountain" was the old Trappen gap, the "first dirt road" a remnant of the old turnpike. Partygoers turned right again off the old pike for Clove Road, the public road unpaved since colonial times. They drove past homes, some abandoned, and past Smitty's Ranch, unaware they were traveling through a hamlet only lately called the Trapps.[2]

It was too late for the Trapps, but not too late for the mountain. The time came when the preservation of great estates, such as those at Mohonk and Minnewaska, was not guaranteed. In 1963, Smiley family members of the Mohonk Mountain House and friends established a land trust called The Mohonk Trust as a means of preserving much of Mohonk's outlying property.[3] Smiley Brothers, Inc. at Mohonk Lake initiated a series of land transfers to The Trust, which in 1964 received charitable tax status. In 1969 the former Enderly farm and mill site in the Trapps was transferred to The Trust as part of a larger Mohonk parcel.[4]

During the 1960s several people passed away who had once called the Trapps home, including Winslow Van Leuven, Anna Enderly Hayes, and Charlie Lyons.[5] In winter 1970 the New Paltz School Board voted to sell an abandoned, rundown building in its district—Schoolhouse Number 8 in the Trapps. A legal notice in the local papers stated that bids on the .27 acre would be accepted until March 30. "Anyone wishing to examine the building," the notice ended, "may do so by picking up the key at the School District Office, 196 Main Street, New Paltz, N. Y." It was the same quarter-acre lot Ben and Laney Fowler sold in 1867 for $1 for a school.

As it turned out, a couple who lived in the former hamlet owned the lot. In April the School Board rejected all bids and in August sold the building to them for $500. The

Left: Ray Otis of the Trapps harvests ice on Lake Minnewaska in January 1956. Modern refrigeration shut down this century-old winter industry in the Shawangunk Mountains. Courtesy of Ray's son, Don Otis.

Above right: Wickie Wackie Country Club advertised in The New Paltz News *in September 1962. The "First Dirt Road" was Trapps Road, a remnant of the old turnpike and still unpaved. The club also advertised in* The New York Age, *a daily African-American newspaper, and was listed with the Clove Valley Dude Ranch in the 1956* Negro Travelers' Green Book. *Courtesy of HHHC-EML.*

---

couple moved away and later privately sold their parcel with mountain school.[6]

For some years Ray and Agnes Otis had moved up to Lake Minnewaska for the winters, returning each summer to their Trapps home. With Ray's retirement in 1972, the couple moved to New Paltz and off the mountain for good. Ken and Lucille Phillips threw a retirement party for Ray at Ski Minne, attended by his family, including his mother, Minnie, then ninety, recalled her grandson Don Otis.[7]

At Mohonk Lake in 1972, Daniel Smiley, his brother Keith, and Keith's wife Ruth interviewed long-time employees, including Arthur Coddington, seventy-seven, one of the last Shawangunk millstone cutters. Art's interviews are some of the few remaining records describing this now-lost skill. Two years later The Mohonk Trust hired Bob Larsen, an ex-climber and ex-Manhattanite who had moved to the upper Clove in the 1960s. Bob joined Thom Scheuer at The Trust as second ranger.[8]

The 1970s marked the passing of: Anna Lyons, German immigrant and widow of Charlie; Minnie Otis, widow of Jake; and Leonard and Ethel May George, former Trapps teachers. Minnie had spent winters of her last years with daughter Myrtle or son Ray, recalled Ray's son, Don. Yet every summer she returned to her Trapps home. Minnie stayed there alone, checked on by her family, the last resident of the old Trapps hamlet, until she left for good a few years before her death at age ninety-three in 1975. Her home, owned by Lake Minnewaska, was left unoccupied and burned down in the mid-1970s, about the same time that fire also took the schoolhouse.[9]

In April 1977, Diane Carlson reprinted in a New Paltz paper the section on the Trapps and Clove from Kenneth E. Hasbrouck's 1953 history of Gardiner. (The second edition came out in 1978 for that town's 125th anniversary.) Did the Trapps School or Laurel Inn still stand, she asked? The Smileys of Minnewaska had dismantled the inn in the 1940s; Carlson missed seeing the school by only a few years. Sometime about 1980, Cliff Inn, the old gas station built by Sheridan Burger, was gone, too, dismantled and carted away.[10]

The Trapps hamlet seemed to be disappearing board by board. Yet a rosier future for the northern Shawangunks was unfolding. "By 1980," Bradley Snyder wrote, "over 30 square miles of the ridge had been set aside for public benefit as parkland and preserve." The Mohonk Trust had reorganized, renaming itself the Mohonk Preserve. Another series of land parcels was transferred from Mohonk Mountain House to the preserve, which then protected over 5,000 acres, including part of the old Trapps hamlet.

Late in 1977 and again in mid-1986, first Cliff House and then Wildmere succumbed to fire. The Lake Minnewaska property—a one-time economic mainstay of the Trapps people—ultimately became Minnewaska State Park Preserve and the people of New York State the new owners of part of the former hamlet.[11]

The 1980s marked the passing of Ray Otis of the Trapps and Daniel Smiley II of Mohonk Mountain House. Ray had played a key role in maintaining the mountain hotels at Lake Minnewaska. Daniel had established a research facility that bears his name. The Daniel Smiley Research Center, part of the Mohonk Preserve, is a repository for much material about the Trapps.[12] In 1994 the preserve hired retired New York State surveyor

Author Bob Larsen, right, and a fellow Mohonk Preserve ranger repair a bridge in 1980 over the Coxing Kill in the former Trapps hamlet. The bridge is part of the extensive carriage road network constructed by the Smiley families in the northern Shawangunk Mountains during the nineteenth and twentieth centuries. Courtesy of DSRC-MP.

Norman Van Valkenburgh who, with ranger Bob Larsen, spent many a day "walking the lines" of the preserve's fifty-plus miles of boundaries. Norm pored over chains of title, deeds, wills, and maps for his own survey maps and what Bob calls "meticulous reports" on his findings.

"Norm was an expert at finding the historic boundaries of land grants," Bob said of their work together. "The Mohonk Preserve's boundaries are based a great deal on these old land grants." When Norm uncovered a name, Bob tried to tie it to his genealogy research. In the field, Norm came across stone rubble—a foundation, boundary marker, root cellar, or quarry—that led the two into a new investigation. How was the structure once used? Could they confirm it on a map? How did the place connect to the people who once lived there? Norm and Bob's work became critical to the ongoing preservation of the Trapps mountain hamlet.

Also in 1994, ten private and nonprofit agencies formed the Shawangunk Ridge Biodiversity Partnership. Members manage and protect more than 40,000 mountain acres of the Shawangunks; contained within are parts of the former Trapps mountain hamlet. Partnership members include: the Mohonk Preserve; Friends of the Shawangunks; Open

Space Institute; New York State Office of Parks, Recreation, and Historic Preservation; New York State Department of Environmental Conservation; and The Nature Conservancy, which that same year designated the entire Shawangunk range as one of the "Last Great Places" on Earth.

In 1995, Black Dome Press published Marc B. Fried's *The Huckleberry Pickers*, the result of Fried's more than ten years' research. The book followed his 1982 *Tales from the Shawangunk Mountains*. Both contain interviews with the last generation of pickers, including former Trapps residents Bertha Van Leuven and her children, Roger, Doris, and Ethel.

Bertha Van Leuven, ninety-one, had passed away in January 1990. Two days after her death, columnist John Conway interviewed Bertha's nephew, Ray Veres, seventy-two, for a story about a "lost hamlet" called the Traps, spelled the old way. Ray described the Trapps people for modern readers: "They made their living farming, hunting, trapping, picking huckleberries, cutting cordwood for the hotels at Mohonk and Minnewaska, and by moonshining. They were a tough, colorful, resourceful breed of people, and extremely reclusive."

Yet the last generation was dying "one by one," Ray said. Conway wrote, "In fact, Veres fears that he is one of the few remaining who remember the hamlet at all." Ray had just attended the funeral of his Aunt Bertha. Conway wondered how a place could disappear so completely. "I suppose it was like [any place] else," Ray said. "The kids went off to school, decided to look for the better things in life, and never returned. The parents died off and moved away. Before long, the whole town had vanished." Ray Veres, who had penned his own reminiscence for the *Huguenot Herald* in New Paltz a decade earlier, passed away in 1992.[13]

Some of Winslow and Bertha's children and other cousins then found themselves advisors to an unlikely but welcome preservation project. In 1969 the Mohonk Preserve had become owner and caretaker of a cabin owned previously by Mohonk Mountain House. Originally made of plank and white in color, the cabin had been resided with horizontal clapboard and painted red. It was a crash pad for climbers and Appalachian Mountain Club members. Preserve rangers tore off the enclosed porch, not original, then asked if the little building was worth saving. The Appy Cabin, as it was known, needed at minimum a new roof and at maximum major funding to restore it. The rangers proposed tearing it down or letting the Gardiner Fire Department burn it down.

One ranger turned the discussion upside down by suggesting a third option that none had seriously considered. Bob Larsen had learned a thing or two about local history since coming up from New York City. He had spoken with Irv Van Leuven in the flickering

This plank house, one of the last remaining structures in the Trapps, was in dire need of repair by the later 1990s when the Mohonk Preserve began restoration efforts. The house was renamed the Eli Van Leuven Cabin for its last owner. Courtesy of DSRC-MP.

lights of climbers' campfires. Bob lived in the upper Clove house of onetime stonecutter Chris Schoonmaker and had donated Chris's stonecutting tools and gear left there to the Barn Museum at Mohonk Mountain House. As trail keeper of the Mohonk Preserve's boundaries, Bob had stumbled across many a remnant of human habitation while walking the lines and studying old deeds. He had seen the schoolhouse before it had burned down. He knew families had once lived there.

At the prospect of razing the Appy Cabin, Bob's preservation instincts went into high gear. He urged the preserve to view the cabin from a historical and cultural perspective. The cabin itself might have dated back to the late 1800s, and its restoration was an opportunity that could not and should not be missed. The preserve realized that it was about "to wipe out [a] last remnant" of a settlement dating back about 200 years, reported Deborah Medenbach. From the start the preserve's executive director, Glenn Hoagland, championed the project, which took off with the renaming of the cabin for Eli Van Leuven, its last resident.

The Mohonk Preserve's goals were twofold—to obtain funding to restore the cabin and to secure recognition for the hamlet as a historic site. The New York State Environmental Protection Fund awarded the preserve a 50 percent matching grant for restoration through the state's Office of Parks, Recreation and Historic Preservation (O.P.R.H.P.). Once matching funds were met, the preserve began restoring the cabin. The O.P.R.H.P. approved the Trapps Mountain Hamlet for listing on the State Register of Historic

Places—a first for a vanished subsistence village of New York. This designation applied to a district that included lands of the Mohonk Preserve and Minnewaska State Park Preserve. Once the state approved, the hamlet was nominated for listing on the National Register of Historic Places.

News stories sprouted across the region in tandem with the spring 1997 groundbreaking of the preserve's new visitor center. Reporter Bond Brungard interviewed Bob Larsen and Thelma Atkins, eighty, of Marlboro, in April 1997. A great-granddaughter of Hiram "Daddy Hype" Van Leuven, Thelma recently had found his grave up on the mountain and recalled a trip made almost seventy years earlier to see her older cousin, Irv Van Leuven, a great adventure for a ten-year-old girl. Bob admitted feeling "daunted" by the prospect of preserving the hamlet's history, which included restoring the cabin and creating an interpretive trail with accompanying guidebook. Bob added, "It's kind of like a murder mystery, putting all this together."

Research by Bob and other Mohonk Preserve staff soon revealed that the cabin did indeed date back to ca. 1889. William Hagen constructed the little plank home when Benjamin Harrison was the 23rd U.S. president. Hagen sold it to Eli Van Leuven, who sold it to the Smiley family at Lake Mohonk.

The Mohonk Preserve continued to gather private and public support, including from the Palisades Interstate Park Commission, which oversees Minnewaska State Park Preserve. The new interpretive trail to the cabin would cross over Minnewaska land and pass by the Fowler Burying Ground, also located on Minnewaska property. Both agencies lent full support to the project, giving approval for the section of trail crossing Minnewaska land.

In late November 1997 the Mohonk Preserve hosted a planning meeting followed by a tour of the Van Leuven Cabin for members of the O.P.R.H.P. and of the Preservation League of New York State. Also invited were Roger Van Leuven, his sister Doris Hall—among the last to have been born and raised in the Trapps—and their cousin, Melvin Sharpe, Jr., a Van Leuven descendant whose mother grew up in the hamlet. They and other former residents or descendants became key figures during restoration, invaluable as advisors and contributors. Reporter Richard A. D'Errico, considered them "the lifeblood of the project because they tell the tales behind the architecture and land."

The meeting and tour sparked many memories—living without central heat or electricity, cutting wood to burn for heat, drinking water from the Coxing Kill—recalled Roger and Mel. Doris recalled swimming in "the brook," as she called the stream.

Mel and his cousin Thelma Atkins penned letters of support to the preserve. Mel wrote: "Most of us wait too late to delve into our past generations and a lot of the information is either buried deep or lost forever. Therefore, anything that can be done to preserve this site and its history is important to me as a descendant." Thelma echoed his feelings: "The Trapps and the Van Leuven Cabin hold a special place in my heart and in my family's history. Their preservation is not only important to me, but to the many other Hudson Valley families whose ancestors lived there, as well."

Roger, Doris, Mel, and Thelma were joined by cousins Harold Van Leuven and Evelyn Coddington DeWitt. All were great-grandchildren of Hiram and Anna Maria Van Leuven except for Evelyn, who was a grandchild. Others came forward one by one: Clarence Kidney, great-grandson of Calvin and Rachel Burger; Don and Joe Otis, grandsons of Jake and Minnie; and Joan Hayes Wustrau, descendant of both the Enderlys and Gorslines. Roger, Doris, Harold, Don, and Joe all had once called the hamlet home.

For the November 1997 meeting and tour, the Mohonk Preserve mounted its first exhibit on the Trapps, largely made up of photographs family descendants and others had given to Bob. Bob's very first photograph was a copy of the Enderly mill given to him by Daniel Smiley II of Mohonk Mountain House before his death in 1989.

Bob also began to hand-draw family trees from information gathered from descendants and public records. Bob recalls the first time he saw a map of the 1799 Nineteen Partners Tract marking many landmarks—the Roosa homestead, the "Foot Path to Shawangunk," and the "Public Road" through the Clove. By then Bob was moving more fully into research, pouring over church and census records, land deeds, and news articles. A vanished village was coming to life before his eyes.

It was then that Bob first collaborated with Robi Josephson, who was the Mohonk Preserve's contract publications editor. The two worked on articles for the preserve newsletter and on the interpretive guide (now a brochure) for the preserve's new path to the Van Leuven Cabin from the West Trapps Trailhead. Robi was dipping her toes into the layered sands of Trapps history. Little could she guess that soon she, too, would be engulfed by the same wave that Bob had surfed for more than a decade.

The day came on April 27, 1999, when a group gathered at the Eli Van Leuven Cabin for its official groundbreaking: Roger Van Leuven, his sister Doris Hall, and their cousins Mel Sharpe and Evelyn DeWitt; Bernadette Castro, commissioner of the New York State Office of Parks, Recreation and Historic Preservation (yes, the child star of her family's convertible sofa commercials); Kenneth Hewes Barricklo, project architect from Kingston; and preserve staff, including Glenn Hoagland and Bob Larsen.

"All the cousins agreed," reporter Nancy Ward wrote, "that this gathering was like a family reunion for the Van Leuvens."

After the groundbreaking, work commenced on the little cabin. Doris Hall told reporter Jesse Nason she was pleased that her ancestor's home would be saved. "Times weren't easy in those days," Doris said. "It was the old-fashioned way. We used to have a lot of fun though because we were together. ... It was a great life. You lived more with nature in those days. Now people live in the malls."

Historic restoration contractor Zeke Boyle told reporter Jeremiah Horrigan in July 2000 that he could "almost feel the presence of past generations in the frameless plank-and-batten walls." Boyle uncovered some surprising insulation within the walls—newspaper, flowered wallpaper, and green gingham cloth. Some of the newspaper dated back to McKinley's days as an Ohio congressman. "These people might not have had a lot for themselves," Boyle added, "but they used everything they had." Boyle restored interior

Rick Zirpoli, left, and Gerry Stokes work under the direction of Zeke Boyle, far right hidden by the ladder, to restore the Van Leuven Cabin as a museum for the Mohonk Preserve. Courtesy of DSRC-MP.

Right: Newspaper, wallpaper, and cloth were cheap and readily available insulation. All were discovered inside the walls of the Van Leuven Cabin, constructed about 1889 in the Trapps hamlet. William McKinley, then a Republican congressman from Ohio, was the main supporter of the Tariff Act of 1890. Photograph by Robi Josephson.

walls to display the original rough planks and the later milled paneling and plaster, with the original insulation protected under Plexiglas.

Restoration work on the cabin took more than a year to complete. Thursday, October 12, 2000, was the long-awaited dedication. About 100 people gathered on a pleasant, mild day: Mohonk Preserve staff, including Glenn Hoagland and Bob Larsen; the restoration team; state dignitaries; former Trapps residents Roger Van Leuven, Doris Van Leuven Hall, Harold Van Leuven, and brothers Don and Joe Otis; and Trapps descendants Thelma Atkins, Evelyn DeWitt, and Joan Wustrau.

"The project cost $120,000," Deborah Medenbach reported, "which is a fair sum for a building that memorializes the hand-to-mouth subsistence lifestyle of the people on the Shawangunk Ridge's western escarpment." Bond Brungard, who had also followed restoration efforts since they began, reported that the cabin now had "a bright new future." Visitors following the new interpretive trail to the cabin passed an old stone quarry, an old bridge abutment, and the old Fowler Burying Ground. At the cabin they peered into wall panels revealing the inner layers of life once endured by the Trapps people.

The preserve also opened the new path to the Van Leuven Cabin from the West Trapps Trailhead on Route 44/55. Along the way visitors can explore an abandoned stone quarry and the Fowler Burying Ground. At the old quarry, stone giants lay untouched for a century, flawed in some way, perhaps discarded by famed Trapps stonecutter Hiram "Daddy Hype" Van Leuven himself.

In December 2000 a long-awaited letter arrived on Bob Larsen's desk from the New York State Office of Parks, Recreation and Historic Preservation. The subject was the Trapps Hamlet Historic District. Ruth L. Pierpont, director of the Historic Preservation

Field Services Bureau, wrote, "I am pleased to inform you that the above referenced property has been listed on the National Register of Historic Places."

After those words had sunk in, Bob made many calls over the next few days. Best of all was calling those who loved most the little cabin and mountain hamlet—Roger, Doris, Melvin, Harold, Thelma, Evelyn, Don, and Joan. Their family histories would now be preserved for the next generations. It was a unique and hard-won recognition for a vanished, hardscrabble community.

Yet the Mohonk Preserve's work was not finished. Over the next two years, rangers cleaned up stone foundations at the former Enderly mill and farm on the Coxing Kill. While hauling away brush they discovered remains believed to be the blacksmith shop, root cellar, and spring house. Along the path to Split Rock, the preserve installed a kiosk explaining the site's role as an early industrial center of the Trapps.

Top Left: Pole rafters at the loft of the restored Van Leuven Cabin were left with the bark on. A loft or attic provided a bedroom for children. Top Right: Bob Larsen gives a tour at the Van Leuven Cabin in the former Trapps hamlet. The Mohonk Preserve restored the walls to reveal original plank wall construction. During its earlier ownership, Mohonk Mountain House added more modern paneling, seen at right, and rented the cabin as a summer cottage during the 1920s and later. Photographs by Robi Josephson.

Bob Larsen speaks during ribbon-cutting ceremonies to celebrate the opening of the Eli Van Leuven Cabin, restored by the Mohonk Preserve. From left: Mary Jane and Harold Van Leuven; Thelma Atkins and Evelyn DeWitt, both seated; Annie O'Neill behind Evelyn; Ruth Dylewski; and Julia Stokes, Deputy Commissioner for Operations of the NYS Office of Parks, Recreation and Historic Preservation. Harold, Thelma, and Evelyn were Trapps descendants. Photograph by Veronique Dietrich. Courtesy of DSRC-MP.

Rangers also improved a section of the hamlet's most historic route, the path over the mountain once traveled by Native Americans and mountain families. The same path that cut through the old Trappen gap once connecting the Rondout and Wallkill Valleys. The same path that served, in part, as the boundary line between lands of the Grote Transport and Nineteen Partners Tract. As early as the 1970s, Bob worked on the path's lower section, its course changed slightly mainly to prevent erosion. The path's continuing history is a good illustration of how roads and paths can alter over time.[14]

In 2004 gravestone rubbings were the topic of the day. As the preserve's cultural historian, Bob was anxious to have rubbings done of gravestones in the Trapps and Clove before the elements eroded them further. Bob brought in artist Suzanne Caporael, author Bruce Murkoff, and retired art teacher Barbara Rubin, who rubbed stones with special tracing paper and black crayon at the Davis and Fowler grounds, among others. Over the winter of 2006–07, Bob and Barbara mounted an exhibit of the rubbings at the preserve's visitor center. "Grave rubbings are better documentation than photographs," Bob said. "We wanted to display them as another important way to interpret the history of the hamlet for visitors."

During this period, Mohonk Preserve research staff located with global positioning several hundred cultural sites in the Trapps and northern Shawangunks, including: cellar holes of former homes and farmhouses; foundations of outbuildings, such as barns and shops; and charcoal pits. As new sites are discovered, this project enlarges our understanding of how people once survived in the Trapps and on the mountain.

On Wednesday evening, June 7, 2006, local filmgoers packed the Shadowland Theater in Ellenville, New York, to see the world premiere of *The Huckleberry Pickers: A Vanished Shawangunk Culture*. Produced by The Nature Conservancy and Open Space Institute, the film documents the story of twentieth-century berry picking around Sam's Point. The last generation of pickers was interviewed for the film, and most were there to talk about a way of life since gone.

Also that year, New York State designated the Shawangunk Mountains Scenic Byway, thanks to efforts spearheaded by Al Wegener. Today, one section of the eighty-eight-mile loop takes drivers along Route 44/55 through the heart of the old Trapps hamlet. As well, New Paltz high school senior Annie Slingerland restored the Enderly Ground on Mohonk Preserve land. Assisted by her mother, Rae, Annie reset and cleaned headstones, cleared away brush, planted bulbs, and built a cedar rail fence enclosing the burying ground. For her efforts Annie received the Gold Award, the highest honor in girl scouting. In 2008 the preserve replaced the cedar fence with a stone one.

In September 2010, George Black, the boy once called G.G., returned to the Trapps. He explored his boyhood haunts and spoke kindly of its people—Oscar and Charlie Lyons, Tillie Dilschmann, the Otis family. He had been back before, but this time, for the first time in about sixty years, he grasped the hands of childhood neighbors Don and Joe Otis. Memory after memory spilled out from the three during a day spent touring the former hamlet. For George those early days in the Trapps were a time when dandelion rings and necklaces belonged to "carefree" summer days. A time not too long ago when mountain men still hunted game and spun tales for eager audiences. A time when a young boy did not share every mountain adventure with his parents.[15]

As the past is being restored and remembered, those with deep ties to the hamlet continue to pass on—Agnes Bouton Otis, Harold Van Leuven, and Thelma Atkins in the

Gravestone rubbings, such as these done in 2004, help preserve stones being eroded by time. Albert Smith and his wife, Cornelia Ann Slater, are buried in the Davis Burying Ground. Four young children of George and Rachel Coddington were laid to rest in the Fowler Burying Ground. All were Trapps residents. Photographs by Robi Josephson. Grave rubbings by, and courtesy, of Barbara Rubin and Suzanne Caporael.

first decade of the new century. Thelma's family honored her in her obituary: "Through her family research, she was an instrumental part of the Preservation of the Van Leuvan [*sic*] Cabin at the Trapps Hamlet at Mohonk Preserve." Never missing an event, Thelma had sat with her cousin Evelyn on the large boulder next to the cabin at the ribbon cutting. The same boulder bearing a work shelf cut out by a local stonemason for his family. The same boulder that kids of another era clambered over in play.

Final respects were also paid to Ken and Lucille Phillips of Lake Minnewaska, who employed the last generation of Trapps residents and their descendants, as well as to Keith and Ruth Smiley of Mohonk Mountain House. Keith and his brother, Daniel, had worked as youths under the tutelage of Irv Van Leuven, and the brothers had remained keenly involved in the cultural history of the mountain. The two cofounded The Mohonk Trust, later the Mohonk Preserve. Ruth, a botanist and photographer, shared the same commitment to preserving the past.

Farewells were bid as well to Mel Sharpe, Jr. and his older cousin, Evelyn Coddington DeWitt. Their mothers, Elizabeth Van Leuven and Elizabeth's great-aunt Laura Van Leuven, respectively, had grown up in the Trapps. Evelyn became a major contributor of information about the Trapps, the Clove, and Lawrence Hill.

In her last years Evelyn had a visitor knock on her door almost weekly. When a week or two would go by without a visit from Bob Larsen, she would call him up and make sure everything was all right. "She adopted me," Bob said. Right up until her death at age ninety-eight, she was able to recall her life for Bob.[16]

We cherish the friendships made while researching the hamlet's history and working on this book. Yet, these friendships have become bittersweet. Joe Otis, seventy-six, passed away most recently, in March 2013. Joe missed seeing the book in print by months, but he had read and was proud of the many contributions he and his brother Don had made. We both now have bid "fare thee well" to individuals who openly shared their memories with us and who became dear friends. As the years and the Trapps people pass, they become even dearer, forever remaining "my people," as Bob calls them.

Our people.

• 15 •

Remembering the Past

Summer 2000. Harold Van Leuven drives up the mountain to the old Davis Burying Ground. He parks off Route 44/55 on a road that was once an unpaved turnpike and walks into the cemetery to one grave in particular.

He tends the grave of his ancestor John D. Van Leuven, Civil War veteran and survivor of Libby and Andersonville Prisons.

After tidying up, Harold sets a small American flag into the ground beside the headstone. He straightens up and salutes his elder cousin and fellow veteran three generations back. Harold returns to his car and drives around the Trapps, a place where he lived as a boy, where he attended the one-room school, where he once called home.

His annual respects paid, Harold heads back down the Minnewaska Trail to his life in New Paltz.

The headstone of Civil War veteran John D. Van Leuven is one of the few remaining today in the Davis Burying Ground. As of 2012 a small American flag stood beside John's headstone to honor his military service. It likely was one of the last flags placed there by Harold Van Leuven before his own death in October 2007.[1]

It has been our great honor and good fortune to become acquainted with and, in some cases, dear friends with former Trapps residents, their descendants, or those with close ties there. This book could not have been possible without the following: Thelma Atkins; George Black; Harold Bittner; Evelyn Coddington DeWitt; Doris Hall; Clarence Kidney; Don & Flo Otis; Joe Otis, Don's brother; Melvin Sharpe; Alfred B. Smiley; Grace C. Smith; Harold Van Leuven; John Van Leuven, Jr.; Marie & Roger Van Leuven, Doris's brother; Marian Whitford, Evelyn's daughter and Grace's cousin; and Joan Hayes Wustrau. Their many contributions are found in the pages of this book. Below are excerpts from lengthier interviews with Don, Doris, and Roger, plus excerpts of newspaper reminiscences by Ray Veres—these four among the last to grow up or live in the Trapps hamlet.

• • • • • •

It was mid-March 1937 and a big day for two youngsters growing up in the Trapps. That morning Agnes Otis bundled up her two oldest children, Don and June, and sent them off for their first day of school. Don had just turned six and June four a few weeks earlier. Agnes watched them go, two small figures trundling along the dirt road, then returned to caring for her youngest, Joe, eight months old.

That first day, however, Don and June did not head to the schoolhouse. Instead, they trudged across the paved highway to the George house. On very cold days, school was held in the warmth of the teacher's home rather than in an icy school. Schoolteacher Leonard George was an accomplished artist who impressed Don with landscapes of his native Oregon and a portrait of a young June. (Whether her portrait still exists is unknown.) Mr. George and his wife Ethel May, also a teacher, often team-taught at the Trapps School. Don called school days in the Trapps "a great learning experience." No matter their ages, students worked together on various lessons. The Georges set up an old cash register inside one corner of the school for the students to study banking. A huge sandbox inside also aided the study of geography.

Trapps neighbors gather ca. 1939 at the Otis home in the Trapps. From left: an unidentified woman; Alice Grimm; "Don't-Mess-With-Me" Joe Otis, about three years old; Joe's older brother Don; their grandmother Minnie Otis; and their mother Agnes Otis in the doorway. Courtesy of Don Otis.

Mr. George's lessons continued outside, where he and the students built a see-saw, a trapeze five feet off the ground, and a merry-go-round. Other projects remain firmly etched in Don's memory seven decades later. One year they dammed up a seasonal creek near the school for a small pond, where they skated in winter and played Terry and the Pirates in summer. The year Don started school, the class built replicas of the George Washington Bridge and the Golden Gate Bridge. (The George Washington Bridge, opened in 1931, was the world's longest suspension bridge until the Golden Gate Bridge broke the record in May 1937. The Golden Gate's opening in San Francisco was big news that year.)

The two bridges were constructed over a drainage ditch near the school. Each one had a three-foot-wide deck of rough-cut pine boards, rose about four to five feet above the water, and extended some thirty feet from end to end, sizable enough to span heavy spring runoff. Towers of wooden poles were set about four feet into the ground and reached about four feet higher than the bridge itself. Mr. George found cables, perhaps cast off from the construction of Route 44/55, for the class to recreate the suspension system. The larger "main" cables were about one inch in diameter, the smaller "suspension" ones about 1/8 inch. "We were able to walk on them, sit on them when we had lunch, and play on them," Don recalled of the bridges that stood until the school closed for good. Town of Gardiner road crews likely took them down one year while cleaning out the ditches.

With winter came sled rides and other adventures. The Town of Gardiner never wanted to plow the roads in the Trapps during winter. Don's father, Ray, and a Trapps neighbor, Randolph Taylor, mail-ordered a snow plow from Sears Roebuck and attached it to Mr. Taylor's Packard automobile. One day Don, June, and Joe were walking back to school after lunch and started to roll a snowball down the middle of the road. Before they reached the school, the icy mass had gotten so big and heavy that they couldn't budge it. Ray and Mr. Taylor tried to move the snowball with the plow, but instead it broke the plow to pieces. So much for the Sears plow.

Coal heated the school. Herbert Phillips of Napanoch drove an old dump truck to pick up the coal in Honesdale, Pennsylvania. Herb was the father of Ken Phillips, later owner of Lake Minnewaska. Coal was stored in a red coal shed about 100 feet from the school. The kids hauled in the coal, which in winter was kept in a small bin inside the school. As the school had no running water, the boys had the job of filling five-quart lidded containers, similar to milk cans, with water from the Coxing Kill and lugging them up the hill to the school. There were two outhouses, one for each sex, but without any fence between them (as state law required). Local men, such as Albert Smith of the Trapps, were hired to maintain the school and grounds.

Don and June graduated sixth grade from the Trapps School in 1943. Their departure left only one student, their younger brother, Joe, not quite seven years old, and the school shut its doors. After the school closed, Mr. George was awarded the contract to transport the students down to school in New Paltz. For a while it was just Don, June, and Joe. Mr. George had a chauffeur's license and bought a new Chevy for the route. He was required to have the car inspected by an agency to ensure its safety for transporting children. Mr. George stopped driving the route in the later 1940s. Minnewaska picked up the service using guest limousines, arranging for a young Trapps resident to drive the school run. Don, sixteen, had taken driver training in high school and, with a new license in hand, became the driver. During his senior year Don started the run at Minnewaska to pick up one of the Phillips kids, then stopped at his own house to pick up June and Joe. Don drove the limo to school, kept it there all day, then returned it each afternoon to Minnewaska, getting a ride home sometimes with his father. After he and June graduated in 1949, Minnewaska hired other drivers, including Joe, to transport mountain students to school in New Paltz.

Don, June, and Joe left the Trapps after high school but, all in all, Don said it was a "great childhood" on the mountain.[2]

• • • • • •

Doris Hall and her brother, Roger Van Leuven, share many good memories of life on the mountain. The Van Leuven family spent summers at the berry camps, but also picked huckleberries "up behind" Lake Minnewaska, a wagon ride from their Trapps home. After finishing one day, the family gathered in the wagon for the ride back. With a tug on the reins the horse began to trot, the wagon bouncing behind. At one point Roger looked back to see "Doris sitting in the middle of the road." The kids often sat on the end of the wagon with their legs dangling, and Doris had fallen off. Doris said, "I thought they were going to leave me!" Big brother Roger kidded her that they should have made the horses trot faster, a comment causing both to laugh about a long-ago memory.

Doris's godmother was Trapps neighbor Theresa "Tillie" Dilschman, whom her mother, Bertha, knew from Brooklyn and whose daughter, Tessie, had been Bertha's maid of honor when she married Winslow. Roger and Doris said that Tillie would walk up to Lake Awosting to pick berries, then trudge home to the Trapps carrying a full ten-quart pail in each hand. "I don't know how she did it," Doris said. Tillie also walked several miles to pick up her mail. Mail delivery stopped short of the Trapps, ending down on Clove Road where the road to Rock Hill veered off; the hamlet's mailboxes were later moved to the base of the mountain on the old Gardiner Road (today U.S. Route 44/45) below the intersection with present-day Route 299. Roger added that "we all got our mail" that way then.

Winslow and Bertha Van Leuven pose ca. 1932–33 at their Trapps home with four of their five children. The children, from left, are Doris, Ethel, Alice, and Roger. Their youngest brother, Richard, was not yet born. Courtesy of DSRC-MP. Donated by Roger Van Leuven.

Come summer, Winslow and Bertha Van Leuven, with five children in tow, journeyed up to Four-Mile Post on the old Smiley Road above Ellenville for the berry picking season. The Van Leuvens sold and delivered berries to the mountain houses at Minnewaska and Mohonk. Berry money paid the $6 a month rent to Minnewaska for their Trapps house, including rent owed from the previous winter on credit extended to them by the Smiley family at that lake.

The Van Leuvens spent Sundays down in Ellenville, where they shopped at the fivc-and-dime store. The kids used their berry money for 15¢ sundaes but had to save some of it to buy clothes. Other memories of Four-Mile Post linger for Roger and Doris. Trucks laden with berries swaying and lumbering down from the camps along the old Smiley Road. At night, tossing horseshoes on the road or playing, by the light of a kerosene lantern, eight-handed pinochle on a big Coca-Cola sign used as a table. Visiting the nearby campfire of a Hurley minister and his family. Bootleg liquor at other camps.

Berry pickers gathered into their own social groups, and the Van Leuvens had theirs at Four-Mile Post. "We knew everybody up there," Doris recalled of a way of life now gone. "We lived in nature, we picked berries, we could hear the birds sing."

It also was the Depression, but they had everything they needed. Back in the Trapps the Van Leuvens went food shopping off the mountain about once a week. In the summer, Bertha sometimes walked with the kids down the old road (the old turnpike) to the Grand Union in Kerhonkson. If they bought enough food, they got a wagon with wooden sides to haul home the groceries. Doris remembered only four cereals available then—corn flakes, puffed rice, puffed wheat, and Shredded Wheat.

The two said that their family made do by "digging up" a small garden. They turned the soil using a fork and grew "all kinds of things," including potatoes, carrots, pole beans, and peas. A few families kept chickens, and the Van Leuvens had two pigs. Their father and older cousin, Irv, slaughtered the pigs in their backyard, and their mother canned the meat. Doris said, "She was from the city [New York] but learned how to do everything up here." Food was stored down in the coolness of an old-fashioned cellar.

When snow piled up three feet high during the winter and Winslow could not cut wood to support his family, he went down to Wallkill for federal relief. As far as Doris remembers, he received only grapefruit juice and rice, but relief in the form of canned meat and milk was also delivered to Trapps families. Each child received a quart of milk, which Roger recalled making into ice cream. The kids brought the canned meat with home-grown carrots and potatoes to school, where they and the teacher cooked on the stove. "By noontime, we had our lunch," said Doris.

She remembered hoboes sometimes traveled through the Trapps. They would stop at the Van Leuven house and ask for food. Whenever one came to the door, Bertha hid her pocketbook before giving them a sandwich. Roger recalled the junk man and the rag man who bought rags for 2¢ a pound. Goldwasser's of Rosendale was one of the grocery trucks making rounds on the mountain, selling "a little bit of everything," Roger said. In summertime, Doris couldn't wait to come out of the woods after picking berries in time to meet Minard's ice cream truck, which came up from the Highland area to make stops at their house.

Roger and Doris played various games as children, including baseball, hide and seek, and duck on a rock. In the last game, a rock (the duck) was placed on a flat rock. The idea was to knock it off with another rock.

Doris had one special toy, a small china doll with moveable joints given to her one Christmas in the 1930s by her Uncle Alex, her mother's older brother, and his wife Aunt Julia. One day Doris dropped the doll and broke it. She decided to bury it near the Trapps house they lived in at the time, and to this day can point out the tree next to where the little china doll still lies.

At strawberry picking time, Roger and other boys went down to Irv Van Leuven's place during school lunch recess. Picking strawberries at the Van Leuvens was a rite of passage for many a Trapps child. One time, Irv's wife Cassie came out toting a pistol and threatened to shoot them the next time she caught them picking her berries. "That got us back to school in a flash!" Roger said.

Winslow picked elderberries, dandelions, and wild grapes, fermenting them in two wooden tubs to make wine. Roger said that one time his father poured the contents of the two tubs together and found a rat inside. "There goes the wine down the hill," Roger recalled, watching his father dump it all out.

Doris said that living in the Trapps, especially during the Depression, was "rough on our mother and father," but the kids did not have the same worries.

Roger, with a twinkle in his eye, enjoys telling two classic lines about the hamlet: "I could never understand why people lived in the Trapps ... it's all rocks!"; and "You know why they called it the Trapps? You got trapped there!"[3]

• • • • • •

Raymond M. Veres came to the Trapps as a boy from Brooklyn during summers of the 1920s. Ray and his wife Ruth made the Trapps their home, but left in 1977 for Modena in the Wallkill Valley. Memories prompted Ray to write about the Trapps of the 1920s for the *Huguenot Herald* of New Paltz, which printed his two-part series in May 1983. Ray began with the story of how his parents, Frank and Mabel, bought a "termite-ridden" house in the Trapps for $600 and continued with many a tale about the hamlet and its people.

Ray recalled the "fun" drive up the mountain to their summer place in Old Liz, a Model T of "vintage unknown" operated by Ray's Uncle Frank:

> A little further on, the road began to wind tortuously up The Trapps Mountain. Here the fun really began. Old Liz would go grinding and rupturing herself up

the hills, getting hotter and hotter, until finally the potato that my uncle used to use in place of a radiator cap would blow off with a loud whoosh. This meant another prolonged stop while Uncle Frank went off to scout more water from a nearby ditch. "I guess I poured too much oatmeal in 'er," he explained. "She don't circ'late good no more."

More often than not Old Liz would abruptly stall just when we'd get to the switch-back part way up the mountain. Everybody, except Uncle Frank, of course, would leap out in a near-panic and wrestle rocks from the nearest stone wall to throw under the wheels. If you weren't agile enough or if your aim was bad you'd likely find Old Liz over a cliff. Happened to quite a few, I'm told.

The meanest part of the safari still lay ahead. Here it became so treacherous and steep that my uncle would make us all get out and walk alongside while

The Veres family of Brooklyn began spending summers in the Trapps hamlet in 1912 to escape the stifling heat of New York City. They pose at their Trapps home purchased in 1920. Courtesy of DSRC-MP. Donated by Veres family relative Harold Bittner.

he nursed the car over the rise. Lots of time he wouldn't have enough gas in the tank, and the infernal thing would cough, sputter, spit and die. When this happened he'd have to back part way down the mountain slope until he could find a place to turn around, then he'd come groaning back around the grade hind-end first.

Ray also fondly recalled the bachelor brothers Jonas and Eli, last names unknown:

This one character up there, Jonas, would periodically and mysteriously vanish into the deep woods for days on end. You wouldn't see his face around these parts until long after his brother Eli had completed burning off a huge charcoal pit down in the hollow below their shack. Most people said that Jonas was the laziest critter in the mountains for ducking out and letting his brother do all the dirty work like that, and they couldn't see why Eli hadn't "knuckled into him" long ago.

But the two of them got along amicably together for bachelor brothers. Every so often they'd hitch up their mare and we'd hear the clomp-clomp of the horse's hooves as she hove into sight, dragging behind her this big wagonload of charcoal. Old Eli and Jonas would be scrunched down low on the seat croaking out a tune like two granddaddy bullfrogs on a lily pad, all the while passing a bottle back and forth between them.

Hours later everyone would be rousted out of bed by the raucous cussin', shoutin', singin', and shootin' they were doing on their way back from town. Folks would spend the remainder of the night gossiping over the way those two had draped themselves over the unfortunate beast in a drunken heap. Sometimes it wasn't the beast they were draped over, either, rather some toothless hag they'd brought home for company. In any case, tongues wagged for days thereafter, and it wouldn't be too long before someone felt it his bounden duty to visit their domicile to quiet a likely case of D.T.s. [delirium tremens]

My father volunteered for this thankless mission once. Just as he stepped through their doorway, old Jonas leaped wildly to his feet and blew a hole as big as an orange through the roof of the cabin with a .30-.30 [rifle]. "Git 'em off'n me," he shouted at Dad. "Damn sneakin' crawlin' varmints."

Well, after a few days of this larking around they'd sober up, but in short order the whole cycle would be in full circle again.

The mystery of their getting along so agreeably was eventually solved the time they went "outside" to High Falls and didn't return to their home until late the following year. It seems that the "Revenoors" had gotten wind of their periodic nocturnal visits to High Falls, and so the Revs decided to be on hand this particular night as a sort of welcoming committee, you might say. They handcuffed the boys to a wagon spoke, then went topside and shoveled all of Eli's charcoal out into the road. Guess what? Underneath this blackened mound of hickory lay forty jugs of "lazy" Jonas's home-distilled mountain "corn." That old coot had been bamboozling everybody all along.

Ray Veres closed his memoirs as follows:

Sure, mister, there were the sinners.

But we had a share of saints, too.

One such latter, certainly, was this elderly duffer going under the name of Huckleberry Charlie. No one knew his real name, or where he hailed from, but nearly every summer morn he'd vanish into the wilderness silent as a ghost, not to reappear again until late in the afternoon. He'd come out of the woods lugging maybe fifty to sixty quarts of blueberries that he'd picked, slung over his back in a huge woven "rush" basket.

Evenings he ran a little general store in one of the rear rooms of his house. While the grownups were being waited upon up at the front counter, we kids would scuttle back to the mint and licorice jars for our treats. He had them neatly set out on a low shelf, and they were the only kind of sweets that he ever carried in stock. You either bought them or you did without altogether. We bought them.

Old Charlie had an equally ancient white horse that he fondly called Mister Sunday. Once a week this oatburner was locked securely in his stall while he, Charlie, went trudging off the eight miles to New Paltz for fresh supplies for his store. He'd come staggering home hours later under this great burden of groceries while the horse luxuriated in his stall munching contentedly on grain. To all horrified remarks Charlie'd say, "Oh, it's too hot for old Mister Sunday." And there you had it.

Well mister, most of them are gone now. But sometimes on a clear night I can hear the ghostly clomp-clomp of the mare's hooves striking flint on the old dirt road leading past our house. Or I can hear the wailing of Jonas as he flits about in everlasting search of his long-gone still. To all such creatures I cast a knowing eye and say, "So long and God bless you, oldtimer," as I turn back with a sigh to the awe of a space age.[4]

• 16 •

Reentering the Wilderness

November 1876. A *New York Times* correspondent makes a two-day tramp in the northern Shawangunks. He arrives late Saturday in the Trapps and boards overnight at the home of George and Sarah Davis. He takes note of the scattered farms and rustic dwellings of the Trapps. His later article juxtaposes the "barren wilderness" of the range with the "mountaintop haven" of the Smiley brothers at Lake Mohonk.

Yet, *The Times* reporter passes the Enderly sawmill Saturday at dusk and the Davis sawmill the next morning with nary a mention. Both likely are idle until Monday. That Sunday the Trapps people tend to farm chores, get ready for church down in the Wallkill or Rondout Valley or, if they can, just sleep in. It is a day of rest.

Had he passed through the Trapps at any other time, he might have returned to New York with a far different impression. On any other day men and women would have been working, their children at school or helping with chores. At any other time of year, he would have seen people engaged in farming, charcoal burning, logging, hoop pole cutting, sawmilling, stonecutting, berry picking, the building of carriage roads and mountain hotels. Families would have been laboring hard on a hard land.

By late November, especially with the early snow the reporter observed, most outdoor labor had ceased. Men were out hunting, but also moving into their shops to shave hoops and to do other indoor work. Our visitor passed through the Trapps during a quiet time, when hardly a creature was stirring.

One hundred years later, these ghostly images only play like an old-time silent movie in our minds. This vibrant place, once populated by a hardy and spirited people, is now silent, its structures for the most part vanished, its farm fields and pastures reverted back to woods, its wagon roads narrowed to paths or lost to time. Most of what was the Trapps is now buried under layers of leaves and new growth, but some leavings lay in wait, ready to whisper their stories to passersby. One day these too, though, will be gone as nature continues to reclaim the hamlet's final flotsam and jetsam.

Yet, if you find yourself up on the mountain standing next to a roaring Peters Kill or gazing at a serene pool on the Coxing Kill, close your eyes and imagine the Trapps people at work long ago. Imagine the sawmills on these streams, those structures of bygone industry roofed over, their inner workings grinding away once more.

Picture the steady stream of traffic on horseback, by wagon, or on foot that once traveled here. Or Trapps children walking to school. Families heading to Sunday services.

Folks going to buy dry goods at the Billingham store or to sell berries to Abe Billingham, the King of the Trapps.

Visit the Van Leuven Cabin or the Enderly Mill Site on Mohonk Preserve lands. Visualize the stonecutter cutting a work shelf in the huge boulder by the cabin doorway or cutting stone at a quarry near his home. The Enderly men setting up a log to be cut

Bob Larsen is dwarfed by the massive stone foundation of the George Davis sawmill, destroyed by fire in 1874. It was one of two sawmills that once operated in the Trapps hamlet. Photographs by Robi Josephson.

by that old up-and-down, slower-than-molasses saw. Their wives keeping house for their families, their children lugging water and tending to other chores.

While out on the Smiley carriage roads, consider how they were built by able hands. Some were so well constructed that sections have lasted in their original state for more than a century. Other grand structures, also built with the strength of Trapps men, stay true. While the mountain hotels at Lake Minnewaska are now gone, Mohonk Mountain House, a *grande dame* of the Victoria era, will celebrate its 150th anniversary in 2019. The New York City aqueducts continue to deliver daily millions of gallons of fresh water to millions of city residents and other communities along their underground waterways. U.S. Routes 44/55 and 52 remain sturdy and graceful escorts for those driving over the mountain.

Opposite page: The Enderly sawmill once operated at this waterfalls on the Coxing Kill. Above: Nothing remains of the mill today except for remnants of the milldam above the falls. Today, the area is a popular spot called Split Rock and is an easy walk from the Mohonk Preserve's Coxing Trailhead. Photos by Ted Reiss.

The people of the Trapps would have been proud of these statements. If they were here today, they would be prouder still that more than a century later many enjoy and marvel at the fruits of their hard labor.

The human history of the Trapps made it "singular" after all, as Philip H. Smith knew it was. The essence of the Trapps was its pioneer spirit, and so it remained until shortly after World War II. It's a story about taking chances, of making a new life for one's family atop a rugged mountain range of New York State called the Shawangunks. Most of all it's a story of hard-working mountain families who scraped by as best they could but whose vigor transformed an unforgiving land into farms, into homes and, most of all, into a community.

We have peered into the past, as if staring at faded or torn photographs of loved ones, and tried to bring them back to life again. In some ways we have succeeded, but local history is often a hit-or-miss business. Stories ever evolve. What we have done here

will be added onto and revised in future. We have only two regrets. Borrowing from one historian, we wish our history could have been longer. As well, William Doughty wrote, "Much of the epic story of the founding and development of Minnewaska will never be told." How true of the Trapps mountain hamlet.

Even so, the Trapps people endure in many ways. One admirer perhaps said it best, "We are still living with them today."[1]

Appendix A:

Time Line of the Trapps Hamlet

1700–1780 Queen Anne conveys Rochester Patent (1703). Town of Rochester trustees carve Grote Transport (1730) and Nineteen Partners Tract (1770) from this patent.

1780–1800 First pioneers, including Teunis Roosa and Peter Aldrich, homestead in the Trapps. The Grote Transport and Nineteen Partners Tract partitioned. The Tract's partition map of 1799 shows the public road through the Clove and Trapps.

1800–1830 Peter Aldrich operates first tavern (believed) in the Trapps. Johannes Enderly operates sawmill on the Coxing Kill on land once owned by Teunis Roosa. Delaware and Hudson Canal opens, providing work to Trapps men and allowing expansion of mountain industries such as stone quarrying and barrel hoop shaving. Burger, Gorsline, and Van Leuven families settle in the Trapps.

1830–1850 Growth of Trapps mountain hamlet continues. Early excursions to Lake Mohonk and environs by landscape artists of the Hudson River School, adventurers, and sportsmen.

1850–1870 One-room Trapps school opens. Davis sawmill operates on Peters Kill. New Paltz–Wawarsing Turnpike built through the Trapps over the mountain. Ben Fowler buys Aldrich tavern and adds small hotel. Trapps men fight in the Civil War.

1870–1890 Smiley families establish hotels at Lake Mohonk and Lake Minnewaska, bringing steadier work to Trapps people. Union Chapel constructed in the Trapps.

1890–1900 Eli Van Leuven buys plank house likely built by William Hagen ca. 1889.

1900–1930 Decline of Trapps population begins and accelerates. Eli Van Leuven and second wife Anna Ostrom Van Leuven sell plank house to the Smileys of Mohonk and leave the hamlet. First paved highway, U.S. Route 44/55, constructed through the Trapps over the mountain.

1930–1960 The Smileys of Minnewaska sell out to the Phillips family. Irv Van Leuven dies in 1956, the last full-time Trapps resident to live the old ways.

1960–1990 The Smileys of Mohonk Mountain House establish The Mohonk Trust, today the Mohonk Preserve. New York State creates Minnewaska State Park Preserve. The two preserves protect much land of the former Trapps hamlet.

1990–Present The Mohonk Preserve restores historically significant sites in the Trapps, including a plank house renamed the Eli Van Leuven Cabin. Through the preserve's efforts, the Trapps Mountain Hamlet is placed on National and State Registers of Historic Places. The all-but-vanished hamlet is the first subsistence community of New York State to receive this designation.

Appendix B:

Selected Family Trees

The Aldrich Family
(Abbreviated)

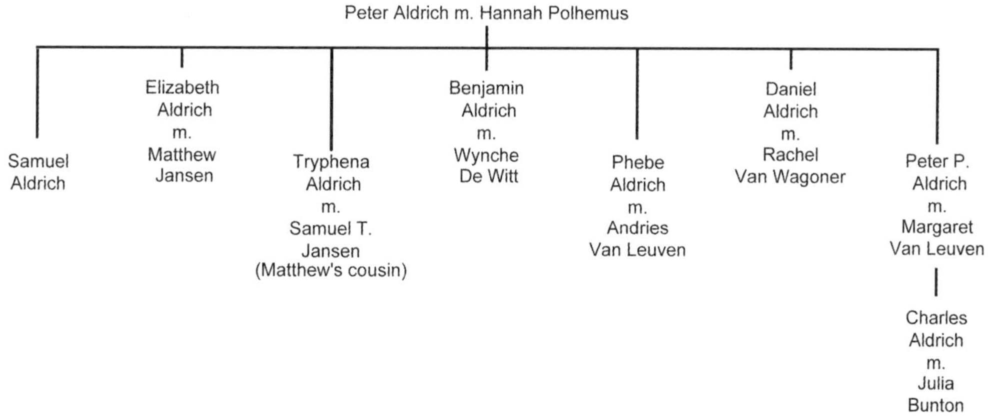

The Burger Family
(Abbreviated)

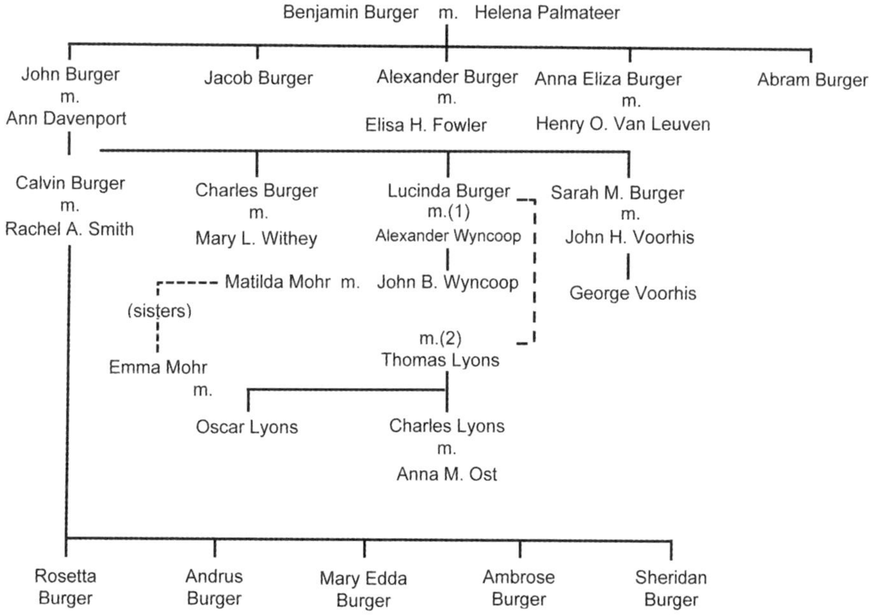

The Enderly Family (Abbreviated)

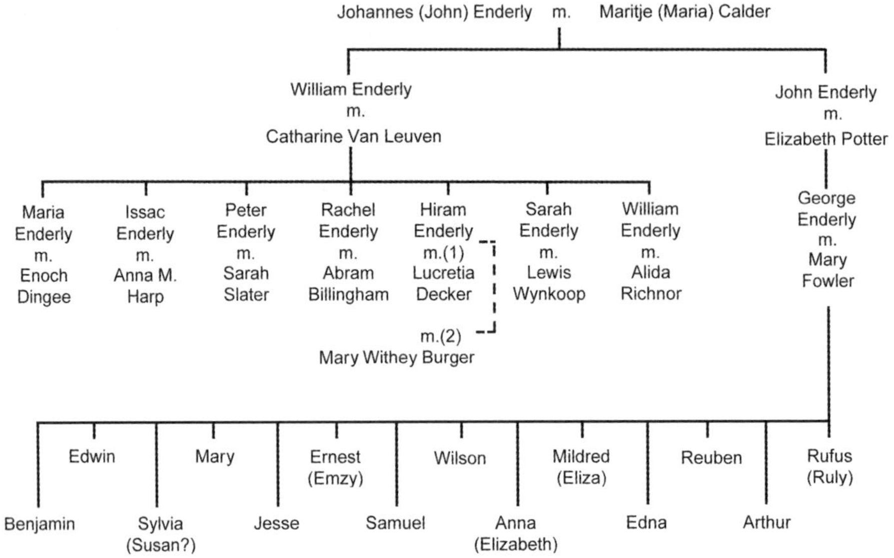

The Harp Family (Abbreviated)

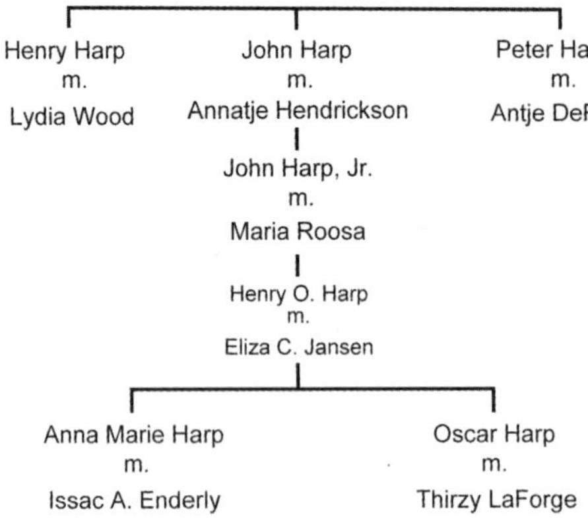

The Otis Family (Abbreviated)

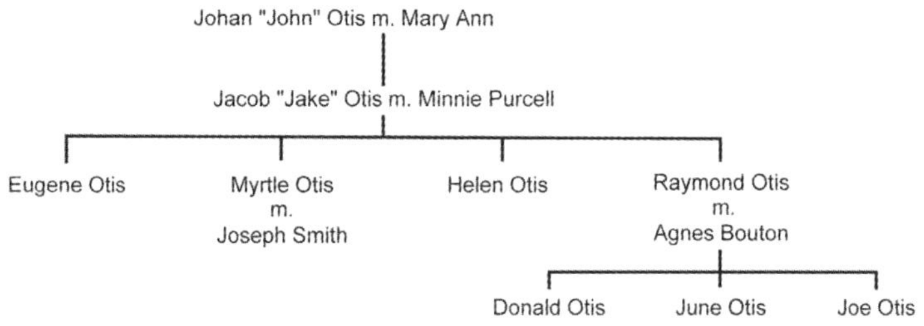

The Van Leuven Family (Abbreviated)

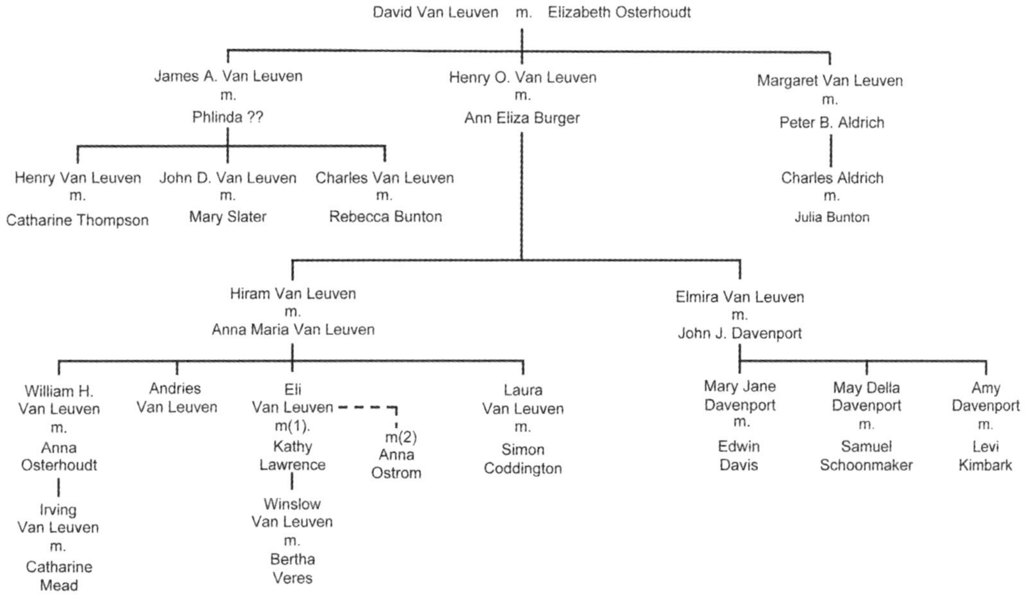

The Wynkoop Family (Abbreviated)

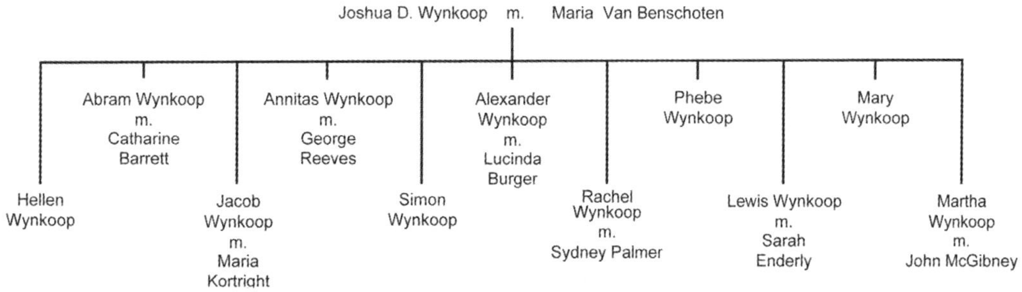

Appendix C:

List and Index of Trapps Family Names

More than 100 family names have appeared in records for the Trapps mountain hamlet from circa 1790 to the mid-1900s. Some lived on the mountain for several generations while others lived there for a short time. Extensive records exist for some, but only a single record for others. The list of names below, with variant spellings, has been compiled from numerous sources. If you have evidence of a name not on this list, contact us at rjhudson@hvc.rr.com. Pages numbers indicate notation in the text.

Appendix D:

Visting the Trapps Today

Heading west from the Wallkill Valley, U.S. Route 44/55 climbs quickly after it intersects with N.Y.S. Route 299 from New Paltz. Route 44/55 passes the Mohonk Preserve Visitor Center on the right, loops into a hair-pin turn, and then crests through a gap in the cliffs.

Drive through the gap and under the steel bridge to the preserve's West Trapps Trailhead parking lot on the right. (Day fees or annual passes available.) From here, visitors can take a short walk back to the steel bridge and follow unpaved carriage roads or hiking paths to see spectacular views of the surrounding valleys.

Starting at the West Trapps Trailhead, visitors can also take a 1 ½-mile, moderate walk, not handicapped-accessible, to the Van Leuven Cabin, a plank house once lived in by a Trapps family and restored by the preserve. Pick up a brochure titled "Take a walk back in time ... along the Trapps Mountain Hamlet Path" at the visitor center or trailhead. This guided path leads to historic sites, including an abandoned stone quarry and the Fowler Burying Ground. The last stop is the Van Leuven Cabin, which is open by appointment or during scheduled tours.

After the West Trapps Trailhead, continue west on Route 44/55 to the first right, Trapps Road. Take the next right fork onto Clove Road to the preserve's Coxing Trailhead parking lot at Split Rock. Explore the former homestead, farm, and sawmill site of the Enderly family. A few steps west of and behind the parking lot is the Enderly Burying Ground.

Signs and maps at the Van Leuven Cabin and Coxing Trailhead interpret these sites for visitors.

Return to Route 44/55 and continue west to the entrance of Minnewaska State Park Preserve on the left. (Day fees or annual passes available.) At the lake explore the area where the two Lake Minnewaska hotels once stood and the Trapps people once worked. A modern stone house, nestled in the cliff across the lake from the parking lot, was built by the Phillips family in the later twentieth century and will be reopened as a visitor center.

Enjoy your walks back in time, and along the way please respect private property.

Notes

Abbreviations:

(ABS) Alfred B. Smiley Interviews and Correspondence
(AMC) Alligerville Methodist Church
(CB) *Commemorative Biography*
(CFI) Coddington Family Interviews
(CR) *Census Records*
(DF) *Daily Freeman*
(DSRC-MP) Daniel Smiley Research Center, Mohonk Preserve, New Paltz, N.Y.
(E&E) Ebert and Ebert
(EJ) *The Ellenville Journal*
(GW) *The Gardiner Weekly*
(HBI) Harold Bittner Interviews
(HH) *Huguenot Herald*
(HHHC-EML) Haviland-Heidgerd Historical Collection, Elting Memorial Library, New Paltz, N.Y.
(J&J) Jandebeur and Jandebeur
(KDF) *The Kingston Daily Freeman*
(KDJ) *The Kingston Democratic Journal*
(LAT) *Los Angeles Times*
(MB) *The Mohonk Bulletin*
(MLCWS) Mohonk Lake Cooperative Weather Station
(MMHA) Mohonk Mountain House Archives, New Paltz, N.Y.
(N&M) Needham and Mussey
(NPI) *The New Paltz Independent*
(NPI&T) *New Paltz Independent and Times*
(NPN) *The New Paltz News*
(NPT) *The New Paltz Times*
(NYT) *The New York Times*
(OFI) Otis Family Interviews
(P&T) Poucher and Terwilliger
(PDE) *The Poughkeepsie Daily Eagle*
(PJ) *Poughkeepsie Journal*
(PSNY) *Poughkeepsie Sunday New Yorker*
(RDC) Reformed Dutch Churches
(R-MP) *Ridgelines,* Mohonk Preserve newsletter
(S-ACI) Smiley "Art Coddington Interviews"
(SF) *Sunday Freeman*
(SUT) *The Southern Ulster Times*
(SW) *Shawangunk Watch*
(TA) *The Accordian*
(THR) *Times Herald-Record*
(TSR) Trapps School Records
(UCCO) Ulster County Clerk's Office
(UCSO) Ulster County Surrogate's Office
(VLFI) Van Leuven Family Interviews
(W-E) (W-F) (W-G) Wustrau-Enderly, Fowler, Gorsline family histories
(WVW) *The Wallkill Valley World*

Note: The citation "#?" indicates a page number missing from a source.

Preface: A Beginning

1. Arthur Friel was an early-twentieth-century adventurer and novelist; his two novels set in the Trapps are discussed in Chapter Twelve. Friel 1923:5.

2. For in-depth discussions of Trapps and other place-names such as Mohonk, Minnewaska, and Shawangunk, see Marc B. Fried's *Shawangunk Place Names* (2005).

Don and Joe Otis, brothers and former residents of the Trapps hamlet, said the gap also was called the Trapps when they grew up there during the 1930s and 1940s. OFI.

3. Child:86; Friel 1923:35.

4. Snyder:7–10; Fagan:1–3.

Birth of a Hamlet, 1730–1830

Chapter 1: Entering the Wilderness

1. A fictional vignette, in italics, opens this and all subsequent chapters. Each is based on real events, some more loosely than others, and some with imaginary dialogue.

2. The original line read, "the least wealthy of the native population." The term "native" then referred to the local inhabitants, not Native Americans.

3. "Mohunk" was an earlier spelling for Mohonk, and "Mohunk Point" an earlier name for Sky Top. "Trapps" with two 'p's became the modern spelling.

4. The headline phrase "End to End" is inaccurate; *The New York Times* reporter walked less than half the range's length. NYT, 24 Dec 1876:4.

Chapter 2: Colonial Land Grant

1. Shawangunk topography determined settlement of the Trapps. Charles E. Stickney cited similar growth restrictions in the range's Minisink Region, today Port Jervis and vicinity: "While the population of Orange county was thus rapidly increasing, and new villages springing up almost weekly within its limits, the history of the Minisink region appears to have been shrouded in darkness. The Shawangunk mountains formed a barrier which for a long time, kept from the outside world a view of the tide of civilization and improvement that was going on, and shut from the page of history much knowledge of the early settlers of that locality." The Trapps, like Minisink, remained isolated well into the 1900s.

The Rochester Patent, like others granted during that period, was immense; a clue to its size is found in Gordon's 1836 Gazetteer, which noted that the Town of Rochester then spanned 37,505 acres, or about 58 square miles, 10 times larger than today's City of Boston. Many land purchases predated the 1703 Rochester Patent, with trade deals a common method of acquiring land from American Indian tribes. The Trapps lay entirely in the Town of Rochester until 1853, when the Town of Gardiner was formed from parts of the towns of Rochester, Shawangunk, and New Paltz; much of the Trapps then became part of Gardiner, with smaller portions in Rochester.

Le Fevre:2; Foote (Rochester):343–44; Gordon:744; Mather and Brockett:181; Stickney:23; P&T:xi.

2. Traps was the standard spelling during much of the 1800s. Two p's occur as early as the 1870s, but only became permanent after about 1930.

The "Foot Path to Shawangunk," noted on eighteenth-century documents such as the Nineteen Partners Tract, referred to the Wallkill Valley town, not the mountain range. The path has acted as a boundary marker into modern times. A June 1928 deed to a mountain parcel noted "the old Shawangunk footpath," as well as "Lot No. 1 of the Grote Transport" and "King's Lane." The last, a wagon road, was another colonial route over the mountain. The footpath and wagon road crossed each other in the Trapps.

Berendsen website; NPI, 17 Apr 1879:3; NPT, 16 Nov 1892:3; UCCO, Liber 532/Deeds:544 (Jun 1928); UCCO, Archives Division, Grote Transport, Liber M:44 (map) and 52 (deed) and Nineteen Partners Tract from 1799 Partition Records, Liber M:274 (deed) and 300–301 (map).

3. "Kloof" in modern Dutch means "a ravine." The place-name Coxing originally referred to "a region of land" in Marbletown, but was transferred to the stream and a mountain pond later renamed Lake Minnewas-

ka, explained Fried. The word has not yet definitively been defined. Variants include "coxen" and "coxsinck."

The names of individual parcel owners in the upper Clove and Trapps, with the dates of acquisition, appear on the 1799 partition map of the Nineteen Partners Tract.

Fried 2005:23–28, 67–69; UCCO, Libers CC/Deeds:130 (Apr 1720), FF/Deeds:17 (Jul 1750); and Smiley and Cross:16–18.

Chapter 3: A Pioneer Settlement

1. The Burger vignette was based on a March 1906 newspaper interview with their daughter, Sarah Voorhis, then aged sixty-one.

Smith:159–61; Bevier:95; NPI, 8 Sep 1905:2, 2 Mar 1906:3, 17 Sep 1912:#?; NPI&T, 30 Sep 1920:1.

2. After 1785 the Grote Transport and Nineteen Partners Tract underwent several partitions. The Grote Transport was divided in three among descendants of the original patentees, then subdivided for a second and final time into twenty-four long, narrow parcels spanning east to west across the Shawangunks. The Nineteen Partners Tract, partitioned in December 1799, was divided among the nineteen partners according to estimated productivity. The 1799 map and an 1875 copy show the subdivisions. A parcel with poor land too rocky to farm received more acres than a parcel with good, arable soil.

3. Wessels:56; van Wagenen, Jr.:27–35.

4. The place-name Anderieses, with variant spellings, appears in early records, including the deed for the 1720 Beatty parcel, which extended from near "Mogonick [Sky Top] ... Runing South ... to a great white oak tree marked with three notches & a Cross on the southeast side of Andriesonkill or Cocksinck kill."

Anjou II:29; UCSO, Liber A/Wills:71; Hoes:610; William Heidgerd, NYS Biographical Dictionary, Harp section; Fried 2005:23–28; UCCO, Liber CC/Deeds:130 (Apr 1720).

5. Teunis Roosa and first wife, Maria, sold the mill site to the Enderlys; the Roosas' later whereabouts are unknown. The deed describes the lot "on both sides of the Coxsink Kill, beginning by a white oak marked, with three notches and cross, above them standing on the East Side of the foot road leading from Rochester to Shawangunk."

The Enderly property totaled 200 acres in the early 1870s.

Anjou I:74–75; CB:944; Banta:13–14; Ruth Heidgerd 1977:86, 203–05; W-G:1–5, W-E 8–18; Town of Rochester Archival Records; Marbletown RDC 1985:9, 81, 112; Rochester RDC:7, 15, 41, 61–83, 232; Guilford RDC, Marriages 1834–1923:29; Clove RDC:2–15; Luke and Donohue Bk I:3, 87, 200, 224–25, Bk III:837; Hoes:276, 309, 578; CR, Town of Rochester, Marbletown (1790–1840); UCCO, Libers 60/Deeds:503 (Jul 1843, noting 1801, 1823, 1839 land acquisitions and transfers of the Enderly family), 88/Deeds:188 (Nov 1852); Child:161; Town of Rochester historical records.

6. CR, Town of Rochester (1800); Ellis:115–16; Sylvester I:214; WVW, 19 Dec 1924:12.

7. In February 1905 a house owned by Peter Aldrich was dated ca. 1805. It is unknown if Peter built this house or if it was his home or tavern. A 1905 deed noted the Yankee Lot of Peter Aldrich. A record of the woodlot Philip DuBois Bevier sold to him in 1801 has been found only on an 1851 deed for a lot sold to Henry O. Van Leuven.

CR, Town of Rochester (1800, 1850–1855), Shawangunk (1800, 1850), Gardiner (1800, 1810, 1855–1865), Marbletown (1880), New Marlborough (1790), Rochester (1850–1870), Newburgh (1790); NPI, 24 Feb 1905:3; UCCO, Libers 78/Deeds:211 (Feb 1851), 385/Deeds:559 (Jan 1905); Marbletown RDC 1985:112; P&T:201; Peck:174–78; Town of Rochester Archival Records.

8. Calvin Burger's interview, published in *The New Paltz Independent* based on font, is dated ca. 1901–04, its page unknown.

The Van Leuven family first arrived in the Hudson Valley during the 1660s, then moved up into the Rondout Valley, arriving in the Trapps after 1800. Henry Van Leuven's middle name (and nickname) Ocea was pronounced "Osha." It likely meant "sea" in Greek, perhaps reflecting the popularity of Greek culture in nineteenth-century America.

CR, Town of Rochester (1790–1850), Marbletown (1790–1830), Wawarsing (1810), Gardiner (1855, 1860); Ruth Heidgerd 1977:27; Clove RDC:6; Marbletown RDC 1985:34, 46, 95–113, 208; Rochester RDC:23, 84–86, 194; Hoes:32, 282, 449; Smith:159; Sylvester I:214; P&T:xi–xii, 90; NPI, 11 Sep 1896:2, ca. 1901–04:#?, 25 Sep 1896:3, 2 Mar 1906:3, 17 Sep 1912:#?; UCCO, Liber 36/Deeds:474 (Dec 1830); Hasbrouck 1978:13–18; Banta:12–15, 23; Thelma Atkins, letter; Ulster County Historical Society website; Brink I:107–10, III:262–265.

9. Sylvester II:198; Clove RDC:1; KDF, 16 Feb 1940:14, 14 Jan 1967:3.

10. New York's Common School Act of 1812 divided the state into more than 11,000 school districts, noted van Wagenen, Jr. Corbett stated, "School districts [were] limited by law to a maximum of three miles in diameter, placing a school within walking distance (1 1/2 miles) of every school age child."

Hough:120–21; French 135–39; *Central Rondout Valley Schools Newsletter*, Nov 1975, 1(2):1–3; Foote (Rochester):345–352; *Rock Hill Bulletin*:4; Whittaker:7; van Wagenen, Jr.:13; Corbett website; James Folts history paper, University of the State of New York website; CR, Town of Marbletown, Rochester (1790–1830).

11. For War of 1812 military claims, the higher the number, the later the claim. Tryphena Aldrich, widow of Samuel T. Jansen (#14,648) of Rochester, received $55. Samuel Fowler (#15,702) of Gardiner received $50. As the Town of Gardiner was not formed until 1853, the Fowler claim likely was not made until that year or later. David Van Leuven (#11,624) of Ulster County received $58.

Sylvester II:202–03, 226; Foote (Gardiner):256; Adams:94–96; A Census of Pensioners:103; White 1989:1075, 1877, 1991 II:1278–79; Young, United States Genealogy Network website.

12. Construction of the D&H Canal brought an unexpected yet critical advancement impacting the Trapps. During construction between High Falls and Rosendale, natural limestone cement was discovered. Rosendale cement was poured into some of America's most important building projects. Cement and many other products were shipped in barrels, their staves secured with barrel hoops. Hoop shaving thus became a major cash industry for Trapps men during the 1800s. See chapter nine, 109–112.

Wakefield 1–31, 79–92, 143, 171–72, 204; Andrist:10–75, 136–38; Whitford, Chapters I, XX, Town of Rochester website; Ries:681; Foote (Rosendale):358–59; Hough:660.

13. James Gorsline settled on the Simon J. Van Wagonen tract (1751) along the Coxing Kill, which James had acquired from Simon or his descendants. James's granddaughter Catharine, or Laney, married Ben Fowler, son of Sam.

Before his death in 1905, Calvin Burger said that his grandfather "Benjamin Burger built his house a short distance east of Coxing Pond, now called Minnewaska Lake." This location is the same area where Ben's son, John, bought the Bevier parcel. In 1896, John's brother Alex died at his home a few feet from the one he was born in.

A 1709 deed records a mill operated downstream on the "Peter's Kill" by Gysbert Van Gorden and his brother-in-law, noted Fried. It is believed the stream was named for Gysbert's son, Peter, Fried explained.

Census records show the existence of slavery in the neighboring Clove during this period of early settlement, but not in the Trapps.

UCCO, Libers 34/Deeds:497 (Jul 1828), 36/Deeds:474 (Apr 1829), 60/Deeds:503 (Jul 1843, noting Jun 1823 Johannes Enderly acquisition of about 111 acres from unknown party); P&T:xi, 28, 201–02; Rochester RDC:67; NPI, 11 Sep 1896:2, ca. 1901–04:#?; CR, Town of Rochester (1850), Gardiner (1855); E&E:42–43; Fried 2005:78–80.

Growth, 1830–1900

Chapter 4: Hearth and Home

1. To add color to this vignette, it was supposed that Thomas Lyons immigrated to America as an adult and therefore spoke with an Irish brogue.

The Trapps once referred to the gap in the cliffs, the mountain hollow, or the community therein; Trapps Valley referred to the mountain hollow. Trapp's (Trap's) Mountain referred to a section of cliffs facing the Wallkill Valley just south of the gap. Today these cliffs are called the Near Trapps, with the cliffs north of the gap called The Trapps.

CR, Town of Gardiner (1870); NPI, 10 Apr 1879:1; William J. Upright, PSNY, 10 Oct 1943:7A.

2. Ireland's Great Famine of 1845 escalated an exodus to North America begun decades earlier. Between 1840 and 1860, the time period during which Thomas Lyons arrived, about 1.7 million Irish came to America. Thom might have first gravitated from New York City to Irish enclaves in the Hudson Valley. McDermott compared their arrival to "a tidal wave … [that] changed the character of Poughkeepsie. Suddenly one of every five individuals was an Irish immigrant or a child of an immigrant Irish family." Poughkeepsie and other "small industrial" cities attracted Irish immigrants, where they found work in a variety of local trades, noted Miller. In Marbletown in 1850, Irish men, many single boarders, worked as laborers, carpenters

and wagon makers, while others worked on the D&H Canal as lock tenders, night watchmen, and overseers. Some Gardiner residents were listed on the 1855 census as two-year residents. This referred to their residence not in the area, but in the Town of Gardiner, formed in 1853.

Griffin:1–12, 39, citing *Historical Statistics from Colonial Times to 1957* (Washington, D.C.:U.S. Government Printing Office, 1957:15–16); Glazier I:vii–xix, II:vii–ix; McDermott:171; Miller:315–322; CR, Town of Rochester (1850), Marbletown (1850, 1855), Gardiner (1855–1865); Child:163; E&E:30.

3. Lucinda's headstone describes her as the wife of Thomas Lyons, but no record exists of their marriage. She and Alexander Wynkoop of the Trapps wed in 1859, but by 1865 had separated. In 1870 their son John B. lived elsewhere and so was not included in the vignette. By 1880, John, nineteen, lived with his mother, stepfather, and half-brothers, Oscar, thirteen, and Charles, four months.

CR, Town of Rochester (1850, 1880), Gardiner (1855, 1860, 1870, 1900); Rochester RDC:146, 208; AMC:7, 31; E&E:35.

4. Some state census years for local towns are unavailable. Fire destroyed most of the 1890 federal census, with only fragments remaining.

No conclusion can be drawn for percentages of log, plank, and frame homes in the Trapps for any given census year. Types of construction were noted only for some years and only for some homes. In general roofing terms, shakes are split, shingles sawn.

The American chestnut was predominant in the eastern United States until blight struck in the early 1900s. This tree was resistant to rot, making it a favored wood for log cabin walls, fence rails and posts, railroad ties, or structures in contact with the ground.

CR, Town of Gardiner (1855–1875), Rochester (1850); Hasbrouck, 1978:16; Smith:123; van Wagenen, Jr.:52 (term "rolled up"); NPI, 5 Jun 1878:3; E&E:42; Guilford RDC, Marriages 1834–1923:29; P&T:202.

5. Lester Walker illustrated the construction of a basic two-story house somewhat similar to "plank-and-beam" houses found in the Trapps. Walker's example had corner posts but no ridge pole. Plank houses on the mountain typically had one room about twelve by eighteen feet, ceilings about six feet in height, and simple gable roofs. An original first floor wall of that height can be seen today at the Eli Van Leuven Cabin, restored by the Mohonk Preserve. Rooms were added onto the basic plank house as families grew. Plank-and-beam construction was also used in two-story homes. One rare example was the two-story plank home of Lewis Wynkoop, built of planks cut long enough to reach the roof line of the second floor. The two-story section was twenty-four by sixteen feet; a smaller one-story kitchen wing, also plank, was perhaps built first.

Marbletown RDC 1985:222; Larsen:13, 17; Walker:70–71; CR, Town of Gardiner (1865); Hayes:4; W-E:17–18, 39–43; NPI, 5 Sep, 1878:4; Phillips and Weintraub:19–41(Fein), 71(Phillips), 77–97(Rhoads), and 112; Fried, Fall 2006:46–47.

6. Junior League 1969:5(25); CR, Town of Gardiner (1855, 1865).

7. NYT, 24 Dec 1876:4; Pike:472.

8. Wash day followed Sunday, the day of rest, for it required, as Lynn Siprelle noted, "a great deal of strength and fortitude to hand-wring clothes and carry big baskets of laundry to the clothesline." Wooden neck yokes for humans were manufactured to fit adults and children in "sizes from one to six, like shirts," Sloane wrote.

NPT 17 Sep 1874:3; N&M:131–33; CFI; OFI; Sears Archives website; Siprelle website; Sloane 1974 (Diary):66, 92.

9. Bray (DeWitt); Hayes:3; *The Saturday Evening Post*, 29 Nov 1924:8 (Friel); William Shakespeare, *As You Like It*, Act II:i:16.

Chapter 5: Artists and Builders

1. "Correspondence ...," 3 Oct 1875.

2. During the twentieth century, climbers and hikers popularized the nickname "Gunks" for the Shawangunks, but Cole's letter to Durand indicates the nickname dates back much earlier. Cole says nothing further about "the Gunks" in his letter.

Smith:159; Vesell (Noble):276–77; Thomas Cole Papers, 1801–1848.

3. P&T:28; Marbletown RDC 1985:112; CR, Town of Marbletown (1850, 1855), Rochester (1860).

4. The New Paltz–Wawarsing Turnpike was so called long after it became public. A July 1921 deed noted the "New Turnpike Road." Housed at Ulster County Archives, the turnpike survey map retains its vibrant colors, notably the bright blue of water, more than 150 years after William Jones drafted the ten plates,

each eighteen by twenty-four inches long.

Marian Whitford said her mother, Evelyn Coddington DeWitt, pronounced Anna Maria as Anna "Mariah."

Le Fevre:197–98; "The New-Paltz and Liberty turnpike company," Apr 1834 Act, HHHC-EML; CFI; Harp:25–26; NPT, 27 Jul 1860:2; Wood:1–23; P&T:27; CR, Town of Gardiner (1850–1860); Hasbrouck 1978:15–16; NPI, 17 Sep 1912:#?; NPI&T, 26 Mar 1925:1; UCCO, Liber 483/Deeds:443 (Jul 1921); UCCO, Archives Division, New Paltz–Wawarsing Turnpike map; KDJ 8 Aug 1860:2.

5. Ruttenber and Clark:111–16; Sylvester I:152, II:169, 174, 289; Hakam and Houghtaling:11; Le Fevre:212–16.

6. Le Fevre:195, 212–16; Harp:16–17.

7. The term "internal revenue" was coined in 1862 when Congress instituted the first federal tax. The federal government previously had raised money through tariffs, but the Civil War proved costly. Cynthia G. Fox explained: "Taxes were levied on residents of all states and territories not in rebellion. ... States that seceded were included in the tax base as soon as Union troops established control. ... Hotels, inns, and taverns were classified according to the annual rent or estimated rent, from a first-class establishment with a yearly rental of $1,000 to an eighth-class hotel with a yearly rental of less than $100, and charged license fees of from $200 to $5 accordingly."

From 1862–64, John Shaffer paid $5 tax as an eighth-class hotel keeper and $20 tax for selling liquor. Taxes rose to support the rising costs of war. By 1865, John paid $35 in taxes; his hotel tax doubled to $10 and liquor tax rose 25 percent to $25.

Gardiner residents could pay taxes at Ben Fowler's store in the Trapps, among ten other sites. Two news briefs of 1871 and 1881 are the only known references to the store's ownership. Philip H. Smith noted the store, not its owner, in his book.

Isaac Dingee owned the first hotel in the Trapps, Hasbrouck noted. Maps of the 1850s show Isaac's tavern at the base of the cliffs near the present-day intersection of Routes 44/55 and 299. The 1858 French map shows Mrs. Jane Hasbrouck, widow of Joseph, Jr., at the site. Smaller hotels or boardinghouses were operated during the 1800s by several Trapps residents, including Henry O. Harp, Samuel M. Decker, Calvin Burger, and the George Davis family.

P&T:xi, 28; AMC:26; E&E:38, 68; CR, Town of Gardiner (1855, 1865), Rochester (1850–1900); Child:231; UCCO, Liber 110/Deeds:302 (Dec 1859), Liber 135/Deeds:385 (Apr 1861); Hasbrouck 1978:17; NPT, 22 Feb 1861:2, 11 Apr 1862:2, 23 May 1862:4, 8; Jul 1864:3, 15 Jul 1880:3; SUT, 26 Jan 1866:2, 2 Feb 1866:2; Internal Revenue Assessment for Ulster County, Town of Gardiner, 1862–1866; IRS website; Cynthia G. Fox, National Archives website; NPI, 21 Dec 1871:2–3, 12 Aug 1881:3; Smith:159.

8. On their 1853 Ulster County map, Tillson and Brink noted Lake Mohonk as the "Pond," Lake Minnewaska was unnamed, and Lake Awosting was called "Big Pond."

NYT, 14 Aug 1882:5; CR, Town of New Paltz (1840), Marbletown (1850); CB:957–58, 1066–67; P&T:90; Beers Atlas:134; Avery and Kelly:10, 128–30, 164–66, 258–59; KDJ, 8 and 15 Aug 1860:2 (EJ article); Minks:17–19; Burgess 1996:11–17, 1969:3–18; Le Fevre:195, 216; Daniel Smiley 1989:15; Buff:3–5, 7; Thyssen:182–83; Guide Book to Mohonk Lake, 1883:5.

NPI: 8 Jul 1869:2, 29 Jul 1869:1, 26 Aug 1869:2, 30 Sept 1869:2, 11 Nov 1869:2, 4 Sep 1873:3, 8 Aug 1884:2, 17 Sep 1912:#?.

NPT: 6 Jul 1860:2, 3 Aug 1860:2, 10 Aug 1860:1, 28 Sep 1860:2, 28 Jun–2 Aug 1861:3, 6 Sep–4 Oct 1861:3, 6 and 13 Dec 1861:2, 3 Jun 1869:2, 1 Jul 1869:3, 2 Sep 1869:3, 23 Sep 1969:3, 7 Oct 1869:3, 16 Mar 1871:2, 4 Sep 1873:3, 7 Sep 1892:3.

9. Burgess 1996:11–20; LaChance Jul 2006:1–4; Partington:15–42; NPI, 4 Aug 1872:3, 21 Oct 1869:2, 4 Sep 1873:3; NPT, 4 Aug 1870:2, 4 Sep 1873:3; P&T:90.

10. Before rail service reached New Paltz, guests journeyed to Lake Mohonk via Hudson River steamboat, the New York & Hudson River Railroad to Poughkeepsie (with ferry across the Hudson), or the Delaware & Hudson Canal. A stagecoach ride to Mohonk followed. Owned by several companies during its history, the Wallkill Valley Railroad reached its terminus at Kingston in 1872. The West Shore Railroad along the Hudson River joined New York with Kingston in 1883; the West Shore was part of the rail system that connected New York and Buffalo later that year. The Ellenville & Kingston Railroad, a New York Ontario & Western line through the Rondout Valley, reached Kingston in 1901 after the canal went out of business. These railroads improved travel to Mohonk and Minnewaska, providing transport for guests from faraway places.

The summer boarding trade blossomed up and down the Shawangunks during this period. One such new destination south of the Trapps was Evansville, which later became one of the earliest art colonies in New York State, known as Cragsmoor.

Arthur G. Adams 1999:101–112; CB:327; Beers Atlas:94, 134; Child:86–87, 133, 141–2, 329; Hakam and Houtaling:8–15; Phillips and Weintraub:50 (Hoopes), 77, 151 (Rhoads); Terwilliger:92–97, 316; Snyder:27; Burgess 1996:67; briefs on John C. Shaffer, NPT and NPI, 1871–1874; CR, Town of New Paltz (1870, 1880); NPT, 22 Dec 1870:4, 26 Dec 1872:4; Sanderson:131–32.

11. A carriage road joining the two lakes was completed in 1879 as part of the opening of the first mountain hotel at Lake Minnewaska. Deeds at the Ulster County Clerk's Office (Liber 216:301–10) describe rights-of-way the Smiley family purchased for $1 each from ten landowners, eight of whom lived in the Trapps and upper Clove.

Knickerbocker wrote that Alfred Smiley bought about 2,500 acres for $4,500, but figures total $2,850 from deeds dated September 30 to October 22, 1875. Alfred's son, Edward, recalled that his Uncle Albert first discovered Lake Minnewaska.

The McGibney Hotel burned down during the 1880s. In March 1888, John McGibney sold out to Cornelius Benjamin "Benny" Walker of Kerhonkson, who rebuilt an inn and tavern called the Laurel Inn on the same site. John later moved to New Paltz. In 1879, the year Alfred Smiley opened Lake Minnewaska, John McElhone, a bank president in Ellenville, proposed a mountain hotel at Long Pond or Lake Awosting, south of Minnewaska. The next year, *The Kingston Freeman* stated, "the more houses [hotels] there are erected on this mountain the more popular it will be, as those who go to one will naturally draw customers to the other." John never realized his dream, and Alfred eventually acquired Awosting as well.

The 1880 *Kingston Freeman* article, date unknown, was reprinted on August 6, 1880 in an Ellenville newspaper (unknown).

NPI, 24 Sep 1874:2, 14 Oct 1875:3, 11 Nov 1875:3, 25 Nov 1875:3, 4 Sep 1879:2, 16 Oct 1879:3, 20 Jan 1881:3, 12 Aug 1881:3, 22 Nov 1889:3, 22 Aug 1902:#?, ca. 1901–04,#?, 17 Sep 1912:#?; NPT, 1 May 1879:2, 15 May 1879:3; Burgess 1969:3–27, 1996:19–31, 119–20; LaChance Jul 2006:1–4; Alfred B. Smiley, interview with Paul C. Huth and John E. Thompson, 26 Aug 1997; Lewit and Phillips, Jr. website; Child:161, 164; AMC:16; Marbletown RDC 1985:112; P&T:28; Hasbrouck 1978:17; Knickerbocker:14–26, 57; "Correspondence ...," Sep–Oct 1875; Daniel Smiley, "Notes on Awosting Lake"; EJ, Aug 1902:3; UCCO, Libers 195/Deeds:485 (Sep 1875), 195:631 (Oct 1875), 186:522 (Apr 1874), 195:632 (Oct 1875), 196:45 (Oct 1875), 276:465 (Mar 1888), 285:148 (Apr 1889); Historians of Ulster County:3; Stromer website; Lake Minnewaska Mountain Houses website; CR, Town of Wawarsing (1860), Gardiner (1865, 1870), Rochester (1860–1880, 1900), New Paltz (1900), Marbletown (1870–1880), Kingston (1870–1880), City of Poughkeepsie (1870–1880).

12. Railroad tunnels through the Trapps were conceived, but realized only on paper. In August 1853 a Kingston paper reported a railroad from Hoboken, New Jersey, to the Trapps was in the works, with surveys being done for a tunnel "through the rocky sides of the Shawangunk Mountain, but if completed will be a grand achievement of enterprise and capital." The next week the paper declared the project a hoax. The paper, chiding itself, stated that the idea for a tunnel through "a mountain like the 'Traps,'" without a proposal, cost estimates, subscriptions sold, or route defined, "is one of the most glaring absurdities we ever saw in a public print."

In 1887 and 1892, the Canadian Pacific Railroad surveyed a route through the Trapps for a new line to Pennsylvania. Plans for a tunnel were changed to a bridge crossing the "chasm known as Polmaghat [Palmaghatt]," if that route was selected. It never was.

Burgess 1969:18–21, 1980:23, 94; Partington (Part Two, Daniel Smiley, Jr. and Albert K. Smiley, Jr.):86–87; NPI, 1 Jul 1880:3, 22 Jul 1887:2, 5 Aug 1887:2, 20 May 1892:3; NPT, 1 Feb 1872:4, 29 Jul 1880:3, 3 Aug 1887:2; KDJ, 3 Aug 1853:2, 10 Aug 1853:2.

13. *Mountain Drives of Mohonk Lake, New York* of ca. 1901 described one drive through the "little valley" of the "Coxing Clove" leading up "the old Public road [today Clove Valley Road] leading up the Clove, past little farms and pasture lands as far as the old sawmill." The old sawmill was the Enderly mill. This brochure noted Trapps farms, but by 1910 newer carriage roads skirted the hamlet.

Guide Book to Lake Mohonk, 1875:47–51; *Guide-Book to Mohonk Lake*, 1888:52–61; *Mountain Drives ...*: 1–20; NPI, 15 Feb 1901:3; Burgess 1969:14.

Chapter 6: School Days and Heydays

1. The common definition of "heyday" is a period of greatest prosperity. While never prosperous, the Trapps experienced its peak years during the 1880s and 1890s, a topic discussed elsewhere. An older meaning of "high spirits" applies to the second half of this chapter.

Wooden barrel hoops were once everyday playthings for generations of American children. As a boy, Don Otis raced hoops with other Trapps kids in the 1930s and 1940s.

TSR; OFI.

2. Sarah Van Leuven's relationship to that Trapps family is unknown.

The 1812 state law "provided that each town in the State be divided into districts—such division to be determined by three commissioners elected by the voters of the town," Ellsworth and Tinney explained. "A further provision was that three school trustees be elected in each district."

Louise Yeaple; Town of Rochester Archival Records; Mary Polhemus:224–25 (Ellsworth and Tinney).

3. In 1871, New York State reported of nearly 11,700 schoolhouses, that 127 were log, 502 stone, 1,162 brick, and 9,904 frame. Numbers of log and stone schools decreased, with frame and brick ones increasing. As Trapps homes were wood, we presume the schools there were, too.

In 1874, New York State passed a law requiring a minimum of 70 days annual attendance for most children. Over time, attendance laws became stricter, and in 1913 the school year was set at 180 days.

Annual school reports, census records, land deeds, and maps help pinpoint where a family lived in the Trapps.

Hough:121; James Folts, University of the State of New York website.

4. Hasbrouck 1978:17; UCCO, Liber 284/Deeds:590 (Sep 1867).

5. Historians of Ulster County:115; Cikovsky, Jr. and Kelly:64–65, 90–91, 394; Beneduce:1–64.

6. Louise Yeaple.

7. Alf Evers edited and republished excerpts from Bide Snyder's articles for his 1987 Woodstock history.

Shear:518–19; *The Woodstock Weekly*, 27 Sept, 4 and 11 Oct 1924.

8. Burgess 1996:57; Lang and Lang:19–27, 201.

9. A one-room frame schoolhouse was built in 1846 in the Town of Esopus, just south of Kingston, at a cost of $1,000, noted Ellsworth and Tinney. Genuine slate blackboards cost $2 total. Eight dollars in taxes were raised in 1886 to pay for fuel.

Spencerian penmanship was taught in the Trapps in the later 1800s. P. R. Spencer introduced this educational craze in the 1840s; he was born in Dutchess County, New York, then moved to Ohio as a boy. With paper rare in the Buckeye wilderness, young P. R. practiced letters on any scratchable surface: "birch bark, sand, ice, snow, the fly-leaves of his mother's Bible, and by permission of a local cobbler, upon the leather in his shop," noted Virtualology.com.

Charles E. Merrill's system of penmanship came into vogue during the later 1800s. The Trapps School switched to the Merrill system for at least the 1899 fall term.

Peter H. Harp, who grew up in New Paltz during the early 1900s, confirmed the use of potatoes as makeshift hand-warmers. He wrote "potatoes were baked and then wrapped in a woolen cloth to keep hands warm on a sleigh ride."

Mary Polhemus:224–25 (Ellsworth and Tinney); N&M:2–3; Palmer:6–7; Spencerian penmanship series; Virtualology website; Harp:60.

10. John Voorhis's job as night foreman with *The Kingston Daily Freeman* was noted in a January 1901 *New Paltz Independent*, where he worked at the time.

CR, Town of Gardiner (1880), City of Poughkeepsie (1870); Guilford RDC, Marriages 1834–1923:33; NPI, 7/28 Jan 1875:2, 4/18 Mar 1875:3, 27 May 1875:3, 4 Nov 1875:3, 9/16 Dec 1875:3, 6 Jan 1876:4, 11 Jul 1878:3, 4 Jan 1901:3, 16 Aug 1901:3.

11. In August 1900, Ralph Le Fevre dated the second Trapps School to 1897–98. He wrote that the new school "speaks well for the interest of the people in education."

Mabee:73; NPT, 27 Aug 1890:2, 3 Sep 1890:2, 10 Sep 1890:4; VLFI; Mary Polhemus:124 (Kelley); Louise Yeaple; NPI, 27 Jul 1876:1, 17 Aug 1876:3, 10 Aug 1900:2.

12. Attendance was taken in a daily register, its size determined by the number of students. The 1889–1890 register of Gardiner District School No. 1, the one-room Kettleborough School, bears a light brown cover, with black print and corner scroll work, and measures about twelve inches square. String runs through five holes up the fold. Its title reads "Daily School Register/Size for Schools Having 30 Pupils or Less in Attendance, Two Terms of Twenty Weeks Each."

World Health Organization website; TSR; The National Arbor Day Foundation website; Gardiner School District #1, Daily School Register, Jul 1889–Jul 1890; CR, Town of New Paltz (1900), Rochester (1880).

13. NPI, 20 May 1875:3, 10 Apr 1879:1, 17 Apr 1879:3, 4 Sep 1891:3; CR, Town of Esopus (1920).

14. NPT, 21 Oct 1869:3, 18 Feb 1885:3; Hasbrouck 1978:50; NPI, 31 Jul 1914:3.

15. Janey and Phebe Van Leuven letter.

16. Hasbrouck 1878:17; *The Woodstock Weekly*, 27 Sep, 4 and 11 Oct 1924.

17. Bray (DeWitt).

18. *The Woodstock Weekly*, 27 Sep, 4 and 11 Oct 1924; CFI.

19. SUT, 2 Feb 1866:2; NPI, 30 Oct 1879:3; NPT, 16 Oct 1873:3, 27 Apr–1 Jun 1892:4, 8–15 Jun 1892:3, 13 Jul 1892:3; CR, Town of Gardiner (1870, 1880); UCSO, Book R/Wills:376 (Stokes).

20. CR, Town of Gardiner (1880), Shawangunk (1870); NPI, 29 Jul 1880:3.

21. TA, Jun 1991:7–9; AMC Records:11.

22. Clove RDC:1 and 6.

Chapter 7: Saints and Sinners

1. AMC:5–6; NPI, 11 Apr 1878:#?

2. *The Rondout Freeman* is not on microfilm for this period at The Kingston Library. Nothing similar appeared for a month previous in *The Daily Freeman*.

Pen names were popular in the press during the 1800s. The following appeared in *The New Paltz Times* for neighborhood columns: Quixote for Gardiner; Alpha and Omega for Libertyville; Punch and Judy for Butterville. Yum Yum appeared, too. Their real names were known locally, but most, including Trapps columnist Wild Boy, are lost to time except when we chance upon their real names. Elting H. Terwilliger revealed that he was Punch in that paper's June 5, 1879 issue. Was Mrs. Terwilliger his Judy?

Criminal cases, including murder, are recounted later in the chapter. Taverns had existed in the Trapps since the early 1800s; the McGibney Hotel was operating at the time of the article, and Ben Fowler was still alive, presumably running his tavern. Victoria Clafin Woodhull was a famous suffragette of the day. A curtain lecture is a private lecture given by a wife to a husband. The term dates back to colonial days when curtains were hung around a bed. Webster's 1913 *Dictionary* provides a quote by Washington Irving: "A curtain lecture is worth all the sermons in the world for teaching the virtues of patience and long-suffering."

HH, 26 May 1983:28; NPI, 4 Mar 1875:3; NPT, 5 Jun 1879, 22 Jul 1880:3.

3. One not-so-veiled comment about the Trapps appeared in July 1876 in New Paltz. Albany correspondent "D." had published an earlier visit to "Lake Miniewasky." "D." and party "slowly rode through the Traps, about which enough is probably known here to make it unnecessary to say more."

The 1850 census listed four Trapps men who were literate, but their wives were not. Two other couples were illiterate. This was fairly typical, especially for rural areas. If one of the couple was literate, it was usually the husband. An exception was taverner Ben Fowler, who made his mark on his 1875 will. His wife Laney could write, and likely it was she who kept track of the tavern's daily accounts. Ben's will instructed that after all debts were paid, Laney inherited his estate as long as she remained a widow. At her death or remarriage, the estate would be divided equally among their four surviving children, all daughters. If one became disabled, lame or unable to help herself, she would receive one-third more than the others. If a woman became incapacitated, marriage for her was unlikely. Knowing this, Ben provided for his daughters in case of adversity.

NPI, 27 Jul 1876:2; NPT, 24 Aug 1860:2; CR, Town of Gardiner (1870, 1900), Rochester (1850); UCSO, Liber S/Wills:242 (1875).

4. Some reports recounted general religious experiences, such as this August 1871 one from New Paltz: "A negro woman picking berries at 'The Traps,' the other day, was so effected by the descent of a large paper balloon, that she picked up her child and rushed frantically into the nearest house, declaring that 'de Lord had come, for sure, and was down thar on de fence.'"

The Guilford Church was located in the Town of New Paltz until 1853 when that section of New Paltz became part of Gardiner. The first Guilford Church, which no longer exists, was located on the west side of the intersection of present-day Albany Post Road with Bruynswick Road and Gibbons Lane, noted Hasbrouck.

History of the Reformed Church of Guilford:6–11, 18, 28–32; Hasbrouck, 1978:49–50; P&T:146; Child:86–87; NPT, 31 Aug 1871:4, 30 Apr 1890:3.

5. AMC:title page, 5–6; Beers Atlas:74–75; NPT, 7 Dec 1871:3; NPI, 17 Feb 1870:3, 14 Dec 1871:2; CR, Town of Gardiner (1910).

6. Sylvester II:198; UCSO, Liber R/Wills:376 (Stokes 1871, will and codicil certified Sep 1873); UCCO, Liber 199/Deeds:203 (Apr 1876); *The Clove Chapel*:1–14; CB:1066–67; AMC:8.

7. The insurance bill is housed today at the Ellenville Library and Museum.

Ulster County Clerk's Office, Liber 255/Deeds:153 (Dec 1884); Hasbrouck 1978:15; Knickerbocker: 54–56; NPT, 1 May 1879:2; NPI, 15 May 1885:2, 10 Aug 1900:2, 31 May 1901:#?; Mabee:60; CR, Kings County, NY (1880); Bray (DeWitt).

8. Church records, both at Guilford and Alligerville, list the Trapps and Clove as congregations and places of burial.

NPI, 25 Jan 1888:3, 24 May 1923:#?

9. Rotating ministers performed baptisms often witnessed by the "whole congregation," church records noted. Records, though, did not always indicate where people lived. Richard L. Rider, Friends of Historic Rochester Genealogy Library, explained that rotating ministers sometimes entered records in their own churches back home. This is why some family members are found in local records while others might be found in church records far from where they lived.

AMC:7–8 and 25; NPT, 25 Jan 1888:3; Richard L. Rider, interview, 12 Dec 2005.

10. NPI, 4 Oct 1889:2, 16 Feb 1900:3, 22 Jun 1900:3; Le Fevre:212–215.

11. In a *New Paltz Times* piece a few days after the oyster supper, Hasbrouck "thanked his mountain friends for the many acts of kindness he had received from them." The paper stated that of the 150 attending "a large number [were] ladies", but made nary a mention of the fisticuff: "Everything passed off pleasantly, and Mr. and Mrs. Shaffer and their assistants are deserving of considerable praise. ... Last, but not least, and what could have been done without them, we mean the girls present. They were pretty, sociable and by their genial disposition, added much to the hilarity of the evening."

An eminent Tuthill businessman, Hasbrouck rode "the old war horse" of state politics. His crowning, yet most controversial, achievement was the creation of the Town of Gardiner in 1853, an act "entirely his work," a New Paltz paper reported in 1871, "and for this the people of New Paltz could never forgive him, although we presume most of the citizens of Gardiner have become reconciled to the same."

NPT, 22 Feb 1861:2, 23 Aug 1861:2; NPI, 8 Jun 1871:2, 29 Jul 1880:3; Friel 1925:31.

12. Josiah Slater likely was related to Trapps farmer Hugo Slater.

NPI, NPT, 16 Sep 1869:2; CR, Town of Rochester (1850, 1855, 1880), Marbletown (1855), Gardiner (1855–1870); Child:164; Rochester RDC:84, 86, 190; AMC:24; Beers Atlas:134.

13. NPT, 5 Aug 1880:3; NPI, 5 Jul 1877:3; P&T:28; CR, Town of Gardiner (1870, 1880).

14. Alonzo A. is not to be confused with a nephew named for him. The elder Alonzo was a son of Isaac and Sarah Dingee. Alonzo was a popular boy's name in the nineteenth century. Kitty Burns Lorey noted "amid the fashion of the time for all things Greek and Roman—in everything from architecture to hairdos to rhetorical styles to city names ... exotic-sounding monikers came into vogue, especially for boys."

Jacob Wynkoop's parents, Joshua and Maria, gave several of their children the middle names of Brodhead, Schoonmaker, or Hasbrouck. Whether either parent was related to these families is unknown, but it was common then to give children first or middle names of prominent families. The father of Peter H. Harp of New Paltz was given the first name Elting "after the rich Eltings."

CR, Town of Rochester (1850–60), Gardiner (1855–70); Sylvester II:209–10; Foote (Rochester):350; Rochester RDC:63, 126, 132, 136, 232; AMC:16, 25; Guilford RDC, Baptisms 1834–1871:17; J&J:198–99; W-E:36–38; Lorey:25; Harp:51; SUT, 27 Jul 1866:2.

15. The McKinstrys were prominent tanners in Gardiner. David McKinstry established the town's first tannery in the early 1800s.

CR, Town of Gardiner (1855, 1880); Child:86, 164; NPI, 11 Nov 1875:2, 11 May 1876:2, 6 Jul 1876:2, 11 Apr 1878:#?, 4 Sep 1896:3, 4 Jul 1919:1; Hasbrouck 1978:30.

16. NPI, 6 Feb 1879:2, 5 Nov 1886:2; NPT, 3 Nov 1886:3; W-E:37; CR, Town of Wawarsing (1900).

17. One or two family histories suggest mixed racial ancestry in the Trapps, but no definitive evidence for this presently exists. After researching African-American presence in the Shawangunks, Reba W. Laks concluded, "what information and stories there were usually had more to do with New Paltz and other valley settlements rather than relating directly to the ridge."

Smith:159–60; NPT, 4 May 1887:2; CR, Town of New Paltz (1850), Lloyd (1880); Laks:1–10.

18. The 1850 census listed Jacob "Gosling" next to William Enderly of the Trapps. Ben married Eliza Dingee, whose first name may have been Maria, suggested Wustrau. Eliza might have been related to the Trapps family, but is not to be confused with a daughter and granddaughter, both also named Eliza, of Isaac and Sarah Dingee.

CR, Town of Rochester (1850); W-G:5–15; J&J:98, 150–51, 198.

19. *The Kingston Democratic Journal* reported that Maria was the daughter of Eliza Cross, a white woman, and an unnamed black man. Eliza was the daughter of Benjamin Bogee, whose wife had frozen to death fifteen or sixteen years earlier. Benjamin swapped Eliza, his only unmarried daughter, for the ex-wife of a man named Cross.

The Ellenville Journal presumably covered the dismissal of Jacob's indictment in April 1854, but that month's paper is not available on microfilm. *The Poughkeepsie Eagle* ran its initial coverage from the April 22 KDJ.

EJ, 22 Apr 1853:2, 29 Apr, 1853:2, 17 Jun 1853:2, 1 Jul 1853:2; *The Rondout Courier*, 22 Apr 1853:2, 6 May 1853:2, 1 Jul 1853:2, 15 Jul 1853:2; *The Ulster Republican*, 27 Apr 1853:2, 22 Mar 1854:2, 26 Apr 1854:2; KDJ, 27 Apr 1853:2, 29 Jun 1853:1, 13 Jul 1853:2; PDE, 30 Apr 1853:3, 21 May 1853:3.

20. E&E:42; CR, Town of Shawangunk (1860), Orangeville, Michigan (1860, 1870), Viroqua, Wisconsin (1870); *Vernon County Censor*, 5 March 1879:#?; *The Kingston Argus*, 23 Jul 1890:8; NPT, 4 May 1887:2.

SURVIVAL, 1830–1900

Chapter 8: Farming and Gathering

1. NPI, 20 Aug 1897:3; CR, Town of Gardiner (1905).

2. Art Lapp, TA, Mar 2007:4; French:660; Hough:662; Arthur G. Adams 1999:99–100; Evers 1972:196–211, 404–427.

3. van Wagenen, Jr.:32; Van Valkenburgh 1992:38.

4. Harp:24.

5. In 1880 one Town of Gardiner farmer advertised: "Girl wanted—one that understands making butter. 7 cows kept. Wages $8 a month." This type of job was a good opportunity for a girl or young woman to earn extra wages for her family.

NPT, 24 Jul 1880:3, 30 Apr 1890:3; N&M:112–118; van Wagenen, Jr.:9–10.

6. Isaac Dingee and George Davis also raised sheep. During the 1870s, Ben gathered up to forty-six pounds of wool; George, up to thirty. We have no figures for Isaac.

Modern cows in the U.S. produce "an average of 6-1/2 gallons of milk per day," notes the Animal Improvement Programs Laboratory (AIPL) website. Cows of the 1800s likely gave less milk without modern feed, supplements, and medical care. Using the AIPL formula of 25 gallons of milk to make 11 pounds of butter, with rounding down to 10, a Trapps farmer perhaps milked 500 gallons to get 200 pounds of butter.

Rochester RDC:84, 86; CR, Town of Rochester, Gardiner (1820–1880); Guilford RDC, Marriages 1834–1923:35 and Funerals 1847–1923:13–18; *History of the Reformed Church of Guilford*:10–11; Clove RDC:21, 23; W-E:18, 33–35; E&E:27, 42, 98; P&T:201; AIPL website.

7. A shallow well, likely used by both farms, remains today up on the saddle between Dickie and Ronde Barres. "Barre" might be French for "bar" and is pronounced "bar." The origin of these place-names is unknown. "Ronde" might refer to "round."

CR, Town of Rochester (1855–1880); Beers Atlas:134; Child:160–65, 229–236.

8. Michael Swift, TA, Oct 1987:2–3; van Wagenen, Jr.:80; Brooks:180; De Lisser:216.

9. In 1864 widow Catharine Enderly produced sixty pounds of honey. Another Trapps beekeeper was Mary E. Dingee, widow of William E., son of Isaac and Sarah. Mary auctioned four skips [skeps] of bees with other items, noted an April 17, 1899 unidentified newspaper. The two women perhaps sold honey or beeswax. Arthur O. Friel provided one anecdote for collecting wild honey in the Trapps in his 1923 novel *Cat O'Mountain*: "honey-hunting, though productive of a passable revenue to the few whose instincts led them to follow it up, was arduous, uncertain, and often dangerous work because of the roughness of the country, its unexpected pitfalls, and its deadly snakes."

CR, Town of Gardiner (1865, 1875); NPI, 29 Nov 1901:3; Friel 1923:177.

10. Wild ginseng was and still is prized over the farmed variety. Stands are rare in the Shawangunks; ginseng prefers deep, moist, shaded woods such as those found in the Catskills. Ginsengers, as pickers were known, worked large stands in the Catskills, where it was big business well into the 1900s. Harvesting meth-

ods, however, doomed the industry. Pickers dug stands without leaving the rootstock or seeds for reproduction.

Jennifer May internet article, Oct 2005 (Beyfuss citation); Macera:30–33; Harris:424–428; NPI, 10 Apr 1903:3, 24 Apr 1903:3.

11. Root cellars, often located away from the house and built into a hill, were natural refrigerators. The surrounding earth kept food cool in summer, but prevented freezing in winter.

12. Daniel Smiley, Mar 1983:1–16; Burroughs 1886 (1895):242–43.

13. The Bascom Maple Farm website states that it takes thirty to fifty gallons of sap to make one gallon of syrup. Most sources list an average of forty gallons.

"[A] good sap bush" might yield "three pounds of sugar or three pints of syrup for each bucket hung," noted van Wagenen, Jr., who added that cooled maple syrup or sugar "poured on snow becomes the 'jack-wax' of blessed memory."

Another maple treat was pulling off sap icicles to eat. When spring ice storms break off small maple branches, sap runs from the wounds and freezes at night into icicles. As a boy in the Trapps, Don Otis recalled how his mother and grandmother forbade, or tried to, the Otis kids from eating them, probably because of the high sugar content.

Evers 1987:268 (Bide Snyder); Bascom Maple Farms website; CR, Town of Gardiner (1855, 1875), Rochester (1920), City of Kingston (1930); van Wagenen, Jr.:174; OFI.

14. The Dennis Williams interviews were published in Wallkill and likely New Paltz, the second published in 1940.

Sewing was another way for women and girls to earn cash. The Gardiner census of 1870 listed Eliza Slater, twenty-three, as a dressmaker living at home in the Trapps with her parents, Hugo and Rachel. In 1880, Sarah Van Leuven, twenty-seven, lived in the hamlet supporting two young children as a seamstress.

CR, Town of Rochester (1870–1880); Child:231–36; Beers Atlas:135; OFI; WVW, 11 Jul 1941:8 (Williams interview); CR, Town of Gardiner (1870–1880).

15. Like wintergreen, trailing arbutus was grown to sell as fragrant garlands.

Sanderson:38–39 (Ruckle); "Wintergreen Industry," unsourced paper, Ellenville Public Library and Museum; NPI, 1 May 1908:3.

16. Huth, Checklist of Trees and Shrubs, 1990; Polhemus and Polhemus:59.

17. Terwilliger (1907–1994) was an authority on Town of Wawarsing history. Her grandfather, Uriah E. Terwilliger, established a mountain hotel called Mount Meenahga in the Shawangunks above Ellenville.

Six articles from *The New York Times* are quoted extensively: 30 Jul 1871:5; 13 Aug 1884:5; 15 Jun 1885:3; 14 Jul 1886:3; 12 May 1889:10; 7 Jul 1897:10. Each will be cited hereafter by month and year; for example, the July 1871 *New York Times.* (The July 1871 article was reprinted from an undated one in *The Ellenville Press*, source of the term "little black treasures.") Other articles are cited fully in the notes.

To clarify usage, Shawangunk berry pickers picked wild blueberries, not huckleberries. Wild blueberries belong to the vaccinium genus; their cousins, the huckleberries, to the gaylussacia genus. Pickers, though, did not use scientific names. In southern New York and along the Appalachian chain, they called both "huckleberries." Both berries grow together, and both were found in berry pails.

Instead, pickers segregated berries as to height—low-bush, up to eighteen inches high, and high-bush, about head high. Shawangunk Mountain huckleberries ripen from July to early September. However, berry pickers in this range called wild ones "huckleberries" and cultivated ones "blueberries," noted Fried.

True (black) huckleberries grow "half-high," between low- and high-bush. These have nutlets, each with a seed inside, which made them unpopular to eat. Seeds stick between teeth, and eating too many can cause a stomach ache. These berries were also called "crackers" (July 1897 NYT) or "crackerberries," noted Fried.

Whortleberry was an earlier term of English origin for huckleberry that held on until the later 1800s. In July 1880, Henry Hess, who lived below the cliffs south of the Trapps, reported a good whortleberry crop.

THR, 18 Oct 1994:22 (Terwilliger obituary); Stephen Hartshorne, SF, 15 Aug 1976:3; Medsger:68–74; Fried 1995:113, 146–47; Wendy McCormick, TA, Jan 1988:3 and 7; CR, Town of New Paltz (1880); NPT, 30 Jul 1871:5, 1 Jul 1880:3, 7 Jul 1897:10.

18. A July 27, 1860, brief in the NPT (page 2) is the earliest printed in that town on berry picking: "The whortleberry crop on the mountain range west of this village, is a very large one. Hundreds of bushels has already been picked and shipped to eastern markets, bringing on the average 10 to 12 cents per quart."

19. Snyder:21; Daniel Smiley, Jan 1986:2.

20. NPI, 2 Aug 1895:3.

21. The Shawangunks are home to only two poisonous snakes—timber rattlers and copperheads. Adders, also mentioned in the 1897 article, were an older name for copperheads, noted De Kay. Wendy McCormick related the comments of one twentieth-century berry picker: "They wanted to get [rattlesnakes] for the rattlers. Great deal taking the rattles off them ... to keep them and make belts with them ... to show off."

De Kay III:54–57; Wendy McCormick, TA, Oct 1987:2.

22. Peddlers were once a common sight across rural America. Paris Gray operated a general store in High Falls and in July 1884 ran "a peddler's wagon" through the Trapps. Similar to Alonzo, Paris was an "exotic-sounding" boy's name, part of the nineteenth-century rage for all things Greek and Roman, noted Lorey.

The Kingston Journal and Weekly Freeman, 3 Jul 1884:1; Beers Atlas:84.

23. NPI, 3 Aug 1906:3.

24. NPI, 25 Feb 1869:2, 8 Aug 1878:3, 6 Aug 1886:3, 8 Jul 1887:2.

25. The Billingham house was divided into two, half for the family and half for the store and small hotel. Don Otis and his family later lived in the commercial half of the house. He described a very large room downstairs with store counter intact, four bedrooms upstairs, and an attached shed, enclosed on three sides, with horse stalls.

Don recalled his grandparents, Jake and Minnie Otis, telling stories about cattle drives through the Trapps into the early 1900s. One path joining the Rondout and Wallkill Valleys crossed over Dickie Barre and cut down through the Billingham/Otis property; this path was only wide enough for foot, horse, or cattle traffic, not for wagons. Drovers used the Billingham hotel and the Laurel Inn (site of the McGibney Hotel) near the Minnewaska entrance for resting and boarding during the drives.

NPI, 25 Jul 1902:3, 20 Aug 1897:3; Unknown newspaper, 12 Feb 1885:#?; CR, Town of Rochester (1900), Gardiner (1870, 1880, 1905), Neversink (1850); Guilford RDC, Funerals 1847–1923:2, 14, Marriages 1834–1923:36, 41, Baptisms 1873–1918:6; NYT, 13 Aug 1890:2; UCCO, Liber 239/Deeds:533 (Mar 1873); Lawrence:251; Harp 26–27; OFI.

26. Berry picking appears as an occupation only periodically in census records. The 1880 Rochester census listed Mrs. Ellen Wynkoop of the Trapps as a berry picker.

Rochester RDC:151; Guilford RDC, Marriages 1834–1923:36; AMC:5 and 11; NPI, 17 Feb 1870:2, 1 Jul 1880:3; NPT, 7 Jun 1877:3, 30 Jul 1890:3; CR, Town of Gardiner (1870–1900), Rochester (1880); E&E:63; Sylvester II:260.

27. Ralph Le Fevre reprinted the middle and last sentences of the July 21, 1887, *New York Times* story in the August 5, 1887:2 edition of NPI.

Arthur G. Adams 1999:106–08; EJ, 20 Sep 1895:2; NPI, 18 Jul 1878:2–3, 5 Sept 1878:3.

28. NPI, 1 Jul 1880:3, 3 Aug 1906:3; NPT, 22 Jul 1880:3; EJ, 27 Jun 1879:2.

29. A June 1871 paper stated, "A large fire has been raging for several days on the other side of the Shawangunk mountain, about two miles south of the Paltz Point Mountain House, burning over thousands of acres and consuming several small houses." The houses were likely located in the Trapps or upper Clove.

NPI, 13 Apr 1871:2, 1 Jun 1971:2.

Chapter 9: Wood and Stone Cutting

1. CR, Town of Rochester (1880).

2. CR, Town of Gardiner (1880); Smiley and Cross:16–18.

3. van Wagenen, Jr.:185; Evers 1972:224, 326–40, 386–87, 1987:163; NYT, 31 March 1856:3.

4. A side is a half hide cut down the center.

Le Fevre:195, 216; Child:86, 164; Williams:12; NPI, 11 Apr 1878:#?

5. NPI, 19 May 1882:2; Knickerbocker:51–52; Guilford RDC, Baptisms 1834–1871:22.

6. The mountains once teemed with bark peelers and stories about them. H. A. Haring amused readers with stories about slime (the tree pitch), that "secret" sticky stuff of Nature and stuff of "endless jesting." Bark peelers, plastered from head to toe with slime and reeking from it, "would get stuck against the tree, against the fresh bark, against the leaves of the underbrush, even against each other!" he wrote. "Then again, the sap dried on their clothes, which,—it has been said,—were never taken off during the two or three months of the peeling season. Their clothes, therefore, 'would stand alone.'"

Sloane noted that leftover tan bark was sometimes "given to the charcoal men, who in these lean times were glad to have anything that might earn them a dollar."

According to van Wagenen, Jr., Pratt tanned more than 60,000 sides of leather a year, for which he cut down "ten square miles of hemlock forest." His average bark yield was 18 cords per acre, with each cord yielding 194 pounds of leather (or 12 to 15 hides).

Haring:83–105; Hoffman:185–91; New York–New Jersey Trail Conference:134–35; Sloane 1974 (Reverence):61; van Wagenen, Jr.:188–191.

7. Tanneries fouled local air and water. Evers wrote, "Oldtimers used to say that 'Woodstock stank' for a period of almost a century during which the tannery was in business and the refuse of the operation oozed into the brook and lay on its banks." If the tanneries stank up Woodstock, so too did the DeGarmo, McKinstry, and other community tanneries foul the air and water around them.

Snyder:20; Evers 1972:326–36, 384–87, 1987:103; NPT, 3 Jan 1868:2; NPI, 23 Jan 1885:2; Hoffman:185–91; Haring:90–95, 102, 108–09; Pike, 12 Oct 1892:483.

8. Mary Polhemus:148 (Wohltman); Polhemus and Polhemus:115–18.

9. Sloane 1974 (*Reverence*):57–59; *The Chirp*, Nov 1970:9 (Smiley); van Wagenen, Jr.:163.

10. William J. Upright, PSNY, 10 Oct 1943:7A; Sanford A. Cross letter; Daniel Smiley, conversation with Paul C. Huth, late 1970s (PCH, interview, 23 Dec 2008); Daniel Smiley, Jan 1986:8, 1989:14, 25; NPT, 16 Nov 1892:3; Hasbrouck 1978:16; NPI, 14 Aug 1896:2.

11. NPI, 24 Sep 1874:2, 28 Jun 1877:3; N&M:18, 97; Pike:472.

12. Less-efficient undershot wheels could be used on more powerful streams, noted Muise, but it is unknown whether George Davis's mill on the more powerful Peter's Kill was an undershot.

Rotary saws replaced up-and-down saws, and steam power replaced water power. William Roe operated a steam sawmill in the Trapps until 1885, when he found "the timber about exhausted" and sold his engine for a proposed berry cup factory in Clintondale. In 1903 a Mr. Cassady ran a steam sawmill at or near the Trapps. Steam sawmills were not permanent, for they did not use a dammed water source. Cassady likely sold or abandoned his mill after the woodlot was exhausted.

Muise:14, 83; N&M:113–14; NPI, 7 Aug 1885:3, 6–27 Feb 1903:3.

13. Daniel Smiley 1989:26; *Lake Minnewaska News*, Spring 1960:3; NPT, 6 Jan 1892:2.

14. In Harp's *Horse and Buggy Days*, "Our mountains" refer to the Catskills, not the Shawangunks. Haring used De Lisser's *Picturesque Ulster* but did not cite him consistently.

De Lisser:167–68; Burroughs 1886 (1895):233; Haring:116; Evers, 1972:442; Michael Swift, TA, Jan 1988:1–2 (Reavin quote); Harp:65–66.

15. GW, 4 May 1882:#?; NPI, 27 Nov 1885:2, 8 Jan 1886:3; NPT, 18 Feb 1885:3, 11 Oct 1893:3, 27 Dec 1893:3.

16. Haring:116; Second Annual Report:139; NPT, 8 May 1889:3.

17. While waiting to be shaved, poles might be submerged into ponds to keep them pliable, Evers noted. As of 1990 the old hoop shop referred to stood on the property of Shirley Madonna of Rochester. In 1941, Sanford A. Cross wrote, "Each locality had its champion cutter and champion shaver, no union, no eight hour law, no social security."

"Hoops," unsourced paper, Ellenville Public Library and Museum (Hoffman); NYT, 30 Jul 1893:16, 11 Apr 1897:17; Haring:111–12; Evers 1972:442; De Lisser:167–68; Shirley Madonna, TA, Oct 1990:7; Sanford A. Cross letter.

18. Michael Swift noted a firkin held eighteen gallons; a hogshead held fifty-four gallons. Hoop ends were hooked together by notches, thus securing the hoops around the barrel staves and the staves themselves to form a barrel.

Haring:110–16; Evers 1972:441–42, 1987:161–62; NYT, ibid.; Michael Swift, TA, Jan 1988:1–2; van Wagenen, Jr.:108.

19. The largest hoop dealer in America was Theodore Wicklow of Ellenville, who shipped 15–18 million hoops a year "all over the west, especially [to] the great flour mills of Minneapolis," Haring wrote. By 1908, Wicklow quadrupled his output to "an astounding 75 million hoops every year," noted John E. Thompson.

In 1887 the Ulster County cement and lime industries needed 30 million hoops to operate. By 1899 the county's annual cement output topped off at 4,689,167 barrels, noted Ries; 4.6 million barrels times 12 hoops per barrel equals 55.2 million hoops used for cement barrels that year alone. That same year New York State led the nation in natural limestone cement production, with the Rosendale region "one of the best producers" in the country, wrote Ries.

King:34; Simpson:59; Knickerbocker:54; The Century House Historical Society website; NPI, 26 Dec 1902; NPT, 9 Feb 1887:3; Haring:110–16; Second Annual Report:141; Thompson, SW, Summer 2011:5, 9;

Ries:680–83, 693–95, 834–35.

20. Evers 1987:160; NPT, 30 Apr 1890:3; Knickerbocker:54–55; Simpson:59.

21. Daniel Smiley, Mar 1983:2, Jul 1985:1, Jun 1987:1–5, Apr 1988:14, 1989:13–14, 29; Huth and Smiley:1–2.

22. Bader:8–9; CR, Town of Rochester (1880); N&M: 35–36, 95–96, 118–22; NPI, 9 Dec 1875:3; NPT, 23 Dec 1875:3.

23. Burroughs 1886:243–244; Guilford RDC, Baptisms 1834–1871:23, Marriages 1834–1923:41; Rochester RDC:202, 224; Richard L. Rider:4–22 (Coddington); E&E:26; NPT, 30 Apr 1890:3, 11 Jun 1890:3.

24. The Enderly barn, with two stories and hay loft, was a model of efficiency. The Enderlys took advantage of the sloping land by digging out a lower level into which wagons were driven. Cattle were stabled on the main floor, with manure shoveled down onto the wagons then taken out to fertilize crops. The barn and farmhouse foundations can be seen at the Mohonk Preserve's Coxing Trailhead near the Split Rock swimming hole.

A gin pole combines a rigid vertical pole with a boom that rotates 360 degrees. Completely portable, it once was an essential part of construction. One stands today in the front courtyard of the Barn Museum at Mohonk Mountain House and is used at times.

A smaller stone addition was built onto the Davis mill, but its use is unknown.

25. The 1799 partition map of the Nineteen Partners Tract shows a public road running through the old Trappen gap toward the Wallkill Valley. As such, the road existed earlier. In 1892, M. H. Pike stayed at Lake Minnewaska while studying flora. Pike wondered how his stage would get up a "thousand feet of stone [that] seemed to rise like a wall before us." He soon knew:

> Certainly it required faith in modern engineering to believe anything but eagles' wings would raise us to the top of this perpendicular; our road began timidly to take hold of its skirts, and climb the way along its side. ... Then we reached a break in the wall, and turned sharply away into a pass that was once an Indian trail, and affords the only natural access to the interior of this range for many miles. ... our horses labored up the steep ascents, straining their great flanks and standing with their legs gathered under them while they took breath a moment on the shelves cut in the road for that purpose, or dragged the coach with the wheels all locked down the steep descents that seemed aimed at the centre of the earth.

NPI, 30 Aug 1889:2, 1 Nov 1889:2; Pike, 28 Sep 1892:460.

26. NPI, 7 Aug 1885:3; Sloane 1976:64–65; A. Keith Smiley:16–17 (Albert K. Smiley quote).

27. Manning:92–124, 193–210; Snyder:31.

28. Bonnie Langston, DF, 20 Feb 2004:D1, 3 (Hirsch); Newland: 39–40; TA, Apr 1988:6–7; Daniel Smiley, 1989:34; NPI, 5 and 26 Feb 1874:3; NPT, 5 Sep 1878:3; Paul C. Huth, interview, 22 Dec 2006 (Daniel Smiley on testing rock).

29. TA, Apr 1988:6–7; Keith LaBudde, SW, Spr 2002:2; Bray (DeWitt); Jim Clark, interview, 15 Aug 2006.

30. Ruth Happel Smiley:9; MB, 2 Aug 1941:2.

31. NPI, 20 May 1880:3; TA, ibid.; TA, Guy Davenport, June 2011:9–10; Keith LaBudde, ibid.; Sloane 1976:36; Ruth H. Smiley, ibid.; MB, ibid; van Wagenen, Jr.:137.

32. Art Coddington told Ruth H. Smiley that stonecutting ran in his family. Art, born in 1895 in Accord, and his younger brother were "brought up into it." They were sons and grandsons of stonecutters.

S-ACI; Flory and Theodore R. Hazen, Jr., Pond Lily Mill Restorations website; NYT, 22 Feb 1888:2.

33. Quarried stone was transported by horse-drawn wagon and stoneboat. Stone wagons were much stronger than hay wagons; their wheels, spokes, and lugs were larger and sturdier to handle greater weights over longer distances. Stoneboats handled lighter loads for shorter hauls. Evelyn DeWitt said that her father, Simon Coddington, used the term "wagon down" when hauling stone to the canal at High Falls.

The term "Esopus millstones" originated at quarries at that town but spread to quarries across Ulster County, noted the July 1873 *New Paltz Independent*.

Snyder:23; Foote (Rochester):351; TA, ibid.; CFI; NPI, 4 Jul 1873:3.

34. One of the stones that crushed William Stevens might have become his headstone. Hazen, Jr. ex-

plained that stones were "lifted and flipped" at the mills with simple wedges and pry bars. On occasion a stone was dropped, crashing to the floor or floors below. If it killed someone, it was believed to have "tasted human blood" and might kill again. To prevent this, the stone was retired, sometimes buried as the headstone of its victim. A killer stone might also be put down as a doorstep in the belief that when people walked on it, they carried its evil away. Retired millstones also ended up in bridge abutments and milldams, noted Hazen, Jr.

Hazen, Jr., Pond Lily Mill Restorations website; TA, ibid.; Newland: 40; S-ACI; MB, ibid.; Knickerbocker:17; Bonnie Langston, DF, 20 Feb 2004:D1, 3 (Hirsch).

Chapter 10: Hunting, Guiding, and Working for the Tourist Trade

1. CR, Town of Gardiner (1900, 1905); TSR; NPI, 9 Jan 1903:3.

2. NPT, 13 Sep 1861:2, 18 Oct 1861:2, 26 Aug 1864:3; Friel 1923:177.

3. NPT, 18 Feb 1885:3, 31 Dec 1890:3; NPI, 3 Nov 1893:3, 18 Jan 1895:3.

4. SUT, 19 and 26 Jan 1866:2; Anthony:137–39.

5. NPI, 9 Jan 1903:3; Ulster County Poorhouse Project website; TSR; Guilford RDC, Marriages 1834–1923:40–42, Baptisms 1873–1918:3–4.

6. CR, Town of Rochester (1870, 1880); UCCO, Liber 186/Deeds:522; Friel 1925:45–46, 73; Harold Harris:39–75, 80–100; Heusser (*Gold*):1–35, (*Legend*):1–26; Smith:60–65; Fried, 1995:101–02.

7. Hasbrouck 1953:14.

8. Heusser (*Gold*):1–2; Harris:89–92; NPT, 7 Feb 1868:2.

9. In May 1865 the *Poughkeepsie Daily Eagle* touted trout fishing in "the wilds" of Ulster County. Local guides, perhaps Cal Burger himself, were presumably hired to take parties from Dutchess County and elsewhere. In May 1890, Hiram Van Leuven, Jr. "caught a fine mess of trout" in the Trapps.

PDE, 6 May 1865:3; NPT, 14 May 1890:3.

10. Cal said that his grandfather, Benjamin Burger, served in the War of 1812 and was stationed on Staten Island, but this is unconfirmed.

Rachel Smith might have been related to Trapps resident Albert O. Smith.

It was once common for husbands to place legal notices that their wives had left them. Sara Bader wrote that "husbands publicly disparaged their wayward wives, while revoking their power to purchase in the community." She cited a July 1785 notice placed in a Pennsylvania paper by a deserted husband and a rebuttal one month later by his wife, who aired much "ill treatment" by her husband. Did Rachel Burger publish her own rebuttal against Calvin?

CR, Town of Gardiner (1870, 1880), Rochester (1850–1880); Marbletown RDC 1985:223; (NPT), 12 Sep 1862:2, 14 Jun 1867:2; NPI, ca. 1901–04:#?; Bader:116.

11. Knickerbocker wrote that Albert K. Smiley and his party stopped at a rum tavern called the Laurel Inn. Yet, at the time of her story, 1875, the McGibney Hotel was operating. Ben Walker of Kerhonkson bought McGibney out and in May 1889 built a new hotel, calling it the Laurel Inn. The place's reputation as a rum tavern spanned more than one owner and thus resulted in the confusion.

Guide Book, Lake Mohonk, 1875:49; "Correspondence ...," 7 Oct 1875; Knickerbocker:19–20; Hasbrouck 1978:17; NPI, 24 May 1889:2, 8 Sep 1905:3; NPT, 7 Jun 1877:3.

12. During his 1892 visit to Lake Minnewaska, M. H. Pike and his party "followed our rapid guide." Was this Calvin Burger, in his early sixties at the time?

NPI, 27 Jul 1876:2; Garden and Forest, 5 Oct 1892:472.

13. This is the sole story regarding snakes for sale. It is unknown if Calvin trapped the snake or caught it by accident, or if Trapps men hunted snakes to sell venom.

The story about Cal's lichen-carved graffiti first appeared in September 1912. Since Cal's day, lichen has declined in the Shawangunks; today, it is considered an important indicator of ecosystem health. As such, this fragile plant should be left alone.

NPI, 8 Jul 1887:2, 25 Sep 1896:3, 8 Sep 1905:3, 17 Sep 1912:#?

14. CR, Town of New Paltz (1870); NPI, 5 Sep 1878:3; Knickerbocker:23–24.

15. CR, Town of Rochester (1880).

16. Farm jobs at Lake Minnewaska were far fewer than on Mohonk Mountain House property, which had six large farms and a number of smaller ones, noted Smiley, Huth, and Smiley. As such, farmwork at Minnewaska was only a minor source of employment for Trapps men.

Daniel Smiley 1989:3, 28–30; Smiley, Huth, and Smiley:1–11; NPI, 3 Jul 1879:3, 27 Jul 1883:3; CR,

Town of Gardiner (1880, 1900, 1910), New Paltz (1860), Rochester (1880); Marbletown RDC 1985:241; Clove RDC:57, 81; TSR; AMC:31; NPT, 29 Jul 1880:3.

17. NPI, 10 Jun 1880:3; CR, Town of Rochester (1880).

18. The mountain houses at both lakes were closed in winter well into the twentieth century, but construction and maintenance took place all year. Mohonk opened to guests in winter beginning in the 1930s, Minnewaska not until the 1960s. In February 1908 a family lived "during the winter in the Cliff house and another in the Wildmere house," noted a New Paltz paper.

Doughty:64; NPI, 20 Jan 1881:3, 21 Oct 1881:3, 2 Oct 1885:3, 22 May 1896:#?, 8 Jan 1897:3, 7 Feb 1908:3; NPI&T, 15 Mar 1923:5; Burgess 1969:27, 46–60, 77–78, 1996:58; Knickerbocker:39–40.

19. Daniel Smiley 1989:4, 26; Partington:139; Daniel Smiley, ice harvest records, DSRC-MP; Paul C. Huth, interview, 13 Feb, 6 Apr 2007; NPI, 23 Jan 1885:3, 10 Jul 1903:3, 23 Jan 1906:3.

20. Knickerbocker:19–20.

Chapter 11: Living on the Edge

1. The fastest way to transmit messages at Minnewaska was via the telegraph.

CR, Town of Rochester (1880); TSR; NPI, 15 May 1879:3, 31 Aug 1900:3.

2. Smith:iii–x.

3. The tavern in the undated Hill story was likely one operated by Isaac Dingee and later Mrs. Jane Hasbrouck. Smith described the tavern as standing but "straggling." Smith:159.

4. CR, Town of Gardiner (1900); NPT, 14 Jan 1899:#?; NPI, 23 Jan 1903:#?; TSR.

5. The following article appeared about Rachel in June 1887: "The wife of Calvin Burger, the well-known Shawangunk Mountain guide, killed a rattlesnake three feet ten inches in length, and having eleven rattles, one day recently, near her house."

Smith:160–61; NPT, 22 Jun 1887:3.

6. NPT, 17 Aug 1892:3; NPI, 1 Aug 1902:#?

7. Smith 161; NPT, 4 May 1887:2; NPI, 31 Jul 1896:3; KDF, 14 Sep 1912:8.

8. "Nameless" stuck at least once. One Virginia-born man who registered for the draft in World War I signed his full name Nameless Henry Stegall.

World War I Draft Registration Cards, 1917–1918, Ancestry.com; Hayes 1–7; Fowler Family Bible; E&E: 42–43; P&T:91; Guilford RDC, Marriages 1834–1923:29; NPT, 28 Jan 1885:3; GW, 30 Jan 1885:#?

9. CR, Town of Gardiner (1855, 1880); NPT, 12 Sep 1862:2, 30 Apr 1890:3; NPI, 29 Jan 1892:2.

10. Siegfried, Northampton County, Pennsylvania, was a manufacturing hub for Portland cement, noted Lehigh University's Digital Library. About 30 carpenters working in January 1900 at Seigfried, Pennsylvania, were paid $2.25 a day. Despite the spelling difference, the two towns might have been the same where George Wixon and George Davis, Jr. worked as carpenters.

NPT, 17 Aug 1860:2, 15 Jul 1880:3; NPI, 5 Jan 1900:2, 31 Aug 1900:3; CR, Town of Rochester (1880); AMC:26; TSR; Lehigh Univ. Digital Library.

11. UCCO, Liber 483/Deeds:443 (Jul 1921, Fowler); NPI, 25 Feb 1869:2, 7 Jun 1877:3, 22 Jan 1892:2, 24 Feb 1905:3.

12. William E. Dingee, thirty, of the Trapps died of unknown causes in December 1888. He belonged to the Gardiner lodge of the Knights of Pythias, a fraternal order promoting universal peace established in 1864 in Washington, D.C., the first such order "established by an Act of Congress." In February 1889 the lodge held a benefit party, raising about $75 for William's widow, Mary Coddington, twenty-nine, and their five children, the youngest three months old.

Priscilla Prutzman, EJ, 5 Aug 1982:6; Friel 1923:177, 1925:301; NPI, 23 Jan 1885:2, 30 Jan 1885:2–3, 22 Feb 1889:2; NPT, 28 Jan 1885:3; CR, Town of Gardiner (1855, 1870); P&T:28, 85; E&E:26; Guilford RDC, Funerals 1847–1923:12, Marriages 1834–1923:39; J&J:123, 198; Knights of Pythias website.

13. NPI, 8 Jan 1892:3, 22 Jan 1892:2.

14. It is unknown why John Enderly did not board with any of his surviving children, including son George, who lived in the Trapps at the time.

Roads often were impassable in winter. Articles of February 1875 and 1901 printed in New Paltz subtly reinforced the perception of the Trapps as "a hard place":

HARD TIME GETTING HOME.—Mr. and Mrs. Abram Constable of Ellenville, visiting our village on

Saturday last—guests of Mr. and Mrs. Geo. Freer. He had a hard time getting home Monday morning, going by way of 'The Traps,' Mr. C. found the snow drifted on the mountain west of New Paltz, and had to 'break track' before getting through. 'Abe' found that 'Jordan was a hard road to travel.'

On crossing the mountain on Monday by way of Minnewaska Geo. Voorhis found the road badly drifted at the Traps, so that people rode through the fields.

CR, Town of Gardiner (1865, 1880, 1900), Rochester (1850, 1900); E&E:35, 42; P&T:28, 202; NPT, 4 Feb 1875:3, 18 Feb 1885:3; NPI, 15 Feb 1901:3; W-E:39–43.

15. The New Paltz poorhouse closed in 1976; the site today is the home of the Ulster County pool and fairgrounds.

Ulster County Poorhouse Project website; UCCO website; Rochester RDC:52, 67; CR, Town of Rochester (1850, 1890), New Paltz (1850, 1880), Gardiner (1860–1880, 1900), Plattekill (1880); Guilford RDC, Marriages 1834–1923:42; Child:232, 242; NPI, 1 Mar 1876:3.

16. Civil War records were gleaned from: Dilts; Plank; Van Santvoord; Wilt; *Roster of the 156th Regiment*; *New York State's Annual Report of the Adjutant General* (ARAG); U.S. Civil War Records, Ancestry.com; National Archives Trust Fund records acquired by Historic Huguenot Street for Enoch Dingee; National Park Service website, "American Battlefield Protection Program" and Civil War Soldiers and Sailors System. Plank:94–99 (Rigby, Jr.), 122; KDJ, 4 May 1864:2; NPT, 24 Jan 1862:2, 5 Dec 1862:2; WVW, 9 Jan 1932:?; Evers 1987:263–265; History of the Reformed Church of Guilford:19.

17. James Davenport, father of John James, was a first cousin of Wessel Davenport, who supposedly gave Coxen Pond the name of Minnewaska; Wessel's sister, Ann, married John Burger.

CR, Town of Auburn (1910), Gardiner (1855–1905), Marbletown (1890), Montgomery (1900), Newburgh (1890–1900), New Windsor (1890), Plattekill (1900), Rochester (1850–1900), Crawford (1910); Child:160–61; P&T:27–28, 188, 207, 330–31, 346; E&E: 26, 30–38, 92–99; Sylvester II:22, 226–227, 291–292; Foote (Gardiner):256; Plank:1–3, 64, 82–84, 94–99 (Rigby, Jr.); Victor Whitman, THR, 19 Jan 2007:7; AMC:5–7, 11, 31; W-E:66–68, W-F 9–10; Breed:90; NPT, 23 Oct 1863:2, 23 Dec 1875:3; NPI, 9 Dec 1875:3, 9 Mar 1894:2, 1 Nov 1895:#?, 18 Feb 1898:3; J&J:98–99, 150–51, 198; Richard L. Rider, interview, 12 Dec 2005; Guilford RDC, Marriages 1834–1923:29, 33–42, Baptisms 1873–1918:6, 23; Clove RDC:12, 21–23; Hasbrouck 1978:17; TSR; Rochester RDC:10, 15, 126, 132, 143, 232; Sylvester II:226–27; Brink III:262–65; Beers Atlas:117–18.

18. George Pepperrell Frost was born into a family of prominent merchants and politicians; his mother's uncle, Sir William Pepperrell, was a hero of the French and Indian War and early governor of the Massachusetts colony. George, about sixteen, enlisted in 1775 to fight against the British and fought in several major battles, including at the Siege of Boston and in Canada. Much information about the lives of George Frost, his first wife Elizabeth, whom he married in Maine, and their daughter Mary was gleaned from an August 1830 letter George wrote to a cousin in Eliot, Maine. This letter is courtesy of Frost descendants Lester E. Frost and William Carson.

Elizabeth Frost and her brother, James Gorsline, are believed to be the children of James Gorsline, a Tory of French Huguenot descent who settled in Dutchess County, noted descendant Joan Hayes Wustrau.

Kenneth E. Hasbrouck wrote that "two families by the name of Gosling" were early settlers of the Trapps. Our records suggest these were Elizabeth Gorsline Frost and her brother, James Gorsline, both of the upper Clove, or two of James's children. James's son Samuel did move up into the Trapps.

Dutch remained a prevalent, though arrested, language in the Hudson Valley until the Civil War era, noted Brink, and could still be heard after that time. "The Ulster County Dutchman loved his mother tongue with an intensity which is hardly conceivable to us," noted Brink. The New Paltz Academy, a preparatory school, taught the Dutch language as late as the 1880s, wrote Burgess. Remote regions, such as the Trapps, provided places of refuge where Dutch people could more easily retain their traditions after the English took political control.

George Frost; Fogg:168–71; Susan Sinnot, interview, 10 Apr 2007; Ruth Heidgerd 1977:97; Froyle.com website; Wikipedia website; UCCO, Revolutionary War Service Affidavits, 4 Jul 1820 (Frost); W-G:1–7, 10; John Aiello, Rootsweb website; Hasbrouck 1978:16; Brink X:111–14; Burgess 1969:27; Thompson:242; P&T:27–28, 201–02; E&E:42–43, 98–99; Rochester RDC:15; Marbletown RDC 1985:206; Guilford RDC, Funerals 1847–1923:2–15; Richard L. Rider, interview, 23 Mar 2007; OFI; CR, Town of Rochester (1790–1810, 1900, 1910), Fishkill (1790).

19. Hitchcock:25.

DECLINE AND RENEWAL, 1900–PRESENT

Chapter 12: Moving On

1. Rachel and Martha recalled the suicide of their father, Joshua Wynkoop.

CR, Town of Gardiner (1910), Montgomery (1910).

2. CR, Town of Gardiner and Rochester (1900–1920); NPI, 10 Aug 1900:2; NPI&T, 29 Sep 1921:5; TSR.

3. Van Valkenburgh 1999:10; NPI&T, 17 Dec 1924:5.

4. CR, Town of Montgomery (1910); NPT, 28 Jan 1891:2; NPI, 31 Jul 1914:3.

5. George Van Leuven returned to the Trapps, but then left for Walden with older brother, Charles, Jr. Isaac Enderly returned twice to the Trapps. Half-brothers John B. Wynkoop and Oscar Lyons married sisters Emma and Matilda Mohr of Gardiner. The two were daughters of Prussian immigrants; their father was a shoemaker.

The electric trolley made travel between Highland and New Paltz so much easier for guests of Lakes Mohonk and Minnewaska that building a trolley to the foot of the mountain was considered in 1897.

The Ellenville & Kingston Railroad, a branch of the Ontario and Western, replaced the defunct D&H Canal. In 1901, Alfred Smiley selected Kerhonkson as Minnewaska's official railroad depot, hiring Levi J. Haines to be a station agent there. Levi moved his family to the Trapps, where they lived for at least ten years, one of the few to do so during this period of decline.

NPI, 27 Aug 1897:3, 27 Jul 1900:2, 26 Jul 1901:3, 3 Apr 1903:3, 14 Apr 1905:3; EJ, 19 Jul 1901:3; UCCO, Liber 398/Deeds:585 (Jun 1906, Enderly), Liber 451/ Deeds:151 (Feb 1915, Enderly); CR, Town of New Paltz (1900–1920), Gardiner (1880–1915), Montgomery (1920–1930), Rochester (1910–1920), City of Poughkeepsie (1910–1920); New Paltz Chamber of Commerce website; Partington:142; Adams 1999:109.

6. The Lawrences, Coddingtons, and Schoonmakers of Rochester were some of the main stonecutting families left in the region. *The Accordian* noted that Gulden's Mustard, not a gypsum company, was "the biggest buyers of grinding stones" then.

TA, Apr 1988:6–7; S-ACI; Inventory, County Archives of NYS:77; Newland:9–13, 39–40; NYT, 9 Dec 1928:E4; NPI, 9 Nov 1900:3, 18 Apr 1902:#?, 23 Jun 1905:3, 20 Jan 1908:3, 14 Apr 1911:3; NPT, 11 Feb 1902:3; CR, Town of Gardiner (1920), Rochester (1900–1930), Wawarsing (1910), Rosendale (1920); KDF, 28 Mar 1903:7, 17 Jun 1905:6.

7. Minnewaska Business Records; William J. Upright, PSNY, 10 Oct 1943:7A; Knickerbocker:54–55; NPI, 9 Nov 1900:3, 14 Aug 1903:2, 27 Sep 1907:3, 26 Dec 1913:2; George H. Millham, interviews.

8. NPI, 11 Aug 1893:3, 28 Aug 1896:3, 8 Oct 1897:3, 15 Oct 1897:2, 12 Nov 1897:3, 24 Nov 1899:3, 8 Jun 1900:2 (reprinted from *The Ellenville Press*), 25 Feb 1901:3, 11 Oct 1901:3, 2 Oct 1914:2; Camp Awosting brochure, 1921, DSRC-MP; Knickerbocker:24–25; CR, Town of Rochester (1900).

9. NPT, 25 Dec 1900:3, 4 Sep 1908:3; NPI, 28 Mar 1902:3; KDJ, 5 Jun 1907:8, 31 Aug 1908:7, 3 Sep 1908:7, 23 Jul 1909:6, 22 Nov 1909: 2, 23 Mar 1910:6, 29 Mar 1910:3, 11 May 1910:5, 29 May 1912:1, 17 Jul 1912:6; NYT, Aug 23–27 1908 (various pages), 5 Sep 1908:1, 22 Nov 1909.

10. The Mohonk Lake Cooperative Weather Station recorded 4.05 inches of rain with hail and thunder for July 24, 1902. The next day had a 1/4 inch more rain. It is unknown which Trapps resident remembered storms over the previous seventy years. Hiram, Sr. was likely the one whose granary held during the storm.

NPI, 1 Aug 1902:2; CR, Town of Rochester (1900), Gardiner (1900); AMC:28; MLCWS, 24 Jul 1902.

11. LAT, 6 Nov 1902:A6; NPI, 7 Nov 1902:3, 30 Jan 1903:3, 6 Feb 1903:3, 13 Feb 1903:2–3, 20 Mar 1903:3; Burgess 1969:85–87; Knickerbocker:40; Doughty:34.

12. *The New Paltz Independent* reported Gene Otis's death on January 22, 1916:

> A terrible accident, causing the death of Eugene Otis and Tracy Van Wagenen, while crossing the Peterskill, near the Laurel Inn, comes to us. They had been at work on the mountain on Saturday chopping wood on the Minnewaska estate. About 3 p. m. on Saturday as they were crossing the Peterskill on a log bridge, it broke and they were carried down the torrent. One man was found an hour or two afterward, still alive, but fast in the ice. He died before he could be extricated. Further search was made, but the other was not found until 9 o'clock the next morning. He was dead.

Tracy, thirty-four, was the one caught "fast in the ice," with three Minnewaska workers "in the icy waters up to their shoulders for an hour chopping away the ice logs in an effort to recover the bodies," reported an Orange County paper.

Don Otis said that his uncle was not logging, but might have been helping Jake to stack wood. Given the lateness of the afternoon, Gene likely was returning home when he crossed the log bridge with Tracy, who came from the Town of Wawarsing.

George H. Smiley purchased the Billingham parcel of about 150 acres in November 1920; the parcel bordered the old turnpike and ran west to the Peters Kill.

NPI, 23 Aug 1901:3, 27 Feb 1903:3, 10 Apr 1903:3, 19 Jun 1903:3, 25 Sep 1903:2, 24 Feb 1905:3, 7–21 Apr 1905:3, 8 Sep 1905:3, 17 Nov 1905:3, 12 Jan 1906:3, 2 Mar 1906:3, 18 May 1906:3; Knickerbocker:14–26, 57–58; OFI; UCCO, Liber 478/Deeds:583 (Nov 1920); P&T:28; CR, Town of Rochester (1880–1910), Wawarsing (1900–1910); Steuding: 19–37, 104–05; UCCO, Liber 407/Deeds:86, Liber 407:90 (both Walker); Town of Gardiner, Death Records; KDF, 29 Nov 1911:8, 11 Jul 1916:6, 29 Jul 1916:3; *Orange County Times-Press*, 1 Feb 1916:3.

13. NPI, 11 Jan 1907:2; CR, Town of Gardiner (1900–1910); Town of Gardiner, List of Enrolled Electors, 1912; NYT, 3 Dec 1912:15, 13 Jan 1913:11; *The Washington Post*, 3 Dec 1912:6; LAT, 3 Dec 1912:11.

14. In 1914 and 1915 the Smileys of Minnewaska bought other produce from Trapps people: Abe Billingham, potatoes and peaches, $12; Hiram Enderly, eggs, 20 to 35¢ a dozen; Howard Decker, 30 dozen eggs, $5.40; Andrew Davis, 758 pounds of apple and crab apple jelly, $98.57; and Oscar Harp, chickens, no price given.

Minnewaska Business Records.

15. Town of Gardiner Death Certificates; CR, Town of Rochester (1900–1910), Gardiner (1900–1930); TSR; EJ, 6 Oct 1910:5, 23 Nov 1911:1–2; KDF, 24 Aug 1908:3, 26 Aug 1908:6, 23 Oct 1908:4, 24 Oct 1911:18, 14 Nov 1911:5, 20–21 Nov 1911:1, 22 Nov 1911:10, 23 Nov 1911:9, 5 Dec 1911:1; unidentified paper, 24 Nov 1911.

16. Harold Bittner remembered the "Dicka-Bar" fire tower standing during the later 1920s. Roger Van Leuven, of school age then, only heard his father, Winslow, speak of the tower and Schoonmakers. The Albert K. Smiley Memorial Tower was constructed as an observation tower on Sky Top at Mohonk between 1921 and 1923. It was not built as a fire tower, but New York State fire observers were stationed there each summer until 1971. Once the Dickie Barre tower became obsolete, it was likely dismantled and gone by ca. 1930. During the 1930s, Don Otis recalled finding only metal pins drilled into the mountain and other wooden remains.

NPI, 24 Oct 1914:2; HBI; VLFI; CR, Town of Gardiner (1905–1915), Rochester (1900), Lloyd (1920); Keith LaBudde, SW, Fall/Wint 2000:2, Spr/Sum 2000:4; NYT, 2 Jul 1916:XX12; Matteson and LaChance:4–13; OFI.

17. W-E:8, 73–79, 92–96, 104–110; Hayes:1–13; CR, Town of Gardiner (1880, 1900); Guilford RDC, Funerals 1847–1923:6–7, 20; Town of Gardiner Death Records; Harp:26; NPI, 19 Jul 1901:3, 2 Aug 1901:2; EJ, 19 Jul 1901:3.

18. Postcards to Anna Enderly, courtesy of Joan Hayes Wustrau.

19. Wustrau speculates that Benjamin died of tuberculosis. Physical characteristics and occupations of Mary Enderly's sons were gleaned from military enlistment records in Wustrau's collections. Mildred's time of death was not recorded.

Peter H. Harp wrote of Dr. Fred Coddington of New Paltz: "During the Flu epidemic of 1918, he had five horses and two drivers. Doc slept between house calls in his wagon. He would give you an examination, a bottle of medicine and an ear full of good advice for a dollar." We imagine Dr. Fuller also sleeping in his wagon as he tended families, including the Enderlys, around the clock.

Dr. George F. Hanker of Ellenville also saw Trapps patients during the early 1900s, his daughter Molly wrote to Norman Van Valkenburgh. Dr. Hanker, a German immigrant, practiced from at least 1900 to 1930. He rode up and down the mountain on horseback, and was injured once when his horse threw him after stumbling "while going down some steep rocks," Molly recalled.

Friel, 1925:30 ("dread spectre"); W-E:73–79, 92–96, 104–110; Hayes:3; CR, Town of Gardiner (1880–1915), Wawarsing (1900–1930); Town of Gardiner Death Records; NPI, 1 Nov 1918:#?; NPT, 30 Oct 1918:1; Guilford RDC, Funerals 1847–1923:31–2; MLCWS, 16 Oct 1918; TSR; PBS website; Van Valkenburgh, handwritten description of Dr. Hanker, 10 Jan 2000.

20. W-E:73–79, 92–96, 104–110; Census Records, Town of Gardiner (1920); NPI&T, 28 Jul 1921:2; Hayes:1–3; Guilford RDC, Funerals 1847–1923:31; KDF, 22 Sep 1920:1, 25 Jul 1921:8, 14 Jun 1922:11, 4

Aug 1922:1; VLFI.

21. By 1930, James and Carrie Mack moved to the Town of Ulster just south of Kingston; next door lived Carrie's mother, Lucretia Enderly, seventy-two, first wife of Hiram.

After 1900 new families arriving in the Trapps were part of the great wave of Europeans who immigrated to America during the later nineteenth century. For some years, Irishman Thom Lyons was the sole known European immigrant in the hamlet until those of other nationalities came: Veres from Austria-Hungary; Dilschman from Germany; and Sondak from Russia. William Sondak was one Russian Jew of thousands who escaped czarist persecution and made their way to new Jewish settlements in America, including Kerhonkson and the upper Clove in Ulster County. See *In the Catskills, A Century of Jewish Experience in "The Mountains,"* Phil Brown, editor, Columbia University Press.

Mack Family Bible; Guilford RDC, Marriages 1896–1916:43; Town of Gardiner, Marriage Records, 1921; TSR; CR, Town of Gardiner (1900–1920), Rosendale (1910), Ulster (1930), Rochester (1920–1925), Borough of Brooklyn (1900–1930); HH, 19 May 1983:16–17, 26 May 1983:5 and 28; HBI; OFI.

22. When Lucretia Enderly sold the mill, she had long been separated from husband Hiram. Hiram had married again, but had died several years earlier. Hiram and George Enderly were first cousins.

Central Hudson built the dam and hydroelectric plant at Rifton.

Hayes:1–7; W-G:17–21; W-E:46–47, 94–110; HBI; CFI; OFI; KDF, 17 Mar 1923:1; NPI&T, 17 Feb 1921:1, 22 Mar 1923:1.

23. KDF, 4 Mar 1923:1; NPI&T, 27 Nov 1924:5; WVW, 29 Dec 1924:17; Art Lapp, TA, Mar 2007:4; CR, Town of Gardiner (1900–1930).

24. Mabel DePuy, NPN, 21 Jun 1956:10; McAllister:1–24; UCCO, Liber 484/Deeds:394 (Sep 1921), 540/Deeds:440 (Sep 1929); Hewitt 1913:30–56, 94, 113–14, 135, 148; NPI, 7 Feb 1908:3.

25. Oscar Lyons lived in Poughkeepsie in 1920 but had returned to the mountain by 1930.

Friel's description of Uncle Eb matched a later one and a photograph of Sheridan Burger. Sheridan was much younger than seventy, but Abe Billingham was about that age.

The New York Times reported a real-life version of the "30 days after death" scam. In 1887, Charles G. Mead of Rochester sued Shawangunk Mountain millstone dealer Will A. Davis, Jr. for an overdue note for $25. Davis, who had written "30 days after death" instead of "30 days after date" on the note, lost in the Town of Marbletown court but won on appeal in county court.

CR, Town of Gardiner (1920–1930), City of Poughkeepsie (1920), Borough of Brooklyn (1920–1930); NYT, 29 Dec 1887:2, 29 Jan 1959:27; John Conway, THR, 24 Jan 1990:5 (UC Neighbors); Friel 1923:5–8, 31–40, 76, 108, 119, 129–35, 164–77, 272–283; Friel 1925:8, 23–31, 45–47, 141; *The Saturday Evening Post*, 29 Nov 1924: 8–9, 43–44, 46; UCCO, Liber 478/Deeds:583 (Nov 1920); Mabee:26; HBI.

26. Dunn dated the Minnewaska powerhouse foundation to 1924. U.S. Routes 52 and 17 were other highways built over the Shawangunks by New York State. Winslow Van Leuven of the Trapps worked as a laborer on Route 52.

Lake Minnewaska Mountain Houses, Season of 1919; OFI; Dunn:166–67; DSRC History Chronology Card File; HBI; GW, 9 Feb 1923:#?; Partington:142; New York State Department of Transportation website; CR Borough of Brooklyn (1920–1930); NP&T, 3 Nov 1921:4, 27 Sep 1923:5, 10 May 1928:1; NPN, 14 Aug 1969:A5 (reprint of 12 Jan 1929 WVW article); Knickerbocker:46–55.

Chapter 13: Fading from Memory

1. George and Lillian Smiley of Lake Minnewaska visited Jake and Minnie Otis at least once a summer; it was the Otis's custom to cane chairs for entertainment, grandson Don Otis said. A caning needle is used to thread cane through the weave; caning pegs keep the weave taut. OFI.

2. Census taker Stanley Gray included in the Trapps the households of Mary Enderly and her daughter Anna Hayes, who lived below the gap.

Several Trapps roads were abandoned over time, including the eastern loop of the Van Leuven Road out to the old James Van Leuven homestead behind Millbrook Mountain. A section of the western loop became known as Lyons Road, named for that Trapps family and still used today.

CR, Town of Gardiner (1925–1930), Rochester (1930).

3. *The Ossining Citizen-Sentinel* reported Abner's place of residence as the Trapps at the time of burglary; his death certificate reported it as Poughkeepsie.

CR, Town of Ossining (1930), Gardiner (1910), Montgomery (1930); *The Ossining Citizen-Sentinel*, 12

Nov 1931:1, 3.

4. Winslow and Bertha Van Leuven also owned a radio in 1930. Their children Roger and Doris listened to *Amos & Andy* and *The Jack Benny Show*. Similar to the Otis family, they also powered the radio with a car battery and three C batteries.

The Otis's wooden box phone had a front-mounted mouthpiece, corded earpiece, and side crank to ring the operator, in this case at Lake Minnewaska; the phone hung on Jake and Minnie's kitchen wall. Electric lines reached the Otis house from Gardiner ca. 1949–50, recalled Don Otis. At some point lines from High Falls reached Oscar Harp's house on Clove Road, which marked the northern end of the Trapps. Other parts of the Trapps, including Irv Van Leuven's house, located between the Otis and Harp homes, never received utilities during the lifetime of the hamlet.

CR, Town of Gardiner (1930), Woodhull, Steuben County (1920); P&T:27; Pat Rowe, SF, 2 Aug 1998:5; OFI; VLFI.

5. CR, Town of Montgomery (1930), Esopus (1930), Ulster (1930); Peter H. Harp 1994:26–27; E&E:63; Guilford RDC, Funerals 1873–1923:32; Lewis Wynkoop obituary, unidentified newspaper, 1936.

6. Oscar bought ten recorded parcels in the Trapps during the 1910s and 1920s while he lived in New Paltz, Poughkeepsie, and then back on the mountain. He bought them from other Trapps residents, including his brother Charlie, and the State of New York, adding to the land inherited from their father, Thomas. Thomas made his mark, an 'x', in October 1898 on deeds for separate parcels given to each son.

Hal Bittner wrote: "Behind Oscar's house there was a crevice path to Lake Minnewaska. In the summer months the boys would climb up this path, take a dip in the lake, and run down this steep path jumping from rock to rock like mountain goats." This was the same path the Trapps people took to work at Minnewaska. Historically, this was similar to the Clove Path, the route that Mohonk employees who lived in the upper Clove and beyond took to work at that mountain house.

CR, Town of Gardiner (1930); UCCO, Libers 389(1898)–538(1929)/Deeds; HBI.

7. Joseph Speer was a printer who, because of poor health, had moved his family from Queens to the Trapps after 1930. Joe, about sixty-four, died there in 1938. His wife and two sons survived him. George, the son who remained in the Trapps, never married. Roger Van Leuven recalled that George always had projects going on. George reworked his car to start with gas, then switch to kerosene, the latter costing only 5¢ a gallon at that time. George also rigged up electricity from a generator and batteries. Utility lines were slow to come to the Trapps, with parts of the hamlet without electricity as late as the 1960s.

The Highland Post, 18 Nov 1932:5; NPI&T, 9 Dec 1937:8, 16 Dec 1937:9, 24 Mar 1938:4; VLFI.

8. Kenneth Hasbrouck wrote, "The house now occupied by Leonard George was one of the Gosling Houses." The house, on the south side of Route 44/55, possibly dates back to Samuel Gorsline or one of his sons in the 1830s or 1840s.

WVW, 25 Jun 1937:1; KDF, 8 Jul 1937:3; VLFI; OFI; HBI; NPI&T, 8 Jul 1937:3, 7 Jan 1943:1; Hasbrouck 1953:16; NPN, 21 Mar 1979:24; PJ, 18 Mar 1979:5D.

9. Carmer drove up into the Shawangunks in search of "a border people." He categorized the "B'ys from the Traps" with other groups of mixed heritage in New York State, such as the Jackson Whites (also called the Jackson Blacks, both outdated terms for the Ramapough Mountain Indians). Following the excerpt quoted from the book, a section on witchcraft in the Trapps reinforced the misconceptions of Carmer's day. Based on our research and interviews, the Trapps people were unrelated to the groups listed above or others called the Kallikaks or Jukes, the last noted by Evers.

Carmer:302–305; OFI; Larsen and O'Neill, SW, Fall/Wint 2000:3; UCCO, Liber 388:78 (Nov 1896); CR, Town of Montgomery (1920, 1930); *Wallkill Valley Cemetery Interments 1867–1998*:272–73, 280–81; George Black, Sum 2010:8, 13, Interviews and correspondence; AFB; Evers 1987:553–54.

10. NPI&T, 27 Nov 1968:8; PJ, 14 Nov 1999:4B; OFI.

11. Ray Otis's first car was a Marquette convertible, which Roger Van Leuven remembered Ray racing up and down the roads in the Trapps.

Years later, Ken Phillips paid homage to the Otis boys, Joe and Don. "They were employees that sort of grew up here. They operated the power house. They did everything. We built a lot of different things. They were two big boys." Don returned the favor years later, calling his former employer "very innovative" when handling maintenance issues and emergencies at the Minnewaska property.

Central Hudson built a power line in 1937 across the Shawangunks that ran across the Palmaghatt, north of a promontory called Gertrude's Nose, part of the cliff escarpment facing the Wallkill Valley.

NPI&T, 19 May 1938:8; W-G:28 (Gorsline); W-E:104–05; Doughty:36–38; DSRC History Chronology Card File; Daniel Smiley, Jan 1986:8; Burgess 1996:73; VLFI; OFI; Oral History Interviews (Phillips):70–86.

12. Mabee:59; Schoonmaker:311; Wendy McCormack, TA, Oct 1987:4, Jan 1988:3, 7.

13. Winslow and Bertha Van Leuven rented several houses in the Trapps, said their children Roger and Doris. One was a Minnewaska rental previously owned by Winslow's older cousin, Charles Van Leuven, Jr.

Roger made a desk at the Trapps School out of two of the dynamite boxes. He placed a 1-x-12-inch board on top, with a box on either end for storage. He brought the desk to their Trapps home, where he used it for a long time before discarding it.

Mabee:59; VLFI; Fried 1982:93–98, 1995:25–26, 47–50, 58–76, 83, 131, 138; CR, Town of Gardiner (1920, 1930), Rochester (1920, 1930); Schoonmaker: 311; NPN, 31 May 1961:1; Priscilla Prutzman, EJ, 5 Aug 1982:6; Open Space Institute/The Nature Conservancy, *The Huckleberry Pickers*, documentary, 2006.

14. The last one-room school in the New Paltz District closed at Tuthilltown in 1959. The Gardiner School added a second room in 1893, noted Letus; it closed in 1981, with the distinction of being "the oldest, continuously operating school in New York State," a 1978 newspaper stated.

During this period, itinerant preachers held services at the Trapps School and stayed about a week. They were "one of the things that used to scare" a young Don Otis, whose mother made her three children attend. When the preachers talked about dying and rising again, Don said, "I always wondered how I was going to get out" of the casket.

Oscar Lyons drove a Packard with an unusual clutch, Don recalled. Oscar never quite got the hang of that clutch, and the car bucked like a bronco whenever he drove it. One day about 1950, Oscar tried to back out of his garage but "somehow got it into forward." He let out the clutch and "away he went." He and the car crashed out the back of his garage and flew through the air down the hill. "It's a miracle he wasn't killed," Don said. The accident shook Oscar up so badly that he never drove again.

Hasbrouck 1978:15–17; NPN, 4 Oct 1978:18A, 16 Nov 1994:27; Dawn Letus, DF, 10 Aug 2001:A3; VLFI; OFI; Betty Burger, interview, 1990s; Jay O'Connor with Suzanne Spear, interview, 23 May 2008; One-CLE, Inc. website; New Paltz Rural Cemetery Records 1860–1962:103; NPI&T, 9 May 1929:1, 10 Feb 1955:7, 17 Jun 1959:1; *Pine Funeral Home Registers*; George Black, Sum 2010:8 and 13, Interviews and correspondence.

15. Hasbrouck wrote "Van Leuvens" plural, referring to Irv and Cassie. Hasbrouck published his history of Gardiner in 1953, but likely gathered his material previous to Cassie's death in 1949. Ken Phillips was born and raised in Wawarsing; Lucille was born in Brooklyn and raised south of Kingston.

UCSO, Book 691/Wills:439, 21 Apr 1942 (Thirzy Harp); Hasbrouck 1953:18; Oral History Interviews (Phillips:70–86); Lake Minnewaska Mountain Houses website; Eleanor Rosakrans, TA, Nov 1999:1–4, Feb 2000:1–4; Knickerbocker:41–42, 46; OFI.

16. CR, Town of Lloyd (1900–1930), Gardiner (1900–1930); TSR; HBI; Mildred Bendix, interview, 9 Oct 1999; OFI; *Pine Funeral Home Registers*; PSNY, 11 Mar 1956:1C; NPN, 3 Jan 1957:1; NPI&T, 14 Feb 1929:1.

17. Besides Irv and his cousin John Van Leuven, a couple named Constable also burned charcoal in the Trapps during this period and sold it to Minnewaska, recalled Don Otis. After they stopped doing so, Don's father, Ray, drove to a factory on Route 17 west of Liberty to buy charcoal. The factory made plywood and particle board, but also used and sold charcoal. Ray took a dump truck and picked up loads of charcoal from there once or twice a summer for the Minnewaska kitchens for broiling steak.

Daniel Smiley, Jan 1986:8, May 1988:3; A. Keith Smiley, interview with Paul C. Huth, 19 Aug 1997; CR, Town of Gardiner (1940); Daniel Smiley, charcoal records, DSRC-MP; William J. Upright, PSNY, 10 Oct 1943:7A; OFI.

18. A flower car refers either to a separate car that carries the flowers to the cemetery or a hearse with a lower rear chassis in which flowers adorn the casket. It is unknown which Irv arranged; regardless, a flower car was an added luxury. *Pine Funeral Home Registers*.

19. Whether a headstone was ever set for Irv's grave is unknown. Andries Van Leuven likely was a cousin of Henry O. Van Leuven, Irv's great-grandfather.

We consider Irv the last Trapps resident to have lived the old ways. His cousin, John Van Leuven, was born in the Trapps and returned later in life, but spent most of his adult life in Dutchess County, said son John, Jr. The senior John, named for ancestor John D. Van Leuven, built and lived in a cabin in the Trapps, where he made charcoal, went hunting with Irv, and sold berries to the stages that came through. John, Jr.

stayed at his father's cabin while visiting him in the Trapps.

PSNY, 11 Mar 1956:IC; NPI&T, 10 Jan 1957:6; NPN, 3 Jan 1957:1; *Pine Funeral Home Registers*; New Paltz Rural Cemetery Records, 1860–1962:11; John Van Leuven, Jr., interviews ca. 2000, Oct 2002.

Chapter 14: Restoring the Past

1. MLCWS, 12 Oct 2000.

2. Ice cutting records of Daniel Smiley show impressive figures during the last decades. In 1939, 1,254 tons of ice were cut; in 1955, over 600 tons, or just under 7,500 cakes. The 10–12-inch-thick cakes averaged 145 pounds each.

Oral History Interviews (Phillips):8; Burgess, 1996:69, 120–21; OFI; Mohonk ice cutting records, DSRC-MP; OFI; NPN, 27 Sep 1962:20.

3. *Mohonk Preserve newsletter*, Summer 1984.

4. Large parcels of outlying Mohonk Mountain House property transferred to The Mohonk Trust ranged up to nearly 1,300 acres. These parcels were composed of a dozen or so smaller ones bought earlier by the Smileys of Mohonk. The parcels were sold to The Trust for $20 an acre, which covered administrative costs, explained Paul C. Huth, director of research emeritus of the Mohonk Preserve. These transfers included lands once part of the Trapps hamlet.

Mohonk Preserve newsletter, Sum 1984; Paul C. Huth, interview, 7–8 Nov 2007.

5. NPN, 31 May 1961:1; NPI&T, 8 Jan 1969:3; W-E:94.

6. NPI&T, 26 Feb 1970:Sect. 2, p. 6; New Paltz Central School District, Minutes of the School Board, 30 Mar 1970, 20 Apr 1970.

7. OFI.

8. S-ACI.

9. NPN, 6 Dec 1972:15A, 21 Mar 1979:24; E&E:38; U.S. Social Security Death Index; PJ, 18 Mar 1979:5D; OFI.

10. AFB; Diane Carlson, HH, 6 Apr 1977:10, 13 Apr 1977:16; Hasbrouck 1953:15–18, 1978:23–25.

11. Snyder:39; NYT, 3 Jan 1978:33; THR, 13 Jun 1986:1, 3, 20.

12. SF, 2 Mar 1986:5; THR, 20 May 1989:8.

13. THR, 20 Jan 1990:24, 24 Jan 1990:5 ("Ulster County Neighbors"), 10 Sep 1992:31; PJ, 14 Nov 1999:4B.

14. Paul C. Huth, interview, 3 Nov 2005; Richard A. D'Errico, THR Sunday Magazine, 30 Nov 1997:7–9; Letters to Glenn Hoagland, Mohonk Preserve, by Melvin Sharpe, Jr. and Thelma Atkins; R-MP, Spr 1999:7; SF, 12 Oct 1997:3; Pat Rowe, DF, 2 Aug 1999:5; Nancy Ward, *Blue Stone Press*, 7 May 1999:6; Jesse Nason, PJ, 4 Oct 1999:1A, 2A; Jeremiah Horrigan, THR, 12 Jul 2000:5; Deborah Medenbach, THR, 12 Oct 2000:11; Bond Brungard, PJ, 9 Apr 1997:1B, 14 Oct 2000:1B, 2B; Ruth L. Pierpont, letter, 1 Dec 2000.

15. Annie and Rae Slingerland, meeting, 5 Nov 2006; Hank Alicandri, R-MP, Spr/Sum 2007:5 (Enderly Ground).

16. THR, 18 Feb 2004:30, 2 Sep 2007:39; NPN, 7 Feb 2001:4; NPT, 2 Aug 2001:22, 14 Nov 2002:33; DF, 25 Oct 2000:A4, 28 Dec 2000:A2; NYT, 15 Dec 2001:C16; PJ, 2 Nov 2006:4B.

Chapter 15: Remembering the Past

1. Harold Van Leuven, interview, 1990s; TSR; CR, Town of Gardiner (1920).

2. Randolph Taylor was an engineer for the Horn & Hardart Automat in New York City, recalled Don Otis. Automats popularized the sale of prepared food from vending machines during the early 1900s. Taylor commuted to his city job, spending weekends in the Trapps with his family. He also worked as a plumber, noted the 1940 census.

Roger Van Leuven was one of the older students when the bridges were built. When asked decades later if he remembered them, Roger remarked, "I mostly built those bridges!" Pupils sat two to each double desk, some of which the students hand-built, he added. Minnewaska provided the stove cordwood, which Roger and other students cut to heat the school.

Alfred B. Smiley, who grew up at Lake Minnewaska, was driven to school in a hotel limousine but noted, "it was a tight fit as often someone rode on my lap to New Paltz."

OFI; VLFI; Mabee:73; CR, Town of Gardiner (1940); AFB.

3. See Marc B. Fried's *Tales from the Shawangunk Mountains*, 90–91, for fuller descriptions of two versions

of duck on a rock. As a kid, Don Otis also played duck on a rock, hopscotch, and hoops and stick. His mother, Agnes, hated duck on a rock because "someone always got hit" with all the rocks flying around.

Don recalled milk and bread delivery trucks, as well as traveling salesmen coming through the Trapps. He called Goldwasser's grocery truck, "a very professional operation" with a cash register and large hanging scales to weigh meat.

During the winter the roads were often impassable, and families were able to get off the mountain only at times. Roger and Doris told reporter Nancy Ward that Minnewaska Trail, still called the old turnpike by some, "was never plowed in the winter. Their mother got a petition together and got the road plowed in the 1930s," Ward wrote. Roger told Ward that when it snowed, "We were on our own."

Cassie also chased Hal Bittner away from the Van Leuvens' strawberry patch when Hal attended the Trapps School in the 1920s.

VLFI; HBI; OFI; Nancy Ward, *Blue Stone Press*, 7 May 1999:6.

4. Ray Veres's father and uncle both were named Frank. Ray's father had a younger sister named Margaret, likely the Aunt Maggie married to Uncle Frank.

The Prohibition Act of 1919 (the Eighteenth Amendment) was repealed in 1933, thus ending the era of bootleggers and federal "revenoors."

Whether Huckleberry Charlie and Charlie the Peddler, mentioned in chapter thirteen, are the same man is unknown.

HH, 19 May 1983:16–17, 26 May 1983:5, 28; CR, Borough of Brooklyn (1900).

Chapter 16: Reentering the Wilderness

1. Shaughnessy:vii; Doughty:25; Paul C. Huth, interview, 29 Oct 2007.

Bibliography

Accordian, The. Newsletter of Friends of Historic Rochester, Inc. Accord, N.Y.: 1987 to present.

Adams, Arthur G. *The Catskills: An Illustrated Historical Guide with Gazetteer*. Albuquerque, N.M.: Sun Books, 1977. Reprint, New York: Fordham University Press, 1999.

Aiello, John. "George Frost." RootsWeb.com. 1998–2006. http://www.rootsweb.com.

Alligerville Methodist Church Records, 1860–1911. Vols. I–II. Compiled and edited by Kenneth E. Hasbrouck. Town of Rochester, N.Y., 1957.

Andrist, Ralph K. *The Erie Canal*. Mahwah, N.J.: American Heritage Library, 1964.

Animal Improvement Programs Laboratory (AIPL) "Kid's Corner: Facts about Dairy Products." http://www.aipl.arsusda.gov/kc/dairy.html.

Anjou, Gustav. *Ulster County, N.Y. Probate Records in the Office of the Surrogate, and in the County Clerk's Office at Kingston, N.Y.* Vol. I. New York: Gustav Anjou, 1906.

Annual Report of the Adjutant General of the State of New York. For the Years 1900, 1901, 1903, and 1905. Albany, N.Y.: James B. Lyon, 1901, 1902, 1904, and 1906.

Anthony, H. E. *Field Book of North American Mammals*. 1928. New York: G. P. Putnam's Sons, 1954.

Asher and Adams' Pictorial Album of American Industry. 1876. New York: Rutledge Books, 1976.

Atkins, Thelma. Letters, to Glenn Hoagland, executive director, Mohonk Preserve, 1 May 1997; and to Bob Larsen, 30 October 1997. Daniel Smiley Research Center, Mohonk Preserve, New Paltz, N.Y.

Avery, Kevin J., and Franklin Kelly, eds. *Hudson River School Visions: The Landscapes of Sanford R. Gifford*. The Metropolitan Museum of Art Catalogue. New Haven, Conn.: Yale University Press, 2003.

Bader, Sara. *Strange Red Cow and other curious classified ads from the past*. New York: Clarkson Potter, 2005.

Banta, Theodore M. *Year Book of the Holland Society of New York*. New York: Knickerbocker Press, 1902.

Barber, Gertrude A., compiler. *Cemeteries in Orange County, N. Y.* Vol. 4. 1930.

Bascom Maple Farms. "Our Products." http://www.sugarmaple.net/index/html.

Beacon Hill Fine Art. *McEntee & Company*. Exhibition catalog. November 25, 1997–January 17, 1998. New York: Beacon Hill Fine Art, 1997.

Berendsen, Bieneke. (2004, August 1). "Plural nouns endings in -en." http://www.dutchgrammar.com and http://www.gnomjas.nl/html/dutch//.

Bevier, Katherine. *The Bevier Family: A History of the Descendants of Louis Bevier ...* New York: Tobias A. Wright, 1916.

Bittner, Harold. Interviews, late 1990s, including 27 May 1998. Maps and Cards 1998. Daniel Smiley Research Center, Mohonk Preserve, New Paltz, New York.

Black, George. "The Fox." *About Town*, Spring 2010:1 and 15.

———. "Oscar Lyons." *About Town*. Summer 2010:8 and 13.

———. Interviews and correspondence. 2010 to present.

Blue Stone Press. Stone Ridge, N.Y.: October 1996 to present.

Bray, Gregory, Brian Cortright, and Matt Dorcas. Untitled videotape footage for "Steps through History: The Trapps Lifestyle." State University of New York at New Paltz, Communications and Media Department, Spring Semester 2000.

Brink, Benjamin Myer. *The Early History of Saugerties 1660–1825*. Kingston, N.Y.: R.W. Anderson and Son, 1902.

———. *Olde Ulster: A Historical and Genealogical Magazine*. Vol. 1–10. Kingston, N.Y.: R.W. Anderson and Sons, 1905–1914.

Brooks, Karl L. "A Catskill Flora and Economic Botany. III. Apetalae ..." Bulletin No. 443. New York State Museum. Albany, N.Y.: The University of the State of New York, The State Education Department, 1980.

Buff, Barbara Ball. "Introduction." *Windy Summits, Fertile Valleys: An Artistic Journey through the Shawangunk Mountains*. Cragsmoor, N.Y.: Cragsmoor Free Library, 1982:3–5.

Burgess, Larry E. *As the Spirit Moves Them: The Smileys (Alfred, Albert, and Daniel Smiley: A*

Biography.) Redlands, Calif.: Larry E. Burgess, Beacon Printery, 1969.

———. *Mohonk, Its People and Spirit: A History of One Hundred Years of Growth and Service.* 1980. Reprint, Fleischmanns, N.Y.: Purple Mountain Press, 1996.

Burroughs, John. *In the Catskills: Selections from the Writings of John Burroughs.* Boston and New York: Houghton Mifflin Co., 1871. Reprint, The Riverside Press, 1910.

———. *Signs and Seasons.* Boston and New York: Houghton Mifflin Co., The Riverside Press, 1886 and 1895.

Carmer, Carl. *The Hudson.* Illustrated Rivers of America Series. 1939. Revised edition, New York: Grosset & Dunlap, 1968.

A Census of Pensioners for Revolutionary or Military Services ... Under the Act for Taking the Sixth Census. 1841. Reprint prepared with a general index by the Genealogical Society of the Church of Jesus Christ of Latter-Day Saints. Baltimore, Md.: Genealogical Publishing Co., 1974.

Century House Historical Society, The. "History's Our Middle Name!" 2005. http://century-house.org/history.html.

Child, Hamilton, compiler and publisher. *Gazetteer and Business Directory of Ulster County, N. Y. for 1871–2.* Syracuse, N.Y.: Hamilton Child, 1871.

Chirp, The. Field Trip Report, Coxing Camp. Newsletter of the John Burroughs Natural History Society. November 1970:9.

Cikovsky, Nicolai, Jr., and Franklin Kelly. *Winslow Homer.* New Haven, Conn.: Yale University Press, 1995.

Citizen Herald. Walden, New York. 1909–1981. Formerly *The Walden Citizen* and *The Walden Herald.*

Clearwater, Alphonso T., ed. *The History of Ulster County, New York.* Kingston, N.Y.: W. J. Van Deusen, 1907.

Clove Chapel, The. Twenty-fifth anniversary booklet. 1901.

Clove Reformed (Dutch) Church, Baptismal, Marriage and Death Records. High Falls, New York. Kenneth Hasbrouck, compiler. New Paltz, N.Y.: Huguenot Historical Society, 1951.

Coddington Family Interviews. Evelyn Coddington DeWitt, Marian Whitford, and Grace C. Smith. Map of Coddington Farm by Mrs. DeWitt, Daniel Smiley Research Center, Mohonk Preserve, New Paltz, N.Y. 1997 to present.

Cole, Thomas. "Letter to Asher B. Durand, 7 August 1838." Thomas Cole Papers 1801–1848, New York State Library, Manuscripts and Special Collections, Albany, N.Y.

Commemorative Biographical Record of Ulster County, New York, Containing Biographical Sketches of Prominent and Representative Citizens, and of many of the Early Settled Families. Chicago: Beers and Company, 1896.

Corbett, Larry. "A Short History of Public Education in Jefferson County, New York 1802–2001." http://www.geocities.com/lrcorbett/schools.html.

"Correspondence between Albert K. and Alfred H. Smiley, 1869–1879." Mohonk Mountain House Archives, New Paltz, N.Y.

Cross, Sanford A. Letter. 1 October 1941. Daniel Smiley Research Center, Mohonk Preserve, New Paltz, N.Y.

Daily and Sunday Freeman. Kingston, New York. Predecessors include *The Freeman*, Rondout, New York (1859–1872) and *The Kingston Daily Freeman* (1878–1969).

De Kay, James E. *Zoology of New York State or the New-York Fauna ... Part I, Mammalia, and Part III, Reptiles and Amphibia.* Albany, N.Y.: White and Visscher, 1842.

De Lisser, Richard Lionel. *Picturesque Ulster: Being a Pictorial Work on the County of Ulster, State of New York.* Kingston, N.Y.: The Styles & Bruyn Publishing Co. 1896–1905.

Dilts, Bryan Lee, compiler. *1890 New York Census Index of Civil War Veterans or Their Widows.* Salt Lake City, Utah: Index Publishing, 1984.

Directory of the New York, Ontario & Western Railroad, Ellenville to Kingston, 1903–1904. Campbell Hall, N.Y.: Breed Publishing, 1895.

Directory of the Wallkill Valley Railroad 1895. Campbell Hall, N.Y.: Breed Publishing, 1895.

Doherty, Frank J. *Settlers of the Beekman Patent, Dutchess County, New York: An Historical and Genealogical Study of all 18th Century Settlers in the Patent. Volume V (Fackert to Haas) of VIII (to Millington).* 1990–2005. Pleasant Valley, N.Y. F. J. Doherty, Vol. V, 1999.

Doughty, William E. *Lake Minnewaska: In the Heart*

of the Shawangunks. 1946. Third edition, New York: George H. Smiley & Son, and Herbert-Spencer, 1960.

Dunn, Russell. *Catskill Region Waterfall Guide: Cool Cascades of the Catskills & Shawangunks*. Delmar, N.Y.: Black Dome Press, 2004.

Ebert, Milford, and Lucille Ebert, compilers. *Town of Rochester Cemeteries, Ulster County, New York*. Accord, N.Y.: Friends of Historic Rochester, 1997.

Eckel, Edwin C. "Cement Industry in New York." *Bulletin of the New York State Museum*. November 1901. Vol. 8 (44):849–955.

Ellenville Journal, The. Ellenville, N.Y.: 1847 to present.

Ellis, David Maldwyn. *Landlords and Farmers in the Hudson-Mohawk Region 1790–1850*. New York: Octagon Books, 1967.

Evers, Alf. *The Catskills: From Wilderness to Woodstock*. Garden City, N.Y.: Doubleday and Company, 1972.

———. *Woodstock, History of an American Town*. Woodstock, N.Y.: Overlook Press, 1987.

Fagan, Jack. *Time and the Mountain: A Guide to the Geology of the Northern Shawangunk Mountains*. New Paltz, N.Y.: The Mohonk Preserve, 1996.

Fogg, William. "Fogg's Early Families." Excerpt from *The Early Inhabitants of Eliot and Kittery, Maine*. 1859. *In Old Eliot*, Historic Press, Eliot, Maine. January 1901. Vol. IV (1):168–176.

Folts, James D. "History of the University of the State of New York and the State Education Department 1784–1996." New York State Education Department, Board of Regents. New York State Library. 1996. http://www.nysl.nysed.gov/edocs/education/sedhist/htm.

Foote, Charles E. *The History of Ulster County, New York*. Alphonso T. Clearwater, ed. Kingston, N.Y.: W. J. Van Deusen, 1907:253–57, 343–53, 354–59, 360–65, 373–79.

Fowler Family Bible. Courtesy of Joan Hayes Wustrau.

Fox, Cynthia G. "Income Tax Records of the Civil War Years." Originally published in *Prologue Magazine*, Winter 1986, Vol. 18 (4). http://www.archives.gov/publications/prologue/1986/winter/civil-war-tax-records.html.

Frame, C. T. "Town of Marbletown." *The History of Ulster County, New York*. Alphonso T. Clearwater, ed. Kingston, N.Y.: W. J. Van Deusen, 1907, 275–86.

French, J. H. *Gazetteer of the State of New York: Embracing a Comprehensive View of the Geography, Geology, and General History of the State ...* . Syracuse, N.Y.: R. Pearsall Smith, 1860.

Fried, Marc B. *The Huckleberry Pickers: A Raucous History of the Shawangunk Mountains*. Hensonville, N.Y.: Black Dome Press, 1995.

———. *Shawangunk Place Names, Indian, Dutch and English Geographical Place Names of the Shawangunk Mountain Region: Their Origin, Interpretation and Historical Evolution*. Gardiner, N.Y.: Marc B. Fried, 2005.

———. "A Tale of Two Houses: The Bruyn and Smedes Homesteads of the Shawangunk Valley." *de Halve Maen*. Fall 2006:43–50.

———. *Tales from the Shawangunk Mountains: A Naturalist's Musings, A Bushwhacker's Guide*. Glens Falls, N.Y.: Adirondack Mountain Club, 1981.

Friel, Arthur O. *Cat O'Mountain*. Philadelphia: The Penn Publishing, 1923.

———. *Hard Wood*. Philadelphia: Penn Publishing, 1925.

Frost, George P. Letter to Joseph Frost of Eliot, Maine. 11 August 1830. Courtesy of Frost descendants Lester E. Frost and William Carson, and William Fogg Library, Eliot, Maine.

Froyle, Hampshire, England. "An American Connection." http://www.froyle.com/amcom/3.htm.

Gardiner School District #1, "Daily School Register, July 1889–July 1890." Haviland-Heidgerd Historical Collection, Elting Memorial Library.

Gardiner (N.Y.) *Weekly, The*. 1882–19??.

Glazier, Ira A. *The Famine Immigrants: List of Irish Immigrants Arriving at the Port of New York 1846–1851*. Vol. I (Jan 1846–June 1847) and Vol. II (Jul 1847–Jun 1848). Baltimore, Md.: Genealogical Publishing Co., 1983.

Gordon, Thomas F. *Gazetteer of the State of New York: Comprehending Its Colonial History ...* Philadelphia: T.K. and P.G. Collins, 1836.

Green-Wood Cemetery, The. "About Us." http://www.green-wood.com/.

Griffin, William D. *A Portrait of the Irish in America*. New York: Charles Scribner's Sons, 1981.

Guide Book to Lake Mohonk, New Paltz, Ulster Co. N.Y. Providence, R.I.: Rhode Island Printing Company, 1875.

Guide-Book to Mohonk Lake and Its Vicinity. Illustrated by E. J. Whitney. Ulster County, N.Y.: Mohonk Lake House, 1888.

Guide Book to Mohonk Lake, Ulster County, N. Y. New York: Chas. M. Green Printing Company, 1883.

Guilford Reformed (Dutch) Church Records, 1832–1930. Original records and Transcriptions by Kenneth E. Hasbrouck and Shirley V. Anson.

Hakam, Margaret, and Susan Houghtaling. *Cragsmoor—an Historical Sketch*. Cragsmoor, N.Y.: Cragsmoor Free Library, 1983.

Hansen, Harry. "History of the Town of Rochester." Town of Rochester, Ulster County, New York, Historic Resources Reconnaissance Survey. 1993. http://www.accord-kerhonkson.com/myweb/history.htm/.

Haring, H. A. *Our Catskill Mountains*. New York: G. P. Putnam's Sons, 1931.

Harp, Peter H. *Horse and Buggy Days: A History of New Paltz, New York*. 1969. Reprint, Wallkill, N.Y.: Wallkill Valley Federal Savings and Loan, et al., 1994.

Harris, Harold. *Treasure Tales of the Shawangunks and Catskills*. 1955. Published in commemoration of the Ellenville Sesquicentennial, 1805–1955. Second edition, Ellenville, N.Y.: Harold Harris, 1998.

Harris, Jennie E. "Stew Me Some Ginseng. I Need Some Strength." *Natural History*. The American Museum of Natural History. November 1948:424–428.

Hasbrouck, Kenneth E. *A Guide to Huguenot Street and Preservation Efforts of The Huguenot Historical Society, New Paltz, N. Y.* New Paltz, N.Y.: Huguenot Historical Society, 1982.

———, compiler. *History of the Township of Gardiner*. 1953. Reprint, 125th Anniversary 1853–1978, New Windsor, N.Y.: Commercial Offset Printers, 1978.

Hayes, Arthur. "Arthur Hayes." Eight-page typed biographical paper, ca. 1987–1988. Courtesy of daughter Joan Hayes Wustrau.

Heads of Families at the First Census of the United States Taken in the Year 1790, New York. Originally published by the Government Printing Office, Washington, D.C., 1908. Reprint, Baltimore, Md.: Genealogical Publishing, 1971.

Heidgerd, Ruth P. *New Paltz Rural Cemetery Records 1860–1962*. Indexed. 1962.

———. *Ulster County in the Revolution (A Guide to Those who Served)*. New Paltz, N.Y.: Huguenot Historical Society, 1977.

Heidgerd, William, compiler. *The American Descendants of Chretien Du Bois of Wicres, France*. New Paltz, N.Y.: Du Bois Family Association, Huguenot Historical Society, 1968.

———. *Black History of New Paltz, Parts I and II*. New Paltz, N.Y.: Haviland-Heidgerd Historical Collection, Elting Memorial Library, New Paltz, N.Y., 1989.

———. *New York Biographical Dictionary*. New Paltz, N.Y.: Haviland-Heidgerd Historical Collection, Elting Memorial Library. Ongoing, no publication date.

Herald, The. New Paltz, N.Y. 1985–1987.

Heusser, George. *Gold, Silver and Other Mines of the Shawangunks*. Ellenville, N.Y.: George Heusser, 1976.

———. *Legend and History and Minerals of the Ellenville Mines*. Ellenville, N.Y.: George Heusser, 1976.

Hewitt, Charles Edward. *The Hand in the Dark*. Albany, N.Y.: Hamilton Printing, 1947.

———. *Hazel*. New York: Broadway Publishing, 1906.

———. *The Outcast*. New York: J. S. Ogilvie Publishing, 1913.

Historic Wallkill and Hudson River Valleys. Originally published as the *Souvenir of the Wallkill Valley, Its Rivers, Its Environments and Its History by the Wallkill Farmers' Association*. Numbers 1–24. Walden, N.Y.: The Wallkill Publishing Association, 1894–1921.

History of the Reformed Church of Guilford, New York. Published by the Ladies of the Congregation. Somerville, N.J.: Press of the Unionist Gazette, 1890.

Hitchcock, Elwood. *Big Hollow: A Mountaintop History, Maplecrest, Windham, N.Y.* Hensonville, N.Y.: Black Dome Press, 1993.

Hoes, Roswell Randall, ed. *Baptismal and Marriage Registers of the Old Dutch Church of Kingston, Ulster County, New York 1660–1809*. New York: De Vinne Press, 1891. Reprint, Baltimore, Md.: Genealogical Publishing Co., 1997.

Hoffman, A. W. "The Passing of the Hemlock." *Picturesque Ulster*. R. Lionel De Lisser. 1896–1905. Reprint, Cornwallville, N.Y.: Hope Farm Press, 1968:185–191.

Hough, Franklin B. *Gazetteer of the State of New York, Embracing a Comprehensive Account of the History and Statistics of the State ....* Albany, N.Y.: Andrew Boyd, 1872.

Hudson River Valley Heritage, HRVH Historical Newspapers. www.hrvh.org.

Huguenot and Highland Herald. New Paltz, N.Y. 1988–1995.

Huguenot Herald. New Paltz, N.Y. 1976–1985.

Huth, Paul C., compiler. "Checklist of Trees and Shrubs of the Northern Shawangunk Mountains." 1990. Mohonk Preserve, New Paltz, N.Y.

———, and Daniel Smiley. "American Chestnut: 1984 Status in the Shawangunks." April 1985. Mohonk Preserve Research Report. Daniel Smiley Research Center, Mohonk Preserve, New Paltz, N.Y.

"Internal Revenue Assessment for Ulster County, Town of Gardiner, Division Eleven." 1862–66. Haviland-Heidgerd Historical Collection, Elting Memorial Library, New Paltz, N.Y.

Internal Revenue Service, United States Department of the Treasury. "Brief History of IRS." http://www.irs.gov/irs/index.html.

Inventory of the County Archives of New York State. *Historical Records Survey of New York (State).* No. 51. Vol. 2. Ulster County. Albany, N.Y.: 1940.

Jandebeur, Harold E., and Jeannette Dingee Jandebeur. *Dingee Family Genealogy*. Walnut Creek, Calif.: Jandebeur and Jandebeur, 1987.

Janey and Phebe Van Leuvan letter, December 10, 1876. Haviland-Heidgerd Historical Collection, Elting Memorial Library, New Paltz, N.Y.

Junior League of Kingston. *Early Architecture in Ulster County*. New York: Hancock, Little, Calvert, 1974.

———. *Survey of Buildings in Historic Ulster County*. Vols. 1–7. New York: Junior League of Kingston, 1969.

Kilpatrick, Susie, ed. "From the Schoonmaker Family Association." *On Huguenot Street*: Newsletter of the Huguenot Historical Society, New Paltz, N.Y. (Fall 2004):3.

King, Jonathan D. "Rediscovered Rock." *Chronogram*, June 2005:34–35.

Kingston (N.Y) *Argus and Ulster Republican, The.* 1861–1905.

Kingston (N.Y) *Daily Freeman, The.* See *Daily Freeman.*

Kingston (N.Y.) *Democratic Journal, The.* 1837–1868.

Kingston (N.Y.) *Journal and Weekly Freeman, The.* 1881–1886.

Knickerbocker, Nan L. *Minnewaska "Frozen Waters": The Intimate Story of the Origin & Growth of the Lake Minnewaska Mountain Houses.* New York: Nan L. Knickerbocker, 1937.

Knights of Pythias. "Welcome" and "About the Order." From "The Pythians, The Order of the Knights of Pythias." http://www.pythias.org/index.htm.

Kraft, Herbert C. *The Lenape: Archaeology, History, and Ethnography*. Newark, N.J.: New Jersey Historical Society, 1986.

———. *The Lenape-Delaware Indian Heritage: 10,000 B.C.–A.D. 2000.* [S.I.]: Lenape Books, 2001.

Lacey, Simpson & Co., pubs. *Ulster County Directory, for 1892–3, Containing a Patron's Business Directory of the City of Kingston*. New York, 1896. http://distantcousin.com/Directories/NY/UlsterCounty.

LaChance, Joan A. "The True Story of Albert and Alfred Smiley and the Founding of Mohonk and Minnewaska." In *Mohonk Bits*, Mohonk Mountain House, New Paltz, N.Y., Summer 2006.

Lake Minnewaska Mountain Houses. E. A. and G. H. Smiley, Proprietors. Season of 1919. Brochure with unnumbered pages.

Lake Minnewaska Mountain Houses Web site. Employee Listing, Photo Gallery. Brochures and Books, and History Timeline. http://www.lakeminnewaska.org.

Lake Minnewaska News. Lake Minnewaska, N. Y. Spring 1960.

Laks, Reba W. "Early Afro-American Involvement with the Shawangunk Mountains." Historical Cultural Note No. 37–96. Mohonk Preserve, New Paltz, N.Y. 1996.

Lang, Elizabeth, and Robert Lang. *In a Valley Fair, A History of The State University College of Education at New Paltz, N. Y.* New Paltz, N.Y.: State University College of Education at New Paltz, N.Y., 1960.

Larsen, Robert. *Trapps Mountain Hamlet: An Interpretive Walk through a Vanished Shawangunk Community*. New Paltz, N.Y.: Mohonk

Preserve, 1999.

———. "The Trapps Mountain Hamlet: A Part of Rochester's History." *The Accordian*. Vol. XII (1). Accord, N.Y.: Friends of Historic Rochester, March 2003:2–6.

———, and Annie O'Neill. "What a Gas Station!" *Shawangunk Watch*, Fall/Winter 2000:3.

Lawrence, D. S. and Co., pub. *Ulster County Directory for 1880–81, Containing a Historical Sketch of the County ...* . Newburgh, N.Y.: Lawrence, D. S. and Co., 1881.

Le Fevre, Ralph. *History of New Paltz, New York and Its Old Families (From 1678 to 1820)*. Second edition, Albany, N.Y.: Fort Orange Press, 1909.

Lehigh University Digital Library. "Geology of Lehigh and Northhampton Counties, Pennsylvania: Historical Notes." From the *Pennsylvania Geological Survey*, 4th Series. http://digital.lib.lehigh.edu/lvgeolog/_1001.html.

Lewit, Sam, and Ken Philips, Jr. "Lake Minnewaska Mountain Houses—Minnewaska History." http://lakeminnewaska.org/.

Library of Congress. "The Grand Army of the Republic and Kindred Societies." Bibliographies and Guides. http://www.loc.gov/rr/main/gar/garintro.html.

Luke, Lorraine M., compiler, and Donald C. Donohue, editor. *The Roosa Family of North America*. Vol. VIII, Books I–III. Hurley, N.Y.: The Ulster County Genealogical Society, 1998.

Mabee, Carleton. *Gardiner and Minnewaska*. Charleston, S.C.: Arcadia Publishing, 2003.

Macera, Lucile P. "The Wonder Root." *The Conservationist*. State of New York, Department of Environmental Conservation. May–June 1982:30–33.

Mack Family Bible, courtesy of Earl Mack.

Manning, Peter Quincy. "An Interpretive Exploration of the Shawangunk Carriage Roads." A Thesis Presented to the Faculty of the Graduate School of Cornell University in Partial Fulfillment of the Requirements for the Degree of Master of Landscape Architecture. January 1995.

Marbletown Reformed Dutch Church Baptismal Records, Marbletown, New York, 1746–1871. Transcribed and indexed by Arthur C. M. Kelley. Rhinebeck, N.Y., 1979.

Marbletown Reformed Dutch Church Records. Marbletown Reformed Dutch Church, Ulster County, New York, 1737–1944, and Stone Ridge Methodist Church, Ulster County, New York. Copied and indexed by Mrs. Jean D. Worden. Franklin, Ohio, 1985.

Mather, J.H., and L.P. Brockett. *A Geographical History of the State of New York ... the Whole forming a Complete History of the State*. Utica, N.Y.: H.H. Hawley, 1848.

Matteson, Benjamin H., and Joan A. LaChance. *The Summerhouses of Mohonk*. Lake Mohonk, New Paltz, NY: Mohonk Mountain House, June 1998.

May, Jennifer. "For the Love of Ginseng." *Chronogram*. Article dated October 2005. Copyright 2006. http://www.chronogram.com/issue/2005/10/communitynotebook/index.php.

McAllister, Michael F. "Charles Edward Hewitt Revisited." May 17, 1979. Paper written for Professor H. R. Stoneback. State University of New York at New Paltz. Haviland-Heidgerd Historical Collection, Elting Memorial Library, New Paltz, New York (HHHC-EML).

McDermott, William P. "The Famine Irish Arrive in Poughkeepsie—1850." *Dutchess County Historical Society Year Book*. Edited by William P. McDermott. Clinton House Museum, Poughkeepsie, N.Y. 1984, Vol. 69:171–83.

Miller, Kerby A. *Emigrants and Exiles, Ireland and the Irish Exiles*. New York: Oxford University Press, 1985.

Millham, George H. Interviews, 24 Mar 2010, 25 Apr 2011.

Minks, Louise. *The Hudson River School: The Landscape Art of Bierstadt, Cole, Church, Durand, Heade, and twenty other artists*. New York: Crescent Books, Crown Publishers, 1989.

"Minnewaska Business Records for Lake Minnewaska Mountain Houses (Cliff House and Wildmere), 1878–1948." HHHC-EML.

Mohonk Lake Cooperative Weather Station, United States Weather Bureau (NOAA). 1896–present. Daniel Smiley Research Center, Mohonk Preserve, New Paltz, N.Y.

Mohonk Preserve Newsletter, New Paltz, N.Y. Under The Mohonk Trust, No. 1–47, Dec 1967–Winter 1980–1981. Under the Mohonk Preserve, No. 48, Spr 1981 to present. *Ridgelines* (Mohonk Preserve), No. 107, Spr 1996 to present.

Muise, Jeff. *Old Bob's Gift*. Wappingers Falls, N.Y.: The Shawangunk Press, 1996.

National Arbor Day Foundation, The. "The History of Arbor Day" and "Arbor Day's Beginnings." http://www.arborday.org.arborday/history.cfm.

Nature Conservancy, The, and Open Space Institute. *The Huckleberry Pickers: A Vanished Shawangunk Culture*. Documentary film, 2006. Produced by The Ellenbogen Group, Video Production & Multi-Media Services.

Needham, Walter, and Barrows Mussey. *A Book of Country Things*. Brattleboro, Vt.: Stephen Greene Press, 1965.

Newland, D. H. *The Mining and Quarry Industry of New York State: Report of Operations and Production During 1910*. New York State Museum Bulletin, Albany, 1911:151:39–40.

New Paltz Chamber of Commerce. http://www.newpaltzchamber.org/pages/New_Resident_Information.php.

New Paltz (N.Y.) *Independent, The*. September 1868–June 1919.

New Paltz (N.Y.) *Independent and Times*. June 1919–June 1972.

New Paltz (N.Y.) *News, The*. 1953–1980.

New Paltz Reformed Dutch Church Records. *Records of the Reformed Dutch Church of New Paltz, N.Y.* Collections of the Holland Society, Vol. III. New York: Knickerbocker Press, 1896; *New Paltz Reformed Church Marriages, 1817–1844*. Copied by Kenneth E. Hasbrouck.

New Paltz Rural Cemetery Records 1860–1962. Indexed by Ruth P. Heidgerd. New Paltz, N.Y.

New Paltz (N.Y.) *Times*. Ulster Publishing. March 2001 to present.

New Paltz (N.Y.) Times, The. September 1868–June 1919. Renamed *The Southern Ulster Times*, September 1865–June 1866.

New York–New Jersey Trail Conference, and The American Geographical Society. *New York Walk Book*. Fourth edition. Garden City, N.Y.: Doubleday/Natural History Press, 1971.

New York State Adjutant General Report. See State of New York.

New York State Assembly. News from Assembly member Kevin A Cahill. "Legislature Advances Designation of Shawangunk Mountains Scenic Byway." http://assembly.state.ny.us/mem?ad=101&sh=story&story=20190.

New York State Conservationist. "Puma" and "Wolf." Feb–Mar 1948:32.

New York State Department of Transportation. "Past & Present." 1999–2007. http://www.nysdot.gov/portal/page/portal/about-nysdot/history/pastpresent.

New York Times, The. 1851 to present.

OneCLE, Inc. "New York Education Law Section 1501-Existing School Districts Continued." 2008. http://law.onecle.com/new-york/education/EDN01501_501.html.

Oral History Interviews. "Kenneth and Lucille Phillips." 1999 and 2000. Vol. 3:70–86. Taped and transcribed by Marion Ryan for the Haviland-Heidgerd Historical Collection, Elting Memorial Library, New Paltz, N.Y.

Orange County Times-Press. Middletown, N.Y. 1906–1927.

"Orphan Train Riders, The." 2001. D. Bruce Ayler, site editor. http://www.orphantrainriders.com/Menu.html.

Otis Family Interviews. Don and Joe Otis. 1998 to present.

Palmer, Sharon S. "The One-Room Schoolhouse." *Columbia County History & Heritage*. Columbia County Historical Society, Spring 2002, Vol. 1 (1):6–7. http://www.cchsny.org/CCHH_Spring_2002.pdf.

Partington, Frank E. *The Story of Mohonk*. 1911. Fourth edition. Part Two, compiled by Daniel Smiley, Jr. and Albert K. Smiley, Jr., 1932; Part Three by Albert K. Smiley, Jr., 1962. Smiley Brothers, Inc. Annandale, Va.: Turnpike Press, 1970.

Partridge, Eric. *Origins: A Short Etymological Dictionary of Modern English*. New York: Greenwich House, 1983.

PBS.org (Public Broadcasting Service). "Influenza 1918." American Experience series. Michael Azevedo. 1995–2006. http://www.pbs.org/wgbh/amex/influenza/index.html.

Peck, I. Heyward. "The Rev. Johannes Theodorus Polhemus and Some of His Descendants." *The New York Genealogical and Biographical Record*. Vol. XCI, No. 3. (July 1960):174–78.

Phillips, Lucille, ed. *Lake Minnewaska News*. May 1, 1958–March 1964.

Phillips, Sandra S., and Linda Weintraub, eds. *Charmed Places: Hudson River Artists and Their Houses, Studios, and Vistas*. New York: Harry N. Abrams, 1988.

Pike, M. H. "Shongum." Parts I–III. *Garden and Forest*. Vol. 5 (240–242) 28 September

1892:459–60, 5 October 1892:471–72, and 12 October 1892:483–84.

Pine Funeral Home Registers, including transcripts from the original register prepared by Linda M. Kelso. Haviland-Heidgerd Historical Collection, Elting Memorial Library, New Paltz, N.Y.

Plank, Will. *Banners and Bugles.* Marlborough, N.Y.: Centennial Press, 1963.

Polhemus, John F., and Richard V. Polhemus. *Up on Preston Mountain: The Story of an American Ghost Town.* Fleischmanns, N.Y.: Purple Mountain Press, 2005.

Polhemus, Mary, ed. *Town of Esopus Story: 3000 B.C.–1978 A.D.* Hannacroix, N.Y.: Hillcrest Press, 1979.

Pond Lily Mill Restorations. 1996. Theodore R. Hazen, Jr. http://www.angelfire.com/journal/pondlilymill/index.html. Florry, Paul B. "Millstones and Their Varied Usage." 2002; Hazen, Theodore R., Jr. "The Art of Millstones, How They Work." 2001.

Poucher, J. Wilson, and Byron J. Terwilliger, collectors and editors. *Old Gravestones of Ulster County, New York.* Vol. 1. Salem, Mass.: Higginson Book Company, 1931.

Poughkeepsie (N.Y.) *Journal*, and its predecessors, *The Poughkeepsie Eagle* and *Poughkeepsie New Yorker.* 1785 to present.

Register of Deaths, Town of Gardiner, County of Ulster, State of New York, From 1910 to 1923. Rochester, N.Y.: Williamson Book Company.

Rider, Richard, compiler. "Descendants of Daniel Coddington." Friends of Historic Rochester Genealogy Library, Accord, N.Y.

———. Ongoing genealogical research at the Friends of Historic Rochester Museum and Library, Accord, N.Y.

Ries, Heinrich. "Lime and Cement Industries of New York." *Bulletin of the New York State Museum.* November 1901. Vol. 8 (44):640–848.

Rochester Reformed (Dutch) Church, The. 1736–1901. Accord, Ulster County, New York. With Additions: Accord Methodist Church. 1871–1924. Copied, typed, and indexed by Mrs. Jean D. Worden. Franklin, Ohio: Mrs. Jean D. Worden, 1987.

"Rock Hill School," District 13, Town of Rochester, Fall 1912 Bulletin.

Rondout Courier, The. Rondout and Kingston, N.Y. 1847–1877(?).

Roster 156th Regiment, New York Vol. Infantry 1862–1865. Jesse Elting DuBois Collection. Haviland-Heidgerd Historical Collection, Elting Memorial Library, New Paltz, N.Y.

Russell, Emily W. B. "Three Centuries of Vegetational Change in the Shawangunk Mountains." Report to The Nature Conservancy. Department of Biological Sciences, Rutgers University, Newark, N.J. May 17, 2001.

Ruttenber, Edward Manning. "The County of Ulster." *The History of Ulster County, New York.* Alphonso T. Clearwater, ed. Kingston: N.Y.: W. J. Van Deusen, 1907:17–203.

———. *History of the County of Orange with a History of the Town and City of Newburgh.* 1875 and ca. 1887. Reprint, Goshen, N.Y.: Orange County Genealogical Society, 2002.

———, and Lewis H. Clark, and a Corps of Biographical Writers. *History of Orange County, New York with Illustrations and Biographical Sketches of Many of its Pioneers and Prominent Men.* 1881. Reprint, Interlaken, N.Y.: Heart of the Lakes Publishing, 1980.

Salerno, Louis, Questroyal Fine Art, LLC, Manhattan, New York. "Literature" on Thomas Cole's painting *Lake Mohonk* for Exhibition Catalogue, *In the Garden of Liberty: Paintings of the Hudson River School.* March 1–31, 2006.

———. "Thomas Cole (1801–1848)." Exhibition catalogue, *Important American Paintings.* Fall 2007.

Sanderson, Dorothy Hurlbut. *Ellenville Days and Ways until 1897.* Ellenville, N.Y.: Rondout Valley Publishing Company, 1968.

Schoonmaker, Alice, ed. *The History of Ulster County with Emphasis upon the Last 100 Years 1883–1983 Compiled by the Historians of Ulster County For the Tercentenary Year 1983.* Kingston, N.Y.: Ulster County Historians, 1984.

Scott, Charles. "Origin and Meaning of the Word Shawangunk." *Collections of The Ulster Historical Society.* Vol. I. Kingston, N.Y.: Hommel & Lounsbery, 1860:229–33.

Sears Archives. "History of the Sears Catalog." http://www.searsarchives.com/catalogs/chronology.htm.

Second Annual Report of the Forest Commission [N.Y.S.]. Albany: Argus Company, 1887.

Sharpe, Melvin, Jr. Letter, to Glenn Hoagland, executive director, Mohonk Preserve. 22 April 1997. Daniel Smiley Research Center,

Mohonk Preserve, New Paltz, N.Y.

Shaughnessy, Jim. *Delaware & Hudson*. Berkeley, Calif.: Howell-North Books, 1967.

Shawangunk Watch. Spring 1995 to present. (Previous to Spring 1995, *Friends of the Shawangunks newsletter*.) Friends of the Shawangunks & The Shawangunk Conservancy, Accord, N.Y.

Shear, S. R. "The Schools of the County." *The History of Ulster County, New York*. Alphonso T. Clearwater, ed. Kingston, N.Y.: W. J. Van Deusen, 1907:517–35.

Simpson, Jeffrey. *The Hudson River 1850–1918: A Photographic Portrait*. Tarrytown, N.Y.: Sleepy Hollow Press, 1981.

Siprelle, Lynn, et al. "A Chore for Every Day: the 'dishtowel' schedule." 1999–2006. http://www.thenewhomemaker.com/node/434.

Sloane, Eric. *Diary of an Early American Boy, Noah Blake 1805*. 1965. Reprint, New York: Ballantine Books, 1974.

———. *Our Vanishing Landscape*. 1955. Reprint, New York: Ballantine Books, 1976.

———. *A Reverence for Wood*. 1965. Reprint, New York: Ballantine Books, 1974.

Smiley, Alfred B. Interviews and correspondence, 1997 to present.

Smiley, A. Keith. *An Anecdotal History of Mohonk*. 1957. Revised and edited by Pril Smiley and Keith LaBudde. New Paltz, N.Y.: Mohonk Mountain House, 1999.

Smiley, A. K., and D., Proprietors. *Mountain Drives of Mohonk Lake, New York*. Brochure published by Mohonk Mountain House, ca. 1910.

Smiley, Daniel. "Maple Trees to Maple Syrup." Transcript of talk, Pioneer Weekend, Mohonk Parlor. 4 March 1983. Daniel Smiley Research Center, Mohonk Preserve, New Paltz, N.Y. (DSRC-MP).

———. Mohonk Preserve Historical/Cultural Notes. DSRC-MP. "Resource Industries of the Shawangunks." No. 16. January 1986; "Wintergreen Oil Manufacture in the Shawangunks." No. 13. 24 August 1985; "Wood as a Shawangunk Resource." No. 25. June 1987.

———. *Mr. Dan's Landmark Walk*. New Paltz, N.Y.: Mohonk Mountain House, 1989.

———. "Notes on Lake Awosting," 29 September 1982. DSRC-MP.

———, and Alice Cross. "Shawangunk Place Names." April 1988. Mohonk Preserve Research Report, DSRC-MP.

———, and Keith Smiley, and Ruth Smiley. "Art Coddington Interviews." 19 June 1972 and 17 July 1972. DSRC-MP.

———, and Paul C. Huth. Daniel Smiley Research Center History Chronology Card File. On-going file of events at the Mohonk and Minnewaska resorts, DSRC-MP.

———, and Paul C. Huth. "Fish Disappearance in the Northern Shawangunks between 1862 and 1992." 1992. Mohonk Preserve Research Report, DSRC-MP.

———, and Paul C. Huth. "Making Do: Living Within an Ecosystem at Mohonk 1870 to 1930." April 1988. Mohonk Historical Report, DSRC-MP.

———, and Paul C. Huth. "Shawangunk Forest Fires 1842–1982." March 1983. Mohonk Preserve Research Report, DSRC-MP.

———, Paul C. Huth, and Sandra Smiley. "Vegetable and Fruit Growing in the Shawangunks." May 1988. Mohonk Preserve Research Report, DSRC-MP.

Smiley, Ruth Happel. *The Mohonk Gardens: A History and Guide*. 1977. Reprint, New Paltz, N.Y.: Mohonk Mountain House, 2004.

Smiley Mountain House Brochures, New Paltz, New York, 1880–1887. Lake Minnewaska House, Alfred H. Smiley, Proprietor; Lake Mohonk Mountain House, Albert K. Smiley, Proprietor; Minnewaska Mountain House brochures, Alfred H. Smiley, Proprietor.

Smith, Philip H. *Legends of the Shawangunk (Shongum) and Its Environs* ... Pawling, N.Y.: Smith & Company, 1887.

Snyder, Bradley. *The Shawangunk Mountains: A History of Nature and Man*. Mohonk Lake, New Paltz, N.Y.: Mohonk Preserve, 1981.

Spencer, P.R. *Spencerian System of Practical Penmanship*. 1874. Reprint, Fenton, Mich.: Mott Media, 1985.

State of New York Annual Report of the Adjutant General Transmitted to the Legislature January 31st, 1868. Vol. 3. Albany, N.Y.: Charles van Benthuysen & Sons, 1868.

Stessin-Cohn, Susan. "Ulster County Poorhouse Project." 2004. Ulster County Information Services. http://www.co.ulster.ny.us/poorhouse/index.html.

Steuding, Bob. *The Last of the Handmade Dams: The*

Story of the Ashokan Reservoir. Fleischmanns, N.Y.: Purple Mountain Press, 1995.

Stickney, Charles E. *A History of the Minisink Region: Which Includes the Present Towns of Minisink, Deerpark, Mount Hope, Greenville and Wawayanda, in Orange County, New York, etc.* 1867. Reprint, Florida, N.Y.: Minisink Valley Historical Society, 1989.

Stromer, Tim. "Glenwood." *Midwest Getaway Travel and Leisure Magazine*. http://www.midwestgetaway.com/glenwood.html.

Sunday Freeman. See *Daily Freeman*.

Sylvester, Nathaniel Bartlett. *History of Ulster County New York with Illustrations and Biographical Sketches of Its Prominent Men and Pioneers*. Parts First and Second. Philadelphia: Everts and Peck, 1880.

Terwilliger, Katharine T. *Wawarsing, Where the Streams Wind: Historical Glimpses of the Town*. Ellenville, N.Y.: Rondout Valley Publishing Company, 1977.

Thyssen, Esther T., contributor. *American Paradise, The World of the Hudson River School*. The Metropolitan Museum of Art. New York: Harry N. Abrams, 1985.

Times Herald-Record. Middletown, N.Y. 1960 to present.

Town of Rochester Archival Records. Rochester Town Hall, Accord, N.Y.

Trapps School Records. State of New York Department of Public Institution, Daily School Register of School District #8, Town of Gardiner, County of Ulster. Copied by the Haviland-Heidgerd Historical Collection at Elting Memorial Library, New Paltz, N.Y., from original registers owned by Kenneth and Lucille Phillips. For the years 1856, 1892–1908, and 1918–1919.

Ulster County Clerk's Office. 2006. Ulster County Information Services. Nina Postupack, County Clerk. http://www.ulstercountyny.gov/archives.

Ulster County Clerk's Office. various documents, including deeds, maps, court records, and Revolutionary War service affidavits. Kingston, N.Y.

Ulster County Directory for the Year 1885 with a Business Directory and an Appendix. Newburgh, N.Y.: Crum & Fowler, 1885.

Ulster County Historical Society. 2004. "About Bevier House: A House through the Ages and History: A Short History of the Ulster County Historical Society." http://www.bevierhouse-museum.org/.

Ulster County Surrogate's Office. Various wills. Kingston, N.Y.

Ulster County Whig. Rondout, N.Y. 1834–1837.

United States Census Records, federal and state, 1790 to 1930.

"Vanishing Profession, A." *The Mohonk Bulletin*. Mohonk Mountain House, New Paltz, N.Y. 2 August 1941:2.

Van Leuven Family Interviews. Doris Van Leuven Hall and Roger Van Leuven. 1997 to present.

Van Santvoord, C. *The One Hundred and Twentieth Regiment New York State Volunteers. A Narrative of Its Services in the War for the Union*. Kingston, N.Y.: The Press of the Kingston Freeman, 1894.

Van Valkenburgh, Norman. Handwritten description of Dr. George F. Hanker's twentieth-century medical rounds based on letter from Dr. Hanker's daughter, Molly. 10 January 2000. Daniel Smiley Research Center, Mohonk Preserve, New Paltz, N.Y.

Van Valkenburgh, Norman J. *Murder in the Catskills*. Fleischmanns, N.Y.: Purple Mountain Press, 1992.

———. *Murder in the Shawangunks*. Fleischmanns, N.Y.: Purple Mountain Press, 1999.

van Wagenen, Jared, Jr. *The Gold Age of Homespun*. New York: Hill and Wang for the New York Historical Association, 1953.

Vernon County Censor. Viroqua, Wisconsin. 1865–1955. LaCrosse Public Library, LaCrosse, Wisconsin. Obituary of Sarah Gorsline, courtesy Sheri Ballard.

Vesell, Elliot S., ed. *The Life and Works of Thomas Cole* by Louis Legrand Noble. 1964. Reprint, Hensonville, N.Y.: Black Dome Press, 1997.

Virtualology.com. "Virtual American Biographies: Platt Rogers Spencer." 2000 by Evisum, Inc. 2001 by Virtualology. Edited by Appleton's Encyclopedia. http://www.famousamericans.net/plattrogersspencer/.

Wakefield, Manville B. *Coal Boats to Tidewater: The Story of the Delaware & Hudson Canal*. 1965. Reprint, Fleischmanns, N.Y.: Purple Mountain Press, 1992.

Walker, Lester. *American Shelter: An Illustrated Encyclopedia of the American Home*. Woodstock, N.Y.: Overlook Press, 1981.

Wallkill Valley Cemetery, Walden, Orange County, NY, Interments 1867–1998. Photocopied from copies belonging to Ruth Feldt, April 2003.

Wallkill (N.Y.) *Valley World, The.* 1924–1935.

Wessels, Tom. *Reading the Forested Landscape: A Natural History of New England.* Woodstock, Vt.: The Countryman Press, 1997.

White, Virgil D., abstractor. *Genealogical Abstracts of Revolutionary War Pension Files.* Vols. I–III. Waynesboro, Tenn.: National Historical Publishing Company, 1990, 1991, and 1992.

White, Virgil D., transcriber. *Index to War of 1812 Pension Files.* Vols. I–III. Waynesboro, Tennessee: National Historical Publishing Company, 1989.

Whitford, Noble E. *History of the Canal System of the State of New York.* Vol. 1. 1905. http://www.history.rochester.edu/canal/bib/whitford/1906/Chap01.html and /Chap20.html.

Whittaker, Norma D. "An Early Mountain Hamlet: A simple story, written for the Twentieth Century Club of Kingston, and based on a few searched-out facts of the life of Adam Yeaple ..." Typed copy, unbound, 1976.

"William Pepperrell." http://en.wikipedia.org/wiki/William_Pepperrell.

Williams, Frank D. *Genealogy of the Williams Family.* December 1938.

Wilt, Richard A. *New York Soldiers in the Civil War.* Vol. I, A–K and Vol. II, L–Z. Bowie, Md.: Heritage Books, 1999.

Wood, Frederic J. *The Turnpikes of New England.* 1919. Reprint, Pepperell, Mass.: Branch Line Press, 1997.

Woodstock (N.Y.) *Weekly, The.* September–October 1924.

World Health Organization (WHO). "Immunizations, Vaccines and Biologicals" and "The History of Vaccination." 2003. http://www.who.int/vaccines-diseases/history/history.shtml.

Wustrau, Joan. Ongoing family genealogies: *Descendants of Jacob Gorsline*; *Descendants of Samuel Fowler*; and *Michael Enderly (1735–1812) and His Descendants.* Interviews and correspondence, 1997 to present.

Yeaple, Louise. *The Adam Yaple Cabin & the Clove Community.* Second edition. Marbletown, N.Y.: Louise Yeaple, ca. 1998.

Young, Holice B., transcriber. "Index of Awards on Claims of the Soldiers of the War of 1812." United States Genealogy Network, Inc. 1998–2004. http://www.usgennet.org/usa/ny/state/1812/.

MAPS

Beers, F.W. *County Atlas of Ulster, New York from Recent and Actual Surveys and Records.* 1875. Reprint; High Falls, N.Y.: Ulster County Genealogical Society, 1975. Haviland-Heidgerd Historical Collection (HHHC-EML), Elting Memorial Library, New Paltz, N.Y.

Billingsley, Paul, Alfred F. Smiley; and J. A. Billingsley. "Topographical Map of the Minnewaska Estate, Shawangunk Mtns." Summer 1906–07. Daniel Smiley Research Center, Mohonk Preserve (DSRC-MP), New Paltz, N.Y.

Brodhead, Conrad, et al. "Map of the Southern portion of the 'GREAT TRANSPORT' From Surveys by Conrad Brodhead, D. T. Van Buren and T. Schoonmaker and The Map of Jas. Oliver and others." 1860. HHHC-EML.

Burr, David H. "An Atlas of the State of New York, No 9, a map of the County of Ulster." 1829. Also "Map of the County of Ulster." 1840. Map 150, Liber 13 of Maps:3. Ulster County Clerk's Office, Kingston, N.Y. (UCCO) and HHHC-EML.

Chandler, Harry A. "Lake Minnewaska Estate on Minnewaska Trail." Brochure titled "Your Trip to Lake Minnewaska over The Minnewaska Trail." Printed by George H. Smiley & Son, Minnewaska, Ulster County, N.Y. Ca. 1930s. HHHC-EML.

Colton & Co., G. W. & C. B. "Colton's Road Map of Ulster County, New York 1891." HHHC-EML.

French, J. H. "Map of Ulster County, New York." Philadelphia: Taintor Dawson & Co., 1858. HHHC-EML.

"Grote Transport or Philip Dubois Tract." 21 February 1800. (UCCO, Map 44, Liber 31 of Maps:2). Also, "Grote Transport or Tract of Philip DuBois Bevier and Heirs of Lodewick Hoornbeck. 1800." Map 134, Liber 3 of Maps:16; Description, Book M:52. UCCO.

Jones, William D. "The Commissioners Location of the Wawarsing & New Paltz Turnpike." Oliver J. Tillson, Daniel Bodley, and Asa S. Wygant, commissioners. 28 June 1856. Map 165. UCCO.

"Map of Minnewaska Estate." ca. 1896–97. DSRC-MP.

Nineteen Partners Partition Maps. "MAP of a Tract in the Town of Rochester Called the Nineteen Partners Tract from Partition Records." Filed January 1800. Partition Records, Court of Common Pleas, Book M:300–301; description, Book M:274. UCCO; Copy by Daniel T. Van Buren, Kingston, N.Y. June 2, 1875. Map # 1017. New York State Department of Environmental Conservation, Region 3, New Paltz and HHHC-EML.

Oliver, Jas. "A map of the land conveyed to Philip DuBois by the Commissioners of the Town of Rochester." Filed May 2, 1810. Contains divisions of lands, February 8, 1800. Map 134. UCCO.

Tappan, Christopher. "New Paltz and Highland Turnpike." 15 September 1807. Liber 10 of Maps:24 and Description 241A. UCCO.

Tillson, Oliver J. and P. Henry Brink. "Map of Ulster County, New York From Original Surveys by Liver J. Tillson and P. Henry Brink." 1853. HHHC-EML.

Van Valkenburgh, N., and R. Larsen. "The Trapps Mountain hamlet circa 1890, A Part of the Grote Transport of 1730 and the Nineteen Partner Tract of 1770 situate in the Towns of Gardiner & Rochester, Ulster County, New York." 17 Feb 1997. DSRC-MP.

Acknowledgments

We welcome your comments. Any errors are our own. For those of you who will find them, we are easy to find ourselves. Please contact us in the spirit of research at rjhudson@hvc.rr.com.

Three facilities located in New Paltz, New York, have been indispensable to our research: the Daniel Smiley Research Center, Mohonk Preserve (DSRC-MP); the Haviland-Heidgerd Historical Collection, Elting Memorial Library (HHHC-EML); and the Mohonk Mountain House Archives (MMHA). These collections are testaments to the dedicated individuals who established them and who maintain them. The New York State Library, Albany, N.Y., the Library of Congress Prints & Photographs Online Catalog, and the University of Michigan Library's Making of America Digital Library were three other important resources.

This book was only possible with the generous and patient assistance of many. We apologize if we have overlooked anyone. We are indebted to former Trapps residents, descendants, family members, and those with close ties to the hamlet: Doris Hall, Roger and Marie Van Leuven, Don and Flo Otis, the late Joe Otis, Joan and Werner Wustrau, George Black, Marian Whitford, Grace C. Smith, Alfred B. Smiley, John Van Leuven, Jr., Mildred Bendix, Loretta Jenkins, William Carson, Lester E. Frost, Earl Mack, June Sharpe, Michael Sharpe, Sheri Ballard, Gerald Haas, and Linda Mackes Bartos.

Also to: Paul C. Huth, John E. Thompson, Shanan Smiley, and Ethan Pierce, DSRC-MP; Glenn Hoagland, Ellen Sticker, Gretchen Reed, Hank Alicandri, Kathy Ambrosini, Patty Murphy, Garrick Bryant, Cathy Shiga-Gattullo, Justin Key, and Nadia Steinzor, Mohonk Preserve; Carol Johnson, Anne Wandres, Marion Ryan, and Rosalie Nelson, HHHC-EML; Nell Boucher and Joan LaChance, MMMA; Pril Smiley, the Smiley Family, and Jim Clark, Mohonk Mountain House; Vivian Yess Wadlin, *About Town* (Ulster County); Eric Roth, Leslie LeFevre-Stratton, and Susan Stessin-Cohn, Historic Huguenot Street; Richard L. Rider, Friends of Historic Rochester; Kathleen Sergio, Town of Rochester; Suzanne Hauspurg, Alice Cross, and Jessica Phinney, Ulster County Historical Society; Marion Ehrhardt, Camille Regholec, Pat Christian, and Rita Helgesen, Ellenville Public Library & Museum; staff at The Kingston Library, Stone Ridge Public Library, Middletown Thrall Library, and the New York State Library; Ken Gray, Ulster County Clerk's Office; Robin Walsh and staff, Macdonald DeWitt Library, Ulster County Community College; Bev Sickler and Susi Popieluszko, New Paltz Central School District; Susan Sinnot, William

Fogg Library, Eliot, Maine; Jay O'Connor and Suzanne Spear, New York State Education Department; Peter Osborne, Minisink Valley Historical Society; Betsy Jacks, The Thomas Cole National Historic Site; Lori Childers, *Blue Stone Press*, Stone Ridge; Gary Coutu, Gwen Ahlers, and Mike Pogue, New York State Department of Environmental Conservation, Region 3, New Paltz; Stuart Shinske, *Poughkeepsie* (N.Y.) *Journal*; Sharmin Kashem, PARS International Corp.; Geddy Sveikauskas, Ulster Publishing; Ronald Hussy, Claire Posner, and Ellen Faux, Houghton Mifflin Harcourt; Bob Anderberg, Open Space Institute; Cara Lee, Shawangunk Ridge Program of The Nature Conservancy; Rick Remsnyder, Ulster County Tourism; Robert LaPolt; Gail Yeaple; Norman Van Valkenburgh; Suzanne Caporael; Zeke and Ginny Boyle; Robert Miraldi; Mary Beth Pfeiffer; Jeff Salt; Marc B. Fried; Elaine LaFlamme; G. Steve Jordan; Linda Tantillo; and George Millham.

To Jane K. Taylor for the main title suggestion, *An Unforgiving Land*, a gift that came just when we needed it. Our special thanks to Jack and Lin Fagan. Our heartfelt appreciation for your charming illustrations, Jack. Each one, so carefully rendered, captures an aspect of Trapps life for us and our readers.

A very grateful nod to our contributors Glenn Hoagland, Bob Anderberg, and Cara Lee. We blush at your accolades. Our sincerest gratitude to our reviewers for their welcome and insightful comments and generous editorial assistance: Paul Huth, Carleton Mabee, and notably to Janet Mattox for volunteering double duty so graciously. All of you took time out of your busy schedules to help us. How pleased and humbled we were when you said yes without hesitation.

To Steve Hoare of Black Dome Press, as well as Debbie Allen now retired, deepest thanks for making this dream of ours—and so many others—come true. Especially to Steve, for shepherding, at times under tremendous pressure, every aspect of this book to its beautiful reality, and to Black Dome's designers Ron Toelke and Barbara Kempler-Toelke for their most artful touch. Our profound admiration.

Last, but first in our lives, to our spouses, Ted Reiss and Barbara Rubin, who journeyed with us every step of the way, sweetest embraces for their ardent support, encouragement, and enthusiasm. To both of you, all our love, with gratitude to Ted for navigating, once again, the deep waters of Photoshop and, let's be frank, help with *all* computer work. And I couldn't help but add a personal thanks to Bob, my mentor and friend, who has shared it all with me and taken me on the most wonderful of journeys—and full of wonder it continues to be.

General Index

About the Authors

Robi Josephson and Bob Larsen at the restored Van Leuven Cabin, Mohonk Preserve, Gardiner, New York. Photograph by Ted Reiss.

Robi Josephson is the author of *Mohonk: Mountain House and Preserve* (Arcadia Publishing, 2002). She earned her bachelor's and master's degrees in English from SUNY New Paltz, writing her master's thesis on John Burroughs in and around the Shawangunks. Before writing full-time, Robi worked as a freelance consultant in grant writing, publications editing, and collections cataloguing, the last two for the Mohonk Preserve.

Bob Larsen is the author of *Trapps Mountain Hamlet: An Interpretive Walk through a Vanished Shawangunk Community* (Mohonk Preserve, 1999). Bob joined the Mohonk Preserve (then The Mohonk Trust) in 1974 as second ranger. He designed and installed many trails at the preserve. Bob also assisted the surveyor in locating and marking over fifty miles of boundary lines. In recent years he has added cultural historian to his title.

Robi and Bob collaborated for ten years on this book, which is based on Bob's groundbreaking research of Trapps history over a period of forty years. Both live in Ulster County, New York.